S0-EJV-157

UCLA Symposia on Molecular and Cellular Biology, New Series

Series Editor, C. Fred Fox

RECENT TITLES

Volume 33
Yeast Cell Biology, James Hicks, *Editor*

Volume 34
Molecular Genetics of Filamentous Fungi, William Timberlake, *Editor*

Volume 35
Plant Genetics, Michael Freeling, *Editor*

Volume 36
Options for the Control of Influenza, Alan P. Kendal and Peter A. Patriarca, *Editors*

Volume 37
Perspectives in Inflammation, Neoplasia, and Vascular Cell Biology, Thomas Edgington, Russell Ross, and Samuel Silverstein, *Editors*

Volume 38
Membrane Skeletons and Cytoskeletal—Membrane Associations, Vann Bennett, Carl M. Cohen, Samuel E. Lux, and Jiri Palek, *Editors*

Volume 39
Protein Structure, Folding, and Design, Dale L. Oxender, *Editor*

Volume 40
Biochemical and Molecular Epidemiology of Cancer, Curtis C. Harris, *Editor*

Volume 41
Immune Regulation by Characterized Polypeptides, Gideon Goldstein, Jean-François Bach, and Hans Wigzell, *Editors*

Volume 42
Molecular Strategies of Parasitic Invasion, Nina Agabian, Howard Goodman, and Nadia Nogueira, *Editors*

Volume 43
Viruses and Human Cancer, Robert C. Gallo, William Haseltine, George Klein, and Harald zur Hausen, *Editors*

Volume 44
Molecular Biology of Plant Growth Control, J. Eugene Fox and Mark Jacobs, *Editors*

Volume 45
Membrane-Mediated Cytotoxicity, Benjamin Bonavida and R. John Collier, *Editors*

Volume 46
Development and Diseases of Cartilage and Bone Matrix, Arup Sen and Thomas Thornhill, *Editors*

Volume 47
DNA Replication and Recombination, Roger McMacken and Thomas J. Kelly, *Editors*

Volume 48
Molecular Strategies for Crop Protection, Charles J. Arntzen and Clarence Ryan, *Editors*

Volume 49
Molecular Entomology, John Law, *Editor*

Volume 50
Interferons as Cell Growth Inhibitors and Antitumor Factors, Robert M. Friedman, Thomas Merigan, and T. Sreevalsan, *Editors*

UCLA Symposia Board

Viruses and Human Cancer

Viruses and Human Cancer

Proceedings of a UCLA Symposium
Held in Park City, Utah,
February 2–9, 1986

Editors

Robert C. Gallo
Laboratory of Tumor Cell Biology
National Institutes of Health
Bethesda, Maryland

William Haseltine
Dana-Farber Cancer Institute
Boston, Massachusetts

George Klein
Department of Tumor Biology
Karolinska Institute
Stockholm, Sweden

Harald zur Hausen
German Cancer Research Center
Heidelberg, Federal Republic of Germany

Alan R. Liss, Inc. • New York

Address all Inquiries to the Publisher
Alan R. Liss, Inc., 41 East 11th Street, New York, NY 10003

Printed in the United States of America

Library of Congress Cataloging-in-Publication Data

UCLA symposium on Viruses and Human Cancer (1986:
Park City, Utah)
Viruses and human cancer.
(UCLA symposia on molecular and cellular biology;
new ser, v. 43)
Includes bibliographies and index.
1. Viral carcinogenesis—Congresses. 2. Retroviruses
—Congresses. 3. Oncogenic viruses—Congresses.
I. Gallo, Robert C. II. Title. III. Series. [DNLM:
1. Neoplasms—etiology—congresses. 2. Oncogenic
Viruses—congresses. W3 U17N new ser. v.43 /
QW 166 U17v 1986]
RC268.57.U25 1986 616.99'4071 86-27526
ISBN 0-8451-2642-3

Contents

Contributors

Mordechai Aboud, Department of Medicine, Division of Hematology-Oncology, UCLA School of Medicine, Los Angeles, CA 90024; present address: Department of Microbiology and Immunology, Faculty of Health Sciences, Ben Gurion University of the Negev, Beer Sheva, Israel **[109]**

George Acs, Departments of Biochemistry and Pediatrics, Mount Sinai School of Medicine, New York, NY 10029 **[467]**

James C. Alwine, Department of Microbiology, University of Pennsylvania School of Medicine, Philadelphia, PA 19104-6076 **[269]**

Rita Anand, AIDS Program, Centers for Disease Control, Atlanta, GA 30333 **[297]**

Minas Arsenakis, The Marjorie B. Kovler Viral Oncology Laboratories, The University of Chicago, Chicago, IL 60637 **[427]**

Larry O. Arthur, Program Resources, Incorporated, NCI-Frederick Cancer Research Facility, Frederick, MD 21701 **[451]**

Philip J. Barr, Chiron Research Laboratories, Chiron Corporation, Emeryville, CA 94608 **[29, 205]**

Hubert Bayer, Abteilung Virologie und Immunologie, Deutsches Primatenzentrum, D-3400 Göttingen, Federal Republic of Germany **[319]**

Phillip Berman, Department of Molecular Biology, Genentech, Inc., South San Francisco, CA 94080 **[191]**

Robert J. Biggar, Environmental Epidemiology Branch, National Cancer Institute, National Institutes of Health, Bethesda, MD 20892 **[93]**

Beverly J. Blake, New England Regional Primate Research Center, Harvard Medical School, Southboro, MA 01772 **[451]**

William A. Blattner, Environmental Epidemiology Branch, National Cancer Institute, National Institutes of Health, Bethesda, MD 20892 **[93]**

Walter Bodemer, Institut für Klinische Virologie der Universität, D-8520 Erlangen, Federal Republic of Germany **[443]**

Dani Bolognesi, Department of Surgery, Duke University Medical Center, Durham, NC 27710 **[15]**

Carolina Braga, Department of Obstetrics and Gynecology, University of California School of Medicine, San Francisco, CA 94143 **[387]**

The numbers in brackets are the opening page numbers of the contributors' articles.

Ralph L. Brinster, Laboratory of Productive Physiology, University of Pennsylvania School of Veterinary Medicine, Philadelphia, PA 19103 **[493]**

A. Burny, Department of Molecular Biology, University of Brussels, 1640 Rhode-Saint-Genèse, Belgium **[345]**

Alan J. Cann, Department of Medicine, Division of Hematology-Oncology, UCLA School of Medicine, Los Angeles, CA 90024 **[109, 221]**

Susan Carswell, Department of Microbiology, University of Pennsylvania School of Medicine, Philadelphia, PA 19104-6076 **[269]**

Mark Chamberlain, Central Research and Development Department, E.I. DuPont de Nemours & Co., Inc., Wilmington, DE 19898 **[15]**

Nancy T. Chang, Center for Biotechnology, Baylor College of Medicine, Houston, TX 77030 **[237]**

Irvin S.Y. Chen, Department of Medicine, Division of Hematology-Oncology, UCLA School of Medicine, Los Angeles, CA 90024 **[109, 221]**

Mark Chernaik, Department of Biochemistry, The Johns Hopkins University, School of Hygiene and Public Health, Baltimore, MD 21205 **[415]**

Francis V. Chisari, Basic and Clinical Research, Scripps Clinic & Research Foundation, La Jolla, CA 92037 **[493]**

Judith K. Christman, Departments of Biochemistry and Pediatrics, Mount Sinai School of Medicine, New York, NY 10029 **[467]**

Jeffrey Clark, Environmental Epidemiology Branch, National Cancer Institute, National Institutes of Health, Bethesda, MD 20892 **[93]**

Y. Cleuter, Department of Molecular Biology, University of Brussels, 1640 Rhode-Saint-Gènese, Belgium **[345]**

Christopher P. Crum, Departments of Obstetrics and Gynecology, Columbia University College of Physicians and Surgeons, New York, NY 10032 **[355]**

Raymond L. Cybulski, Biomedical Products Department, E.I. DuPont de Nemours & Co., Inc., Wilmington, DE 19898 **[15]**

Chris Dabrowski, Department of Microbiology, University of Pennsylvania School of Medicine, Philadelphia, PA 19104-6076 **[269]**

C. Dandoy, Institut Pasteur, Lille, France **[345]**

M. D. Daniel, New England Regional Primate Research Center, Harvard Medical School, Southboro, MA 01772 **[451]**

Ronald C. Desrosiers, New England Regional Primate Research Center, Harvard Medical School, Southboro, MA 01772 **[309,319,451]**

S. Dewhurst, Department of Pathology and Microbiology, Molecular Biology Laboratory, University of Nebraska Medical Center, Omaha, NE 68105 **[81]**

Dino Dina, Director of Virology, Chiron Corporation, Emeryville, CA 94608 **[29]**

Donald Dowbenko, Department of Molecular Biology, Genentech, Inc., South San Francisco, CA 94080 **[191]**

Renato Dulbecco, The Salk Institute, The Monoclonal Antibody Laboratory of The Armand Hammer Cancer Center, La Jolla, CA 92037 **[1]**

Anthony S. Fauci, Laboratory of Immunoregulation, National Institute of Allergy and Infectious Diseases, National Institutes of Health, Bethesda, MD 20892 **[59]**

Christopher Fennie, Department of Molecular Biology, Genentech, Inc., South San Francisco, CA 94080 **[191]**

Pierre Filippi, Basic and Clinical Research, Scripps Clinic & Research Foundation, La Jolla, CA 92037 **[493]**

Bernhard Fleckenstein, Institut für Klinische Virologie der Universität, D-8520 Erlangen, Federal Republic of Germany **[443]**

Daniel Friedman, Departments of Obstetrics and Gynecology, Columbia University College of Physicians and Surgeons, New York, NY 10032 **[355]**

Masatoshi Fujishita, Department of Medicine, Kochi Medical School, Nankoku 781-51, Japan **[141]**

Patricia Fultz, AIDS Program, Centers for Disease Control, Atlanta, GA 30333 **[297]**

Gregory Gallo, Department of Microbiology, University of Pennsylvania School of Medicine, Philadelphia, PA 19104-6076 **[269]**

Robert C. Gallo, Laboratory of Tumor Cell Biology, National Cancer Institute, National Institutes of Health, Bethesda, MD 20892 **[xxiii, xxvii, 151]**

Suzanne Gartner, Laboratory of Tumor Cell Biology, National Cancer Institute, National Institutes of Health, Bethesda, MD 20892 **[161]**

Richard B. Gaynor, Department of Medicine, Division of Hematology-Oncology, UCLA School of Medicine, Los Angeles, CA 90024 **[109]**

Hans Gelderblom, Bundesgesundheitsamt, D-1000 Berlin 65, Federal Republic of Germany **[319]**

Carlos George-Nascimento, Chiron Research Laboratories, Chiron Corporation, Emeryville, CA 94608 **[29,205]**

Helen L. Gibson, Chiron Research Laboratories, Chiron Corporation, Emeryville, CA 94608 **[205]**

David W. Golde, Department of Medicine, Division of Hematology-Oncology, UCLA School of Medicine, Los Angeles, CA 90024 **[109]**

Beatrice Hahn, Division of Haematology and Oncology, University of Alabama Medical Center, Birmingham, AL 35294 **[151]**

Robert A. Hallewell, Chiron Research Laboratories, Chiron Corporation, Emeryville, CA 94608 **[205]**

William Haseltine, Dana-Farber Cancer Institute, Boston, MA 02115 **[xxvii]**

Masanori Hayami, Department of Animal Pathology, Institute of Medical Science, University of Tokyo, Tokyo 108, Japan **[333]**

Dorothee Herlyn, The Wistar Institute, Philadelphia, PA 19104 **[257]**

Mary K. Howett, Department of Microbiology, College of Medicine, The Milton S. Hershey Medical Center, Hershey, PA 17033 **[371]**

Gerhard Hunsmann, Abteilung Virologie und Immunologie, Deutsches Primatenzentrum, D-3400 Göttingen, Federal Republic of Germany **[319]**

Ronald D. Hunt, New England Regional Primate Research Center, Harvard Medical School, Southboro, MA 01772 **[451]**

Yoji Ikawa, Laboratory of Molecular Oncology, The Institute of Physical and Chemical Research, Wako, Saitama 351-01, Japan **[177]**

Lucinda Ivanoff, Central Research and Development Department, E.I. DuPont de Nemours & Co., Inc., Wilmington, DE 19898 **[15]**

Klaus-Dieter Jentsch, Abteilung Virologie und Immunologie, Deutsches Primatenzentrum, D-3400 Göttingen, Federal Republic of Germany **[319]**

Elke Jurkiewicz, Abteilung Virologie und Immunologie, Deutsches Primatenzentrum, D-3400 Göttingen, Federal Republic of Germany **[319]**

Janis M. Keller, Department of Microbiology, University of Pennsylvania School of Medicine, Philadelphia, PA 19104-6076 **[269]**

William Kenealy, Central Research and Development Department, E.I. DuPont de Nemours & Co., Inc., Wilmington, DE 19898 **[15]**

R. Kettmann, Department of Molecular Biology, University of Brussels, 1640 Rhode-Saint-Genèse, Belgium **[345]**

Norval W. King, New England Regional Primate Research Center, Harvard Medical School, Southboro, MA 01772 **[451]**

Toshio Kitamura, Virology Division, National Cancer Center Research Institute, Tokyo 104, Japan **[131]**

George Klein, Department of Tumor Biology, Karolinska Institute, Stockholm, Sweden **[xxvii]**

Hilary Koprowski, The Wistar Institute, Philadelphia, PA 19104 **[257]**

Shigemitsu Kotani, Department of Medicine, Kochi Medical School, Nankoku 781-51, Japan **[141]**

John W. Kreider, Department of Pathology, College of Medicine, The Milton S. Hershey Medical Center, Hershey, PA 17033 **[371]**

Kai Krohn, Institute of Biomedical Science, University of Tampere, 33101 Tampere, Finland; present address: Laboratory of Tumor Cell Biology, National Cancer Institute, National Institutes of Health, Bethesda, MD 20892 **[43]**

H. Clifford Lane, Laboratory of Immunoregulation, National Institute of Allergy and Infectious Diseases, National Institutes of Health, Bethesda, MD 20892 **[59]**

Laurence A. Lasky, Department of Molecular Biology, Genentech, Inc., South San Francisco, CA 94080 **[191]**

Diana Jin Lee, Chiron Research Laboratories, Chiron Corporation, Emeryville, CA 94608 **[29]**

Chun Ting Lee-Ng, Chiron Research Laboratories, Chiron Corporation, Emeryville, CA 94608 **[205]**

Norman L. Letvin, New England Primate Research Center, Harvard Medical School, Southboro, MA 01772 **[451]**

Paul H. Levine, Environmental Epidemiology Branch, National Cancer Institute, National Institutes of Health, Bethesda, MD 20892 **[93,105]**

Jay A. Levy, Chiron Research Laboratories, Chiron Corporation, Emeryville, CA 94608; present address: Department of Medicine, Cancer Research Institute, University of California, San Francisco, San Francisco, CA 94143 **[29]**

Ruey Liou, Center for Biotechnology, Baylor College of Medicine, Houston, TX 77030; present address: Biotech Research Laboratory, Inc., Rockville, MD 20850 **[237]**

Paul A. Luciw, Chiron Research Laboratories, Chiron Corporation, Emeryville, CA 94608 **[29, 205]**

M. Mammerickx, Department of Virology, National Institute for Veterinary Research, 1180 Uccle, Belgium **[345]**

G. Marbaix, Department of Molecular Biology, University of Brussels, 1640 Rhode-Saint-Genèse, Belgium **[345]**

Joseph B. Margolick, Laboratory of Immunoregulation, National Institute of Allergy and Infectious Diseases, National Institutes of Health, Bethesda, MD 20892; present address: Department of Environmental Sciences, The Johns Hopkins University School of Hygiene and Public Health, Baltimore, MD 21205 **[59]**

Tom Matthews, Department of Surgery, Duke University Medical Center, Durham, NC 27710 **[15]**

Harold McClure, Department of Pathology and Immunobiology, Yerkes Regional Primate Research Center, Emory University, Atlanta, GA 30322 **[297]**

Alan McLachlan, Basic and Clinical Research, Scripps Clinic & Research Foundation, La Jolla, CA 92037 **[493]**

F. Merino, Departamento Medicina Experimental, Instituto Venezolano de Investigaciones, Caracas 1010A, Venezuela **[81]**

David R. Milich, Basic and Clinical Research, Scripps Clinic & Research Foundation, La Jolla, CA 92037 **[493]**

Elizabeth T. Miller, Chiron Research Laboratories, Chiron Corporation, Emeryville, CA 94608 **[29]**

Gregory Milman, Department of Biochemistry, The Johns Hopkins University, School of Hygiene and Public Health, Baltimore, MD 21205 **[415]**

Masanao Miwa, Virology Division, National Cancer Center Research Institute, Tokyo 104, Japan **[131]**

Isao Miyoshi, Department of Medicine, Kochi Medical School, Nankoku 781-51, Japan **[141]**

Shridhara Murthy, New England Regional Primate Research Center, Harvard Medical School, Southboro, MA 01772 **[309]**

Hans Helmut Niller, Institut für Klinische Virologie der Universität, D-8520 Erlangen, Federal Republic of Germany **[443]**

Nikolaus Nitsche, Institut für Klinische Virologie der Universität, D-8520 Erlangen, Federal Republic of Germany **[443]**

Victor Nizet, Department of Medical Microbiology, Stanford University School of Medicine, San Francisco, CA 94305 **[387]**

Wylla Nunes, Department of Molecular Biology, Genentech, Inc., South San Francisco, CA 94080 **[191]**

Gerard Nuovo, Department of Pathology, Columbia University College of Physicians and Surgeons, New York, NY 10032 **[355]**

Kazue Ohishi, Laboratory of Molecular Oncology, The Institute of Physical and Chemical Research, Wako, Saitama 351-01, Japan; present address: Division of Life Sciences, Tonen R & D Laboratories, Saitama 354, Japan **[177]**

Yuji Ohtsuki, Department of Pathology, Kochi Medical School, Nankoku 781-51, Japan **[141]**

Yuko Ootsuyama, Virology Division, National Cancer Center Research Institute, Tokyo 104, Japan **[131]**

Jing-hsiung Ou, Department of Biochemistry and Biophysics, Hormone Research Institute, University of California, San Francisco, San Francisco, CA 94143-0534 **[479]**

Joseph S. Pagano, Department of Medicine, University of North Carolina at Chapel Hill, Chapel Hill, NC 27514 **[397]**

Joel Palefsky, Department of Medical Microbiology, Stanford University School of Medicine, Stanford, CA 94305 **[387]**

Richard D. Palmiter, Department of Biochemistry, University of Washington School of Medicine, Seattle, WA 98195 **[493]**

Debbie Parkes, Chiron Research Laboratories, Chiron Corporation, Emeryville, CA 94608 **[29, 205]**

Rick L. Pesano, Lineberber Cancer Research Center, University of North Carolina at Chapel Hill, Chapel Hill, NC 27514 **[397]**

S.R. Petteway, Jr., Central Research and Development Department, E.I. DuPont de Nemours & Co., Inc., Wilmington, DE 19898 **[15]**

Jane Picardi, Department of Microbiology, University of Pennsylvania School of Medicine, Philadelphia, PA 19104-6076 **[269]**

Carl A. Pinkert, Laboratory of Productive Physiology, University of Pennsylvania, Philadelphia, PA 19103 **[493]**

Kimber Lee Poffenberger, The Marjorie B. Kovler Viral Oncology Laboratories, The University of Chicago, Chicago, IL 60637 **[427]**

Mikulas Popovic, Laboratory of Tumor Cell Biology, National Cancer Institute, National Institutes of Health, Bethesda, MD 20892 **[161]**

D. Portetelle, Department of Molecular Biology, University of Brussels, 1640 Rhode-Saint-Genèse, Belgium **[345]**

Michael D. Power, Chiron Research Laboratories, Chiron Corporation, Emeryville, CA 94608 **[205]**

Peter M. Price, Departments of Biochemistry and Pediatrics, Mount Sinai School of Medicine, New York, NY 10029 **[467]**

Iris Puchtler, Institut für Klinische Virologie der Universität, D-8520 Erlangen, Federal Republic of Germany **[443]**

Annamari Ranki, Department of Dermatology, Helsinki University, 00170 Helsinki, Finland; present address: Laboratory of Tumor Cell Biology, National Cancer Institute, National Institutes of Health, Bethesda, MD 20892 **[43]**

Elizabeth Read-Connole, Laboratory of Tumor Cell Biology, National Cancer Institute, National Institutes of Health, Bethesda, MD 20892 **[161]**

Kevin J. Reagen, Biomedical Products Department, E.I. DuPont de Nemours & Co., Inc., Wilmington, DE 19898 **[15]**

Donna Reed, Central Research and Development Department, E.I. DuPont de Nemours & Co., Inc., Wilmington, DE 19898 **[15]**

Marjorie Robert-Guroff, Laboratory of Tumor Cell Biology, National Cancer Institute, National Institutes of Health, Bethesda, MD 20892 **[93]**

L. Rodriguez, Departamento Medicina Experimental, Instituto Venezolano de Investigaciones, Caracas 1010A, Venezuela **[81]**

Bernard Roizman, The Marjorie B. Kovler Viral Oncology Laboratories, The University of Chicago, Chicago, IL 60637 **[427]**

Joseph D. Rosenblatt, Department of Medicine, Division of Hematology-Oncology, UCLA School of Medicine, Los Angeles, CA 90024 **[109, 221]**

Lucia D. Rossoni, Department of Medicine, Division of Cancer Pharmacology, Dana-Farber Cancer Institute, Harvard Medical School, Boston, MA 02115 **[283]**

Rüdiger Rüger, Institut für Klinische Virologie der Universität, D-8520 Erlangen, Federal Republic of Germany **[443]**

Ruth M. Ruprecht, Department of Medicine, Division of Cancer Pharmacology, Dana-Farber Cancer Institute, Harvard Medical School, Boston, MA 02115 **[283]**

William J. Rutter, Department of Biochemistry and Biophysics, Hormone Research Institute, University of California, San Francisco, San Francisco, CA 94143-0534 **[479]**

Elizabeth A. Sabin, Chiron Research Laboratories, Chiron Corporation, Emeryville, CA 94608 **[205]**

Noriyuki Sagata, Laboratory of Molecular Oncology, The Institute of Physical and Chemical Research, Wako, Saitama 351-01, Japan **[177]**

Ray Sanchez-Pescador, Chiron Research Laboratories, Chiron Corporation, Emeryville, CA 94608 **[205]**

W. Carl Saxinger, Laboratory of Tumor Cell Biology, National Cancer Institute, National Institutes of Health, Bethesda, MD 20892 **[93]**

Josef Schneider, Abteilung Virologie und Immunologie, Deutsches Primatenzentrum, D-3400 Göttingen, Federal Republic of Germany **[319]**

Gary Schoolnik, Department of Medical Microbiology, Stanford University School of Medicine, Stanford, CA 94305 **[387]**

Mary Ann Sells, Department of Biochemistry, Mount Sinai School of Medicine, New York, NY 10029 **[467]**

Neil P. Shah, Department of Medicine, Division of Hematology-Oncology, UCLA School of Medicine, Los Angeles, CA 90024 **[221]**

G. Shaw, Division of Haematology and Oncology, University of Alabama Medical Center, Birmingham, AL 35294 **[151]**

Chip Shearman, Center for Biotechnology, Baylor College of Medicine, Houston, TX 77030; present address: Centocor, Malvern, PA 19355 **[237]**

Hiroshi Shima, Virology Division, National Cancer Center Research Institute, Tokyo 104, Japan **[131]**

Nobuaki Shimizu, Virology Division, National Cancer Center Research Institute, Tokyo 104, Japan **[131]**

Kunitada Shimotohno, Virology Division, National Cancer Center Research Institute, Tokyo 104, Japan **[131]**

Masanori Shimoyama, Drug Therapy Division, National Cancer Center Research Institute, Tokyo 104, Japan **[131]**

Saul J. Silverstein, Department of Microbiology, Cancer Center, Columbia University College of Physicians and Surgeons, New York, NY 10032 **[355]**

F. Sinangil, Department of Pathology and Microbiology, Molecular Biology Laboratory, University of Nebraska Medical Center, Omaha, NE 68105 **[81]**

Dennis J. Slamon, Department of Medicine, Division of Hematology-Oncology, UCLA School of Medicine, Los Angeles, CA 90024 **[109]**

A. Srinivasan, AIDS Program, Centers for Disease Control, Atlanta, GA 30333 **[297]**

Chuck Staben, Chiron Research Laboratories, Chiron Corporation, Emeryville, CA 94608 **[29]**

Bruno Starcich, Laboratory of Tumor Cell Biology, National Cancer Institute, National Institutes of Health, Bethesda, MD 20892 **[151]**

Kathelyn S. Steimer, Chiron Research Laboratories, Chiron Corporation, Emeryville, CA 94608 **[29, 205]**

James C. Stephans, Chiron Research Laboratories, Chiron Corporation, Emeryville, CA 94608 **[29, 205]**

Takashi Sugimura, Virology Division, National Cancer Center Research Institute, Tokyo 104, Japan **[131]**

Hirokuni Taguchi, Department of Medicine, Kochi Medical School, Nankoku 781-51, Japan **[141]**

A. Tartar, Institut Pasteur, Lille, France **[345]**

David E. Tribe, Central Research and Development Department, E.I. DuPont de Nemours & Co., Inc., Wilmington, DE 19898 **[15]**

John Trimble, New England Regional Primate Research Center, Harvard Medical School, Southboro, MA 01772 **[309]**

Radonna Tritch, Central Research and Development Department, E.I. DuPont de Nemours & Co., Inc., Wilmington, DE 19898 **[15]**

Atsumi Tsujimoto, Virology Division, National Cancer Center Research Institute, Tokyo 104, Japan **[131]**

A. Van Den Broeke, Department of Molecular Biology, University of Brussels, 1640 Rhode-Saint-Genèse, Belgium **[345]**

Gary A. Van Nest, Chiron Research Laboratories, Chiron Corporation, Emeryville, CA 94608 **[29]**

D. J. Volsky, Department of Pathology and Microbiology, Molecular Biology Laboratory, University of Nebraska Medical Center, Omaha, NE 68105 **[81]**

William Wachsman, Department of Medicine, Division of Hematology-Oncology, UCLA School of Medicine, Los Angeles, CA 90024 **[109, 221]**

Shaw Watanabe, Epidemiology Division, National Cancer Center Research Institute, Tokyo 104, Japan **[131]**

Stanley Weiss, Environmental Epidemiology Branch, National Cancer Institute, National Institutes of Health, Bethesda, MD 20892 **[93]**

Irene Wendler, Abteilung Virologie und Immunologie, Deutsches Primatenzentrum, D-3400 Göttingen, Federal Republic of Germany **[319]**

John Whitbeck, Department of Microbiology, University of Pennsylvania School of Medicine, Philadelphia, PA 19104-6076 **[269]**

L. Willems, Department of Molecular Biology, University of Brussels, 1640 Rhode-Saint-Genèse, Belgium **[345]**

Jan Williams, Department of Medicine, Division of Hematology-Oncology, UCLA School of Medicine, Los Angeles, CA 90024 **[221]**

Barbara Winkler, Departments of Pathology and Obstetrics, Gynecology, and Reproductive Sciences, University of California School of Medicine, San Francisco, CA 94143 **[387]**

F. Wong-Staal, Laboratory of Tumor Cell Biology, National Cancer Institute, National Institutes of Health, Bethesda, MD 20892 **[151]**

Makoto Yamashita, Department of Medicine, Kochi Medical School, Nankoku 781-51, Japan **[141]**

Kenji Yamato, Department of Medicine, Kochi Medical School, Nankoku 781-51, Japan **[141]**

Teruo Yasunaga, Laboratory of Molecular Oncology, The Institute of Physical and Chemical Research, Wako, Saitama 351-01, Japan; present address: Computation Center, The Institute of Physical and Chemical Research, Wako, Saitama 351-01, Japan **[177]**

Shizuo Yoshimoto, Department of Medicine, Kochi Medical School, Nankoku 781-51, Japan **[141]**

Arthur Zelent, Department of Biochemistry, Mount Sinai School of Medicine, New York, NY 10029 **[467]**

Harald zur Hausen, German Cancer Research Center, Heidelberg, Federal Republic of Germany **[xxvii]**

Introduction

For the past sixteen years my co-workers and I have been engaged in tumor virology and the relevance of viruses to human cancer. Looking back, I am struck by how the field was so much in decline in the seventies and how much it has taken off in the eighties. The International UCLA Symposia on human tumor viruses, that now may occur with regularity, are one indication of how the field has matured. If we were to label the various periods of tumor virus research, we might say that the period beginning from the turn of this century until around 1970 could be called the era of animal tumor virology, the period from 1970 to 1980, the era of molecular biology (replication and mechanisms of effects); and from 1980 onward, the era of human tumor virology. This is illustrated nicely with the retroviruses. Although discovered in chickens between 1908 and 1911, it was not until four decades later that they were first defined in mammals and really linked to disease (leukemia) induction. This was chiefly due to the pioneering work of Ludwig Gross. His mouse leukemia virus was soon verified by the important work of many others including Charlotte Friend, Dick Rauscher, John Moloney, and Werner Kirsten. It was accompanied by the independent work of Bittner and Furth on the breast tumor retrovirus (mouse mammary tumor virus) of mice. The sixties saw the discovery of the feline leukemia virus by William Jarrett in Glasgow, and ultimately, its link to the cause of leukemias, lymphomas, aplasias and an AIDS-like disease in cats. This was historically important not only because it showed that the same virus could cause different kinds of abnormalities of cell growth, but also because it defined an infection as the cause of a *naturally* occurring cancer. This work was a major stimulus to our search for a human retrovirus which we initiated in 1970.

The period of 1970 to 1972 was important to retrovirology for three reasons: bovine leukemia virus (still another retrovirus) was discovered as the cause of bovine leukemia; the first primate leukemia virus was found (gibbon ape leukemia virus) by Thomas Kawakami and his co-workers at Davis, California, and reverse transcriptase was discovered, verifying Howard Temin's novel and ingenious concepts on the replication cycle of retroviruses. By this time, the National Cancer Institute's Virus Cancer Program (created by Rauscher and Moloney during the sixties) was at its peak, as was support for tumor virology in general. Ironically, however, work on putative human retroviruses had declined and there was little interest and little credibility given to the very notion that these viruses might be present in humans. This was due to decades of failure in finding these viruses. The general view held that, if they existed, they would be easy to find (as they then were in chickens, mice and cats), because extensive virus replication was known to occur in the animal models under study. Nonetheless, we decided to intensify our efforts by combining two approaches: 1) sensitive tests for viruses based on refining the assays for reverse transcriptase and using this enzyme assay as an indicator for the presence of these viruses; and 2) development of systems for the appropriate long-term growth of human blood cells. In this respect, we were fortunate to find T-cell growth factor (interleukin - 2 or Il - 2) in 1976. Growth of human T-cells with Il - 2 helped in some areas of cell biology and immunology. It also led to the isolation of the various human retroviruses. There also were some unexpected negative results in human tumor viruses research during this period. For example, in 1970, we expected that by the eighties some human adenoviruses and some of the SV40 polyoma-like human viruses such as BK and JC viruses would have been implicated in human cancer. In fact, they were the major candidates for the first real human tumor viruses. Interest in the adenoviruses was stimulated by the capacity of some strains of adenoviruses to transform cells in culture and suggestions that specific strains might be implicated in some human tumors. The evidence was even stronger for BK and JC viruses. Not only could they transform cells in culture, but closely related viruses could produce tumors in animals. In early reports, BK and JC viruses were found in some human cancers. However, the subsequent work on these viruses has not linked them to any human cancer.

The direct link of viruses to human cancer was stimulated in the eighties by the discovery of the first human retroviruses (HTLV-I and HTLV-II) between 1978 and 1982. These viruses are associated with T-cell malignancies (leukemias and lymphomas). There are hundreds of separate isolates of HTLV-I, but only a few of HTLV-II. The former is linked to an aggressive T-cell malignancy called Adult T-cell lymphoma/leukemia (ATL), while it is too soon to state the situation for HTLV-II. The role of HTLV-I, in these leukemias, is indicated by: 1) the epidemiology (whenever ATL is present, HTLV-I is present); 2) the *in vitro* studies (HTLV-I and -II) immortalize primary human T-cells in culture, and the cells resemble cells of tumors caused by the virus; 3) the molecular biology (the DNA provirus is present in a clonal fashion in all tumor cells of an ATL patient); and 4) animal studies (STLV-I African green monkeys, a virus ninety-five percent homologous to HTLV-I, produces lymphoproliferative disorders when inoculated into some rhesus monkeys).

Studies with HTLV-I and -II also led us to find novel mechanisms for the transformation of normal cells into immortalized cells lines *in vitro* and probably mechanisms involved *in vivo* in the viral-induced tumors. M. Yoshida's sequencing of HTLV-I led to the finding of a sequence called X. Work by W. Haseltine and co-workers and by my colleague, F. Wong-Staal, showed that the X sequence contains a novel gene called *tat*. Later, this was shown to encode a protein which acts at a distance. This trans-acting protein activates expression not only of viral genes, by binding to the viral LTR, but also by activating cell genes important to T-cell proliferation, such as Il - 2 and its receptor. Although this mechanism was long suspected, it was directly demonstrated by W. Greene, et al., only this past year. The total work on the human retroviruses has progressed relatively rapidly and we have learned a considerable amount about their mechanisms. Moreover, it may be that HTLV-I has a more direct role in human cancers than some of the DNA viruses, such as EBV, for the following reasons: 1) the ratio of malignancy to virus infection is far higher than for the ubiquitous DNA viruses such as EBV. Thus, whereas it has been estimated that about one in every few hundred HTLV-I-infected people develop an HTLV-I-related malignancy, only one in about one-hundred thousand to one million EBV-infected people will develop an EBV-related neoplasm. The rate of hepatoma following infection by hepatitis virus is variable, but highly significant. The rate of genital-cervical cancer development following infection with the appropriate strain(s) of papilloma virus, is to my knowledge, unknown. It is now clear that the leukemogenic potential HTLV-I is not dependent on some additional environmental co-factor. For instance, an infected child in Jamaica appears to have the same chance of developing ATL whether growing up in the Carribbean or migrating elsewhere, such as the United States. There is, in short, no HTLV-I geographic "belt" as there is an EBV "belt" which appears to correspond to areas of holoendemic malaria, and there is no chemical carcinogen requirement as there appears to be for papilloma viruses. Hepatitis virus and liver cancer may be closer to the HTLV-I/ATL situation in this respect. There are other similarities between HTLV-I and one or more of the DNA viruses in several important features. Like EBV, HTLV-I has the capacity to immortalize its target cell *in vitro*, whereas, hepatitis virus and papilloma viruses fail to transform cells *in vitro*. Another important element regarding our view of HTLV-I as a human tumor virus is the inescapable evidence that similar retroviruses are important causes of similar malignancies in numerous animal systems. Data implicating naturally occurring papilloma viruses with cervical or genital cancers in animals are indeed scarce. For EBV (herpes viruses), the one example known is Marek's disease of chickens, although lymphomas can sometimes be induced in the laboratory by introducing a herpes virus from some species to a different species. There also are some recent exciting animal models for hepatitis viruses and liver cancer. HTLV-I may have more of the "classical criteria" for a tumor virus than any of the DNA viruses. On the other hand, some of the DNA viruses are far more important in terms of their involvement in numbers of human cancer. If hepatitis virus causes liver cancer, and strains of papilloma virus are involved in the causes of cervical, genital and possibly other epithetical cancers, as we suspect, then the importance of these DNA viruses cannot be overemphasized.

Features common to all the viruses implicated in a human cancer include: 1) transmission by blood, blood products, and by intimate contact, with the exception of EBV which can be more casually transmitted; 2) long latency; 3) ability to induce only one of the few genetic steps apparently needed for the development of a malignancy; and 4) may play a causative role in other (non-malignant) diseases. Thus, EBV is the cause of the self-limiting lymphoproliferative disease known as infectious mononucleosis. Hepatitis virus causes hepatitis and papilloma viruses cause benign tumors. HTLV-I may induce a benign polyclonal T-cell proliferation, slight impairment of immune function leading to an increase in infectious diseases, and recent studies suggest it may be involved in some central nervous system diseases. Regarding the latter, Japanese workers have suggested the name "HAM" (HTLV-I associated myelopathy) for this disorder, which is said to resemble the chronic progressive form of multiple sclerosis.

Work with the four categories of candidates for causes of human cancers has had many spin-offs. It has hastened the pace of the epidemiology of all four virus types. It has opened new insights into the origin of cancer, including discoveries of new molecular mechanisms. It has been the area of research that has best exploited the advances made in molecular biology to human medicine. It has contributed to advances in immunochemical methods for detection of proteins. It has stimulated development of new tissue culture systems and led to the discovery of new growth factors. Last but not least, work with one of them, HTLV-I, laid the conceptual and technical groundwork that prepared us for the greatest pandemic of the twentieth century, AIDS.

Robert C. Gallo

Acknowledgments

We wish to thank the speakers, poster session contributors, and the discussants who made this conference a vital experience.

We gratefully acknowledge Hoffmann-La Roche, Inc. for its generous sponsorship of this conference, and the National Cancer Institute for grant USPHS 1 R13 CA41709. An additional gift was received from Cambridge Bioscience Corporation.

Robert C. Gallo
William Haseltine
George Klein
Harald zur Hausen

Viruses and Human Cancer, pages 1–14

A TURNING POINT IN CANCER RESEARCH: SEQUENCING THE HUMAN GENOME

Renato Dulbecco

The Salk Institute
The Monoclonal Antibody Laboratory of
The Armand Hammer Cancer Center
La Jolla, California 92037

ABSTRACT The present state of research in the mechanisms of cancer reflects the strong impact of tumor virology and of the discovery of oncogenes. Available findings afford a satisfactory explanation of the various modalities by which cancers arise. The findings do not however explain progression, which is the process leading to malignancy. Important in progression are rearrangements of the cellular genome, which may be produced mostly by nonhomologous recombination perturbing many cellular genes. If alterations of cellular genes cause progression, it becomes mandatory to study the cellular genome in detail. Sequencing the human genome would allow the identification of genes important in progression of human cancer, opening up new therapeutic possibilities. It would be also essential for progress in many other fields of human biology and medicine.

INTRODUCTION

Ways of thinking about, and experimental approaches to, the nature of cancer have both undergone rapid evolution during the past thirty years. We are now at an important point of this evolution, a turning point where an even more drastic change in our way of thinking and approaching the problem will be needed.

ONCOGENIC VIRUSES AND CANCER

The past evolution of cancer research has taken us over several steps. The first step was the realization that it is useless to study spontaneous cancers in order to understand cancer, except from an epidemiological point of view. This led to the study of induced cancers in animals, mainly using chemical carcinogens. Another step was to study the differences between normal and cancer cells in vitro. This step got away from the complexity of the whole animal, but not from the genetic complexity of animal cells. A major turn took place when studies shifted to oncogenic viruses. Their use circumvented, or so it seemed, the complexity of the cellular genome, replacing it with the extraordinary simplicity of the viral genome. It was reasoned that if an organism with a few genes can do it, cancer must be a simple process, which can be reduced to the action of one, or a few, of those genes. The use of oncogenic viruses became even more attractive when it was moved from the whole animal to cultures of animal cells. The remarkable effects of these viruses on the cultures were termed "transformation".

Progress in molecular biology permitted the clarification of certain points of cell transformation and carcinogenesis. It was established that the viral and the cellular genome become associated in a permanent way, explaining the persistence of the transformed state in a cell clone (1,2). The association was found in most cases to be integration, although in some systems -- Epstein Barr virus (3) or papilloma viruses (4) -- a stable episomal state of the viral genome was also observed. Experiments of both genetic and molecular nature suggested that transformation or cancer caused by viruses are the consequence of the continued expression of viral genes. Finally the viral transforming genes, or "oncogenes", and the proteins they specify were identified. The crowning development was the demonstration that in retroviruses the oncogenes are picked up from the cellular genome during the viruses' most recent history (5). The oncogenes do not perform any function for these viruses, the viruses are essentially carriers of the oncogenes (6). In the DNA viruses the origin of oncogenes is unclear. It is not known whether they originated independently of the other viral gene and were incorporated permanently into the viral genome at some early evolutionary time. In any case, they underwent specialization for the sake of the virus.

A crucial consequence of these findings was that they seemed to exclude once and for all the possibility that viruses may cause transformation or cancer by a "hit-and-run" mechanism in which the virus alters the cell and then vanishes. This point was later supported by studies with other viruses such as HTLV-1, Bovine Leukemia virus, and Herpesviruses, in which the transforming gene specifies transactive activators which activate viral genes and presumably also cellular genes; this mechanism seems to require the continued expression of viral genes for transformation. It would have been a rather dismal outcome if a "hit-and-run" mechanism had been supported by the data, for it would have shown viral oncogenesis to have the complexity of the host genome from which it tried to escape. Instead, cancer seemed locked to the expression of some viral gene.

VARIOUS EFFECTS OF ONCOGENES

The studies from then on concentrated heavily on the oncogenes revealed by the viral studies. As a paradigm, we can focus for a moment on the findings with polyoma virus. This virus possesses two main oncogenes which specify the proteins large T and middle T in addition to small T, the function of which is less well defined. The study of these and other oncogenes suggested that the many changes of transformed fibroblastic cells can be separated into two main effects (7). One effect is cell immortalization, that is abolition of senescence; the other is tumorigenesis, which closely correlates with an ability of the cells to grow in agarose. As the studies of these effects developed, a simple scheme seemed to emerge: polyoma large T causes immortalization, middle T confers the property of tumorigenesis. The combination of the two main functions causes the development of continuously growing tumors. Parallel work with other oncogenes attributed similar effects to them (8). Reflecting the two different roles of the two polyoma proteins was their differing cellular location. Large T is in the nucleus, where it binds to sites in both viral and cellular DNA; middle T is in the cell plasma membrane. This protein binds pp60 c-src, the protein specified by a cellular oncogene, and increases it protein kinase activity (9).

These and many other observations conveyed the following message. Certain oncogenes, some of which are normal components of the cellular genome, either alone or in combination start a process that leads to cell transformation and cancer. They perform this function when they are activated.

Activation may be brought about by their incorporation into a viral genome or, without mobilization, by their alteration, translocation, amplification, or by excessive transcription. In all cases activation probably means unregulated function of the product of the oncogene, through different molecular mechanisms.

These concepts have a powerful unifying value in carcinogenesis. Cancer induction by chemicals can be attributed to an alteration of proto-oncogenes (10). The role of specific chromosome translocations in some cancers is explained by the alteration of proto-oncogenes located near the breakpoints (11). The role of genetic diseases causing DNA damage (12) in eliciting cancer can be attributed again to activation of some proto-oncogenes by the DNA damages. The concepts also unify cancer and development. Normal proto-oncogenes are closely regulated, and tend to be expressed either at special periods of development (13) or transiently when growth is stimulated (14,15). The regulation of oncogenes explains another aspect of carcinogenesis: the hereditary predisposition to certain cancers (16) results from the loss or inactivation of the regulating genes (antioncogenes) (17).

In this picture, oncogenes causing cell immortalization affect growth control, probably by acting on DNA; those causing tumorigenesis affect the peripheral pathways of the control of the cellular genome, causing important changes in gene expression.

PROGRESSION

All that has been discussed so far has to do with the initial events of cell transformation and cancer. But a large number of observations show that additional events play an important role. Cancers slowly evolve towards malignancy through many definable stages, in a process called "progression" (18). The result is a marked heterogeneity of different parts of tumors (19,20) and chromosomal irregularities such as aberrations and variations in number (21). It is important to differentiate between the action of the oncogenes that initiate the neoplastic process and the subsequent events of progression (22).

Progression is still the least understood and most crucial phase in cancer development. The study of viruses and oncogenes begins to throw a little light on progression. Some proto-oncogenes are amplified in advanced cancers (23) but we don't know whether this is cause or consequence of progression. The main finding is that the cells transformed by these agents also undergo progression. For instance, bursal lymphomas induced in chickens by avian leukosis viruses are initiated through the activation of the normal myc oncogene (24). Transformed follicles are formed, some of which evolve into lymphomas (25). In this process deletions appear in the viral genome, and mutations in the myc oncogene (26). Clear evidence of progression is also found in T-cell mouse lymphomas (27) and in leukemogenesis induced by the Friend leukemia virus in cultures of mouse bone marrow cells (28). Examples of progression are the long latency between injection of SV40 into mice and the appearance of tumors, during which many additional events probably occur (29,30) and the discrete steps through which polyoma-infected hamster embryo cell cultures undergo full transformation (31). Striking evidence of progression was observed in a transgenic mouse containing myc and SV40 sequences (32) which developed an intracranial tumor. Fibroblastic cells from a variety of organs of this mouse expressed SV40 T, but were at first normal. Upon continued cultivation they became gradually transformed and capable of forming tumors in nude mice. Evidently the transformation was not caused directly by the presence of the SV40 genome, but by cellular events occurring during the long term growth of the cultures, probably under the influence of the viral genome itself. A general finding is that permanent cell lines more easily reach full transformation than primary cell cultures. This implies that one of the changes needed for progression is already present in the cell lines.

ROLE OF CHANGES IN CELLULAR GENES

All these observations and many others show that the activation of an oncogene is only the beginning of the process leading to cancer or cell transformation. This information does not explain progressions which must involve cellular changes in other cellular genes. We shall concentrate on these changes because they represent the new direction of research to which the previous results point.

The role of cellular genes in cancer is all pervading, including the initiation step. For instance, whether a certain oncogene causes simple immortalization or full transformation depends not only on the nature and state of the oncogene but also on the cells, whether they are primary cultures or permanent lines, and the animal species from which they derived (33,34). Moreover the oncogene products must act on cellular functions which depend on cellular genes; and the transcription-regulating oncogenes presumably control the expression of cellular genes. These observations point to a limitation of the earlier approach to cancer. They show that transformation or cell immortalization are not defined events, because they depend on characteristics of the cells. The differences among cells of different species show that each species is a unique arrangement of genes and control mechanisms, either in a qualitative or in a quantitative way. In different species the same regulatory problem may be solved in different ways, so that a change in the same parameter can have quite different effects. The observations also show that cellular genes must inevitably participate in the transformation process.

Concerning the nature of progressions, an illuminating finding is the heterogeneity of advanced cancers (19,20). Observations with rat mammary carcinomas show that the expression of many genes is altered in different ways in different parts of the same cancer; the changes have a clonal character, being uniform in small parts of the cancer but different in adjacent parts (unpublished observations). The closeness and smallness of the parts makes it unlikely that their differences are due to environmental factors; they are probably generated by changes of the genes.

The chromosomal rearrangements observed in advanced cancers suggests that the gene changes responsible for the heterogeneity are of a structural nature. This possibility is further supported by the finding that each chemically or UV-induced mouse sarcoma expresses a different class I major histocompatibility antigen. It is likely that this heterogeneity is caused by gene rearrangements that are different in every cancer (35). Changes of the cellular genome may therefore have a determining role in the development of malignancy (36). Their role cannot be overlooked any longer: they may hold the key for understanding what an advanced cancer is and how it comes about.

CONNECTION BETWEEN ONCOGENE AND PROGRESSION: NONHOMOLOGOUS RECOMBINATION?

The major gap in our understanding of cancer is how the activation of an oncogene is related to the subsequent profound alterations of the cellular genome which take place during both initiation and progressions. Concerning progression, certain observations suggest that the link may be nonhomologous recombinations. In transformation by SV40 this type of recombination is responsible for integration of the viral genome and for its subsequent extensive rearrangements, including those of its flanking cellular sequences (37). Also extensive abnormalities sometimes found in the chromosomes of SV40-transformed cell (38) in brain tumors arising in SV40-transgenic mice (39), and in hamster cells transformed by polyoma virus (31), may have the same origin. Similar events go on in cells transformed by adenoviruses (40), in cells of human hepatomas induced by hepatitis B virus (41,42), and in virus-induced woodchuck hepatomas (43).

All these events suggests that an abnormally high rate of nonhomologous recombination takes place in the transformed cells causing either activation or inactivation of many cellular genes. Concomitant methylation changes can further perturb the function of the genome (44,45,46). All these events are likely to contribute in an important way to cancer progression.

It is not surprising that retroviruses promote nonhomologous recombination because they have some similarity to bacterial transposons as well as to the related _copia_-like elements of Drosophila and the Ty elements of yeast. Retroviruses share with them the general organization (47), as well as the property of reverse transcription (48). Oncogenic DNA viruses may also promote nonhomologous recombination: for instance, integrated hepatitus B virus genomes, which also replicate via reverse transcription (49), are often flanked by short repeats of cellular DNA (50,51).

One essential characteristic of transposons is to have a gene for a transposase, which carries out the nonhomologous recombination. Retroviruses specify an enzyme needed for integration of the viral DNAs (52,53). The specificity of the enzyme is not unlike that of a transposase, because integration is site-specific for the viral genome, but essentially nonspecific for the cellular genome. The retroviral enzyme has significant homology with the _copia_ enzyme (54). A similar enzyme is probably responsible for the frequent intracellular transpositions of intracisternal A particles (55), and may contribute to the tumorigenic effect of retroviruses (56).

Nonhomologous recombination takes place in animal cells independently of viral infection, for instance in lymphocytes (57). Moreover, nonhomologous recombination occurs at a high frequency among DNA molecules transfected into cells (58,59,60) and among constructs of the SV40 genome incapable of producing virus except after undergoing rearrangement (61). It may be possibly responsible for the rearrangement of telomeres during chromosome duplication (62). We don't know the enzymes involved in any of these events, but a gene related to the retroviral reverse transcriptase is present in avian cells that lack viral genomes (63). Whether this gene specifies a recombinase we don't know.

Is it possible that oncogene proteins causing tumorigenicity also carry out transposition? It is conceivable that oncogenes are related to transposition genes. There is a suspicious relatedness between gene regulation and transposition. Transposing elements were discovered as regulatory elements by Barbara MacClintoch. Transposon insertion is the simplest and perhaps most primitive mechanism of gene regulation. It is displayed by retroviruses when their genomes activate cellular oncogenes by promoter insertion (24). Transpositional inversions are important for regulating gene expression in many organisms including eukaryotes (64). We don't know, however, whether transforming oncogenes specify recombinases in addition to the known regulatory proteins. It might be instructive to look at the base sequences of oncogenes and transposons with this possibility in mind.

Oncogenes may also act indirectly, by activating transposase genes present in the cellular genome. Such activation might be part of the mechanism of control carried out by the oncogenes.

THE MECHANISM OF VIRAL CANCERS: A REVISION

The discoveries about progression force us back to one of the points that seemed to have been solved by past research: whether transformation and cancer can be the result of a hit-and-run action. This is still excluded for the initiation of cancer. But if an oncogene causes malignancy by generating extensive damage of the genome, its action on malignancy is of the hit-and-run type, even if the oncogene persists in the cancer cells. And in fact there are cases in which the neoplastic state of the cells does not change after the oncogene responsible for causing transformation has become completely inactive (65).

Much more than is now known can be found out about these problems. The prediction that the DNA of an advanced cancer is as heterogenous as the phenotype of its cells is amenable to experimental verification. and if the facts support the hypothesis, a new world of cancer research opens up. A new class of oncogenes may be discovered, whose inactivation or activation determine the most outstanding features of malignancy: infiltration and metastasis. It would not be surprising if, in spite of the enormous heterogeneity of cancer cells, these crucial genes were restricted to a small number: much of the heterogeneity of cancers may well be irrelevant to malignancy. The situation would be comparable to that of initiating oncogenes, which are in a small number.

A TURNING POINT: AN ATTACK ON THE CELLULAR GENOME

It seems that we are at a turning point in the study of tumor virology and oncogenes. We have to face the fact that if we wish to learn more about cancer we must now concentrate on the cellular genome. We are back to where we started, but the situation is drastically different because we have lots of knowledge and crucial tools we did not have before, especially DNA cloning.

Our choice on how to approach this problem is determined by an important fact that results from the analysis I made of the actions of oncogenes: that their effects are determined by many circumstances, one of which is the species of the cells on which an oncogene acts. If the primary objective of our endeavor is to understand human cancer, we must study it in human cells. This conclusion is also borne out by the vast experience showing that a carcinogen will act in different ways in different species, and that cancers of the same organ arising in different species are vastly different. Undoubtedly cancers arising in humans are unique in all their characteristics. We don't even know whether oncogenes that are important in animal cancers are important in human cancers, not only because a small proportion yields transforming oncogenes, but also because no normal human cell has been ever transformed in vitro by an oncogene. If we accept that human cancer must be studied, we are immediately faced with the difficulties that have made its study unpopular. Human cancer is not a simple system: the cancers are highly heterogeneous; cultures isolated from a human cancer reflect the properties of only one of the many cellular types present, and moreover these properties are distorted by in vitro cultivation. The approach followed in experimental cancer where the experimenter works with homogeneous cell populations, and where he can manipulate the animal, is not applicable. In the study of human cancer we should aim at the status of the genes characteristics of each individual cell although the genes cannot be directly manipulated. Such a study can be carried out only by knowing first what the genes are then using methods that allow the determination of the status of each gene in any cell. To do so the first thing we need is a complete knowledge of the human genome: we must begin by sequencing the human genome. Having the sequence we can make probes for each gene; we can then classify genes for their expression in each cell type using cytological hybridization. These tools will make it possible to ascertain what changes the genome undergoes during progression. The task of identifying genes that might be involved in progression will be greatly simplified.

Cancer research in humans will receive a major boost by the detailed knowledge of DNA. Because it will be possible to define cancer in molecular terms, the agents capable of inducing it will be identified with much greater certainty by the combination of in vitro and epidemiological studies. Progression will be described in precise molecular terms at the level of individual cells; the genes involved will be directly identified. man will become the preferred experimental species for cancer research using a combination of approaches. The knowledge gained will open new therapeutic approaches, possibly leading to the long sought general cure for cancer.

Knowledge of the genome and availability of probes for any gene would be also crucial to progress in human physiology and pathology outside cancer: for instance in learning about the regulation of individual genes; for the identification and diagnosis of hereditary diseases or of hereditary predisposition to disease; for understanding development; for classifying and defining the functional characteristics of the many classes of cells of the nervous system. The knowledge would rapidly reflect on practical therapeutic applications in many fields.

An effort of this kind could not be undertaken by any single group: it should be an international effort because the sequence of the human DNA is the reality of our species, and everything that happens in the world depends on those sequences.

Many practical and technical problems would have to be solved. A considerable improvement of the technology would be needed in order to shorten the time required. Increasing by 50-fold the present rate of sequencing would make it possible to complete the main task in perhaps five years with adequate manpower. I don't intend to go into these technical problems, but it is clear that by combining a sound theoretical approach, the use of restriction endonucleases, such as eight-cutters, that make large fragments, perhaps the development of micromethods the development of for sequencing an individual chromosome or the genome of an individual sperm, automation of technical procedures, and computerized analysis of the data as they are generated, the technology can be considerably improved in a short time. Because the first goal is to identify the genes, the problem of polymorphism will be left for subsequent studies.

In one generation we have come a long way in our efforts to understand cancer. The next generation can look forward to exciting new tasks, which may lead to a completion of our knowledge about cancer, closing one of the most challenging chapters in biological research.

REFERENCES

1. Sambrook J, Westphal H, Srinivasan P, Dulbecco R (1968). Proc Natl Acad Sci USA 60:1288.
2. Varmus H, Swanstrom R (1982). In Weiss R, Teich N, Varmus H, Coffin J (eds): "RNA Tumor Viruses," Cold Spring Harbor, NY: Cold Spring Harbor Laboratory, p 369.
3. Robinson JE, Miller G (1982). In Roisman B (ed.): "The Herpesviruses, Vol 1," New York: Plenum Publishing Co, p 151.
4. Pfister H (1984). Rev Physiol Biochem Pharm 99:111.
5. Bishop JM (1983). Ann Rev Biochem 52:301.
6. Bishop JM, Varmus HE (1982). In Weiss R, Teich N, Varmus H, Coffin J (eds): "Molecular Biology of Tumor Viruses," Cold Spring Harbor, New York: Spring Harbor Laboratory, p 999.
7. Rassoulzadegan M, Cowie A, Carr A, Glaichenhous N, Kamen R, Cuzin F (1982). Nature (London) 300:713.
8. Land H, Parada LF, Weinberg RA (1983). Nature (London) 304:596.

9. Courtneidge SA, Smith AE (1984). EMBO J 3:585.
10. Zarbl H, Sukumar S, Arthur AV, Martin-Zanka D, Barbacid M (1985). Nature (London) 315:382.
11. Klein G, Klein E (1985). Nature (London) 315:190.
12. Cairns J (1981). Nature (London) 289:353.
13. Muller R, Verma IM (1984). Curr Topics Microbiol Immunol 112:73.
14. Kelly K, Cochran BH, Stiles CD, Leder P (1983). Cell 35:603.
15. Muller R, Bravo R, Burckardt J, Curran T (1984). Nature (London) 312:716.
16. Knudson AG (1985). Cancer Res 45:1437.
17. Murphree AL, Benedict WF (1984). Science 223:1028.
18. Foulds L (1958). J Chron Dis 8:2.
19. Heppner GH (1984). Cancer Res 44:2259.
20. Rubin H (1985). Cancer Res 45:2935.
21. Wolman SR (1983). Cancer Metastasis Reviews 2:257.
22. Duesberg PH (1985). Science 228:669.
23. Yokota J, Tsunetsugu-Yokota Y, Battifora H, LeFevre C, Cline MJ (1986). Science 231:261.
24. Hayward WS, Neel BG, Astrin SM (1980). Nature (London) 290:475.
25. Baba TW, Humphries EH (1985). Proc Natl Acad Sci USA 82:213.
26. Westaway D, Payne G, Varmus HE (1984). Proc Natl Acad Sci USA 81:843.
27. Haas M, Altman A, Rothenberg E, Bogart MH, Jones OW (1984). Proc Natl Acad Sci USA 81:1742.
28. Heard JM, Fichelson S, Sola B, Martial MA, Varet B, Levy JP (1984). Mol Cell Biol 4:216.
29. Hargis BJ, Malkiel S (1981). J Natl Cancer Inst 63:861.
30. Abramczuk J (1983). In Nagley P, Linnane AW, Peacock WJ, Pateman JA (eds.): "Manipulation and Expression of Genes in Eukaryotes," New York: Academic Press, p 355.
31. Vogt M, Dulbecco R (1962). Cold Spring Harbor Symp Quant Biol 27:367.
32. Small JA, Blair DG, Showalter SD, Scangos GA (1985). Mol Cell Biol 5:642.
33. Spandidos DA, Wilkie NM (1984). Nature (London) 310:465.
34. Bouchard L, Gelinas C, Asselin C, Bastin M (1984). Virology 135:53.

35. Philipps C, McMillan M, Flood PM, Murphy DB, Forman J, Lancki D, Womack JE, Goodenow RS, Schreiber H (1985). Proc Natl Acad Sci USA 82:5140.
36. Nowell PC (1976). Science 194:23.
37. Botchan M, Stringer J, Mitchison T, Sambrook J (1980). Cell 20:143.
38. Norkin LC, Steinberg VI, Kosz-Vnenchak M (1985). J Virol 53:658.
39. Brinster RL, Chen HY, Messing A, van Dyke T, Levine AJ, Palmiter RD (1984). Cell 37:367.
40. Sambrook J, Green R, Stringer J, Mitchison T, Hu SL, Botchan M (1980). Cold Spring Harbor Symp Quant Biol 44:569.
41. Koch S, Freytag A, Hofschneider PH, Koshi R (1984). EMBO J 3:2185.
42. Mizusawa H, Taira M, Yaginuma K, Kobayashi M, Yoshida E, Koike K (1985). Proc Natl Acad Sci USA 82:208.
43. Ogston CW, Jonak GJ, Rogler CE, Astrin SM, Summers J (1982). Cell 29:385.
44. Olsson L, Forchhammer J (1984). Proc Natl Acad Sci USA 81:3389.
45. Ramsden M, Cole G, Smith J, Balmain A (1985). EMBO J 4:1449.
46. Hsiao WLW, Gattoni-Celli S, Weinstein IB (1985). Mol Cell Biol 5:1800.
47. Temin H (1980). Cell 21:599.
48. Baltimore D (1985). Cell 40:481.
49. Summers J, Mason WS (1982). Cell 29:403.
50. Yaginuma K, Kobayashi M, Yoshida E, Koike K (1985). Proc Natl Acad Sci USA 82:4458.
51. Dejan A, Sonigo P, Wain-Hobson S, Tiallais P (1984). Proc Natl Acad Sci USA 81:5350.
52. Donehower LA, Varmus HE (1984). Proc Natl Acad Sci USA 81:6461.
53. Panganiban AT, Temin HM (1984). Proc Natl Acad Sci USA 81:7885.
54. Mount S, Rubin GM (1985). Mol Cell Biol 5:1630.
55. Hawley RG, Shulman MJ, Hozumi N (1984). Mol Cel Biol 4:2565.
56. Oliff A, McKinney MD, Agranovsky D (1985). J Virol 54:864.
57. Tonegawa S (1983). Nature (London) 302:575.
58. Winocour E, Keshet I (1980). Proc Natl Acad Sci USA 77:4861.
59. Subramani S, Berg P (1983). Mol Cell Biol 3:1041.

60. Bandyopadhyay PK, Watanabe S, Temin H (1984). Proc Natl Acad Sci USA 81:3476.
61. Roth DB, Wilson JH (1985). Proc Natl Acad Sci USA 82:3355.
62. Horowith H, Thorburn P, Haber JE (1984). Mol Cell Biol 4:2509.
63. Dunwiddie C, Faras A (1985). Proc Natl Acad Sci USA 82:5097.
64. Plasterk RHA, van de Putte H (1984). BBA 782:111.
65. Kettmann R, Cleuter Y, Gregoire D, Burny A (1985). J Virol 54:899.

Viruses and Human Cancer, pages 15–28

IMMUNOLOGICAL CHARACTERIZATION OF HTLV-III RECOMBINANT PROTEINS: POTENTIAL AS DIAGNOSTICS AND VACCINE CANDIDATES

S.R. Petteway Jr.,[1] D. Reed,[1] K. Reagen,[2] T. Matthews[3]
R. Tritch,[1] L. Ivanoff,[1] D. Tribe,[1] M. Chamberlain,[1]
R. Cybulski,[2] D. Bolognesi,[3] and W. Kenealy[1]

[1]Central Research & Development Department and
[2]Biomedical Products Department,
E. I. Du Pont de Nemours & Co., Inc.
Experimental Station, Wilmington, DE 19898 USA
[3]Department of Surgery, Duke University Medical Center
Durham, NC 27710

ABSTRACT Human T-Cell Lymphotropic Virus (HTLV-III) has been implicated as the causative agent of Acquired Immune Deficiency Syndrome (AIDS). Antibodies to HTLV-III have been detected in patients with AIDS and AIDS related diseases. The availability of individual HTLV-III antigens for seroepidemiology studies, the analysis of viral structure, and the development of diagnostics and vaccines has been limited. In an effort to provide sufficient quantities of viral proteins for these studies we have engineered open reading frame fragments of viral DNA into *E. coli* expression vectors. A variety of recombinant proteins representing different regions of the *gag* and *env* open reading frames have been produced and shown to react specifically with AIDS sera. The recombinant proteins were further characterized by their reactivity with monoclonal and polyclonal antibodies to purified viral proteins and their ability to induce virus-specific antibodies. This approach has been used to define more specifically the antigenic regions of the virus. In an effort to analyze the potential of recombinant proteins as diagnostic tools, we have investigated the reactivity of different AIDS sera with individual recombinant antigens. These studies have implications for the development of specific diagnostics and vaccines.

INTRODUCTION

HTLV-III/LAV is a member of the human T-lymphotropic retrovirus family and is implicated as the cause of acquired immune deficiency syndrome (AIDS) (1). Individuals exposed to HTLV-III respond with antibodies specific for viral proteins (2,3,4), some of which neutralize HTLV-III in tissue culture (5,6). Identification of antigenic determinants on viral proteins that are immunodominant, neutralizing, or protective should lead to a better understanding of the role of HTLV-III in the etiology of AIDS. Among the antibodies in AIDS patients, some have been shown to react with the major core protein p24 (7,8) and the envelope glycoproteins gp160, gp120, and gp41 of the virus (9,10). The DNA sequence of several HTLV-III/LAV clones has been determined and the predicted open reading frames corresponding to the *gag* and *env* genes identified (11,12). Evidence has recently been published demonstrating that the *gag* open reading frame encodes the viral p24 and precursor p55 and p70 proteins (10) and the *env* open reading frame specifies the gp160 and gp120 proteins (13).

The availability of individual HTLV-III antigens has been limited. In an effort to provide sufficient quantities of viral proteins for these studies we have engineered viral DNA fragments into *Escherichia coli* expression vectors and have produced recombinant proteins representing different regions of the *gag* and *env* open reading frames that react specifically with AIDS sera. We have further characterized the recombinant proteins by their reactivity with monoclonal and polyclonal antibodies to viral proteins and their ability to induce virus-specific antibodies. This approach has been used to define more specifically the antigenic regions of the virus. The potential of recombinant proteins as diagnostic tools was analyzed by investigating the reactivity of different AIDS sera with individual recombinant antigens. These studies have implications for the development of diagnostics and vaccines.

RESULTS

Expression of HTLV-III Proteins in *Escherichia coli*.

Specific DNA fragments from the *gag* and *env* open reading frames of HTLV-III lambda clone BH-10 (11) were engineered into *E. coli* expression vectors (pENV 7, pENV 9, and pGAG 1) (Fig. 1). The vectors direct the synthesis of recombinant proteins under control of the E. coli tryptophan promotor (trp).

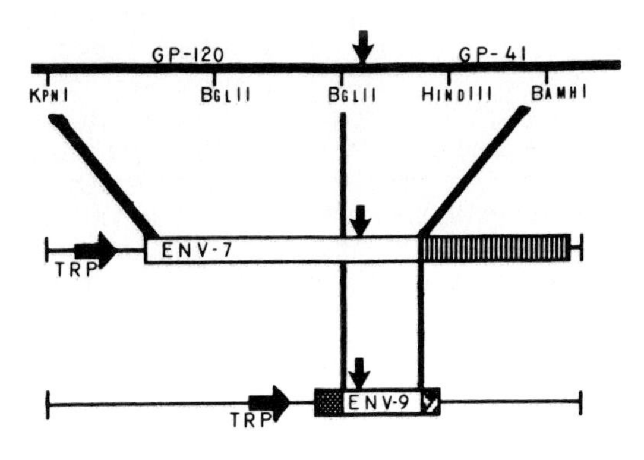

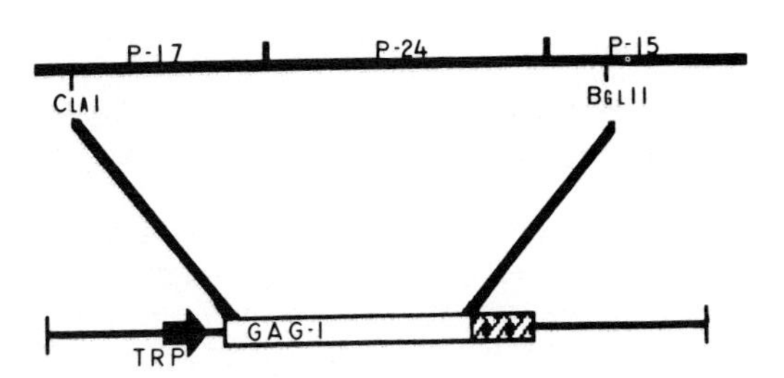

Fig. 1. Construction of vectors for the synthesis of env and gag proteins in E. coli. Horizontal arrows represent the trp promotor; vertical arrows represent the predicted env cleavage site. The β-galactosidase and polioviral protease coding regions are denoted by the lined and stippled boxes respectively. The boxes at the carboxytermini of pENV 9 and pGAG 1 represent the additional out of frame vector sequences.

pENV 7. A DNA fragment between the KpnI and the BamHI sites of the env open reading frame was fused to the lac ΔIZ gene (β-galactosidase) (14) of E. coli. An initiating methionine was provided by inserting an oligonucleotide adapter containing an ATG in frame with the env open reading frame, between the trp promoter and the KpnI site. Western blot analysis (15) of bacterial proteins separated by polyacrylamide gel electrophoresis (16) showed that proteins produced by clone pENV 7 reacted specifically with AIDS sera (Fig. 2A). A reactive band of 190kd corresponds to the expected size of the full length env/lac ΔIZ fusion. Lower molecular weight bands were visible, probably arising from either internal initiations within the HTLV-III sequences or protein degradation. In the recombinant gene the HTLV-III DNA sequences are fused with those of the lac ΔIZ gene, the resultant protein should be reactive with β-galactosidase antiserum. When pENV 7 extracts were analyzed by western blots with β-galactosidase antiserum, reactive protein bands correspond to those seen with AIDS sera (data not shown). These data support the conclusion that clone pENV 7 produced an HTLV-III/β-galactosidase fusion protein.

pENV 9. Vector pExcalibur was constructed to produce high levels of polioviral protease (17). We took advantage of a unique BglII site within the protease coding region to insert the 854bp BglII/BamHI fragment of env, fusing the amino terminus of protease in frame with the env fragment. The construct encodes an aminoterminal fusion of 55 amino acids of protease to 284 amino acids of HTLV-III env, and 33 amino acids downstream from the pExcalibur BamH1 site. Western blot analysis demonstrated that AIDS sera reacted with a 42kd band, which is the expected size for the full length recombinant protein (Fig. 2A). Note that again faster migrating bands appear.

pGAG 1. The ClaI/BglII DNA fragment from the gag open reading frame of BH10 was isolated and modified at the ClaI site with a oligonucleotide adapter placing an initiation ATG in frame with the gag reading frame. This modified DNA was then inserted into an expression vector. Western blot analysis showed a band at approximately 45kd, the expected molecular weight for the full length recombinant gag protein, reacting with AIDS sera. There are also faster migrating bands visible.

Reactivity of Virus Specific Antibodies with Recombinant Proteins.

Both polyclonal and monoclonal antibodies were raised against viral proteins and used in western blot or immuno dot blot analysis to characterize the E. coli derived recombinant proteins. The results of this analysis are presented in Table 1. Rabbit antisera specific for viral p24 was prepared using purified p24 protein. Monoclonal antibodies were generated from Balb/c mice immunized with sucrose gradient purified, detergent disrupted HTLV-III antigen (Biomedical Products Department, E. I. Du Pont & Co.).

TABLE 1
REACTIVITY OF HTLV-III VIRUS-SPECIFIC ANTIBODIES WITH RECOMBINANT PROTEINS[a]

Antibodies	Viral Protein		Recombinant Protein		
	p24	p55	pGAG 1	pENV 7	pENV 9
Monoclonal p24					
BT3	+	+	+	-	-
23D/35.1	+	+	+	ND	ND
23A/2.1	+	+	+	ND	ND
23D/53.1	+	+	+	ND	ND
23D/42.2	+	+	+	ND	ND
23D/15.1	+	+	+	ND	ND
Polyclonal p24	+	+	+	-	-
Monoclonal p41	-	-	ND	+	+

[a]Antibody reactivity to the proteins was determined by western blot analysis for the viral proteins and immuno dot blot and/or western blot for recombinant proteins. The visualization was by second antibody-horse radish peroxidase conjugate systems. ND= Not Determined.

Cells secreting antibodies specific for HTLV-III were selected by reactivity on DuPont HTLV-III Enzyme Linked Immunoabsorbent Assay (ELISA) plates (Biomedical Products Department, E. I. Du Pont Co.), then characterized as p24 reactive by western blot analysis on purified virus. The reactivity of these monoclonal antibodies in ELISA competition studies suggested the existence of three major antigenic sites of viral p24 (designated sites I, II, and III), with III being topographically distinct from I and II (A. Pieper and K. Reagan, manuscript in preparation). Monoclonal antibody BT3 was obtained from Biotech Research Labs, Bethesda, MD (Fig. 2D, lane 2). All these monoclonals

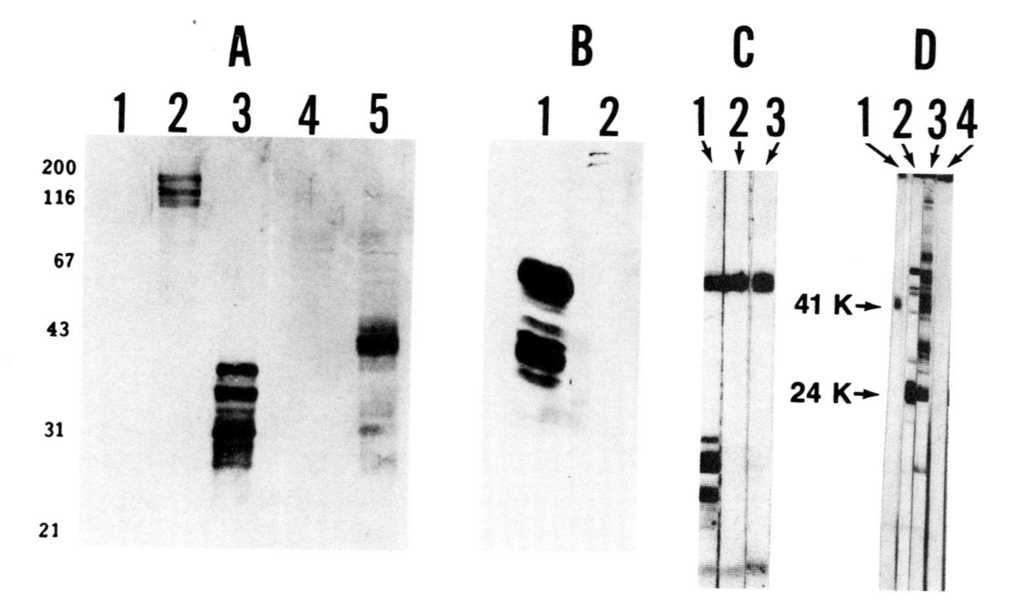

Fig. 2. Gel electrophoresis and western blot analysis of recombinant and HTLV-III viral proteins. Reactive proteins were visualized with second antibody horseradish peroxidase conjugate. A) Extracts of E. coli containing env and gag expression vectors reacted with AIDS serum. 1 & 4 - E. coli controls containing expression vectors lacking HTLV-III specific sequences; 2-pENV 7; 3-pENV 9; 5-pGAG 1. B) Extract of E. coli reacted with monoclonal antibody (BT3) to viral p24 (obtained from Biotech Research Labs); 1-pGAG 1; 2-E. coli control containing vector without HTLV-III sequences. C) Extracts of pENV 9 reacted with monoclonal antibody. 1-gp41 monoclonal antibody to viral gp41 protein obtained from the laboratory of R.C. Gallo. 2-(BT3) p24 specific monoclonal antibody - same as B. 3-E. coli control containing vector without HTLV-III sequences reacted with gp41 monoclonal antibody. D) HTLV-III viral proteins reacted with virus specific antibodies. 1-gp41 monoclonal as in C)1; 2-p24 monoclonal antibody as in B) 1; 3-AIDS sera 4-nonreactive control.

reacted with the p24 core protein, as well as its predicted 55kd precursor (Table 1). The antisera and monoclonal antibodies also react specifically with the recombinant proteins produced by clone pGAG 1 (Fig. 2B, Table 1). HTLV-III gp41 protein has been identified as a major viral antigenic determinant (2). To map the location of the gene encoding gp41 in the HTLV-III genome, we screened our recombinant antigens with a monoclonal antibody specific for gp41 (provided by R.C. Gallo). As shown in Fig. 2D, lane 1, this antibody reacts specifically with the 41kd protein found in purified virus. Significantly this monoclonal antibody also reacted strongly with pENV 9 protein and not with the proteins of E. coli controls (Fig. 2C). Because the monoclonal antibody reacted with denatured protein in the western blot assay, it is likely the epitope recognized is linear. Clone pENV 9 contains DNA sequences mapping between the second BglII and the BamHI restriction sites within the env open reading frame. Therefore, the coding sequence of gp41 maps at least in part between 7199bp to 8053bp in the env open reading frame.

Analysis of Antibodies to Recombinant Proteins.

We took advantage of the insoluble nature of the pENV 7 protein when expressed in E. coli to prepare partially purified protein for preparation of antibody to env derived determinants. Cultures of E. coli containing the pENV 7 protein were disrupted by use of a French press and the insoluble material collected by centrifugation. This partially purified fraction was used to prepare rabbit antiserum. The pENV 7 antisera was tested with pENV 9 protein in western blots and ELISA. Since the pENV 7 and pENV 9 proteins do not share protease or β-galactosidase determinants, a positive reaction would indicate that common env specific antigens were present in both proteins. As presented in Table 2, anti pENV 7 serum reacted with pENV 9 protein.

If the recombinant proteins are to be of value as reagents to produce virus specific antibodies, or as vaccine candidates they must induce antibodies reactive with virus. To confirm this we reacted anti-pENV 7 serum with viral proteins on western blots. As shown in Table 2, antiserum to pENV 7 reacted with viral proteins. Moreover, the reactivity was specific for envelope proteins with no identifiable reactivity to previously identified gag components. Interestingly, the antiserum to the recombinant protein reacts with the predicted gp120 viral protein indicating that exposed protein determinants reside in this highly glycosylated protein. We also observe env specific reactivity with a protein present in purified virus migrating at approximately 90kd. Since the anti-pENV 7 serum is specific for env protein, this would indicate that the 90kd band contains amino acid sequences in common with HTLV-III env proteins.

TABLE 2
REACTIVITY OF ANTI-pENV 7 SERUM
WITH RECOMBINANT AND HTLV-III VIRAL PROTEINS[a]

Protein	Reactivity of Rabbit anti-pENV 7 Serum	
	Western Blot	ELISA
Recombinant		
pENV 9	+	+
pGAG 1	-	ND
Viral		
160kd	+	ND
120kd	+	ND
41kd	+	ND
90kd	+	ND
55kd	-	ND
24kd	-	ND
Whole Virus		+

[a]ND= Not Determined

Fractionation of AIDS Sera With Recombinant Antigens.

In an effort to identify antibody populations in AIDS sera reactive to different envelope protein regions and further characterize recombinant antigens, AIDS serum was fractionated by immunoadsorption chromatography. Antigens pENV 7 and p121 (a recombinant protein mapping to the env open reading frame between the proposed protein cleavage site and the HindIII restriction site, provided by Centocor Inc., Malvern, Pa. (4)) were individually coupled to cyanogen activated Sepharose 4B as described elsewhere (18).

One ml of serum was applied to the column and the 280nm absorbing material that passed through the column was collected as effluent. The resin was washed with PBS and the specifically bound antibody was eluted with 3M guanidine hydrochloride in PBS (pH 7.4). The effluent was again passed over the column to remove residual antibody and the column washed and eluted as before. The final effluent and the combined eluants were adjusted to a 1:5 dilution relative to untreated serum.

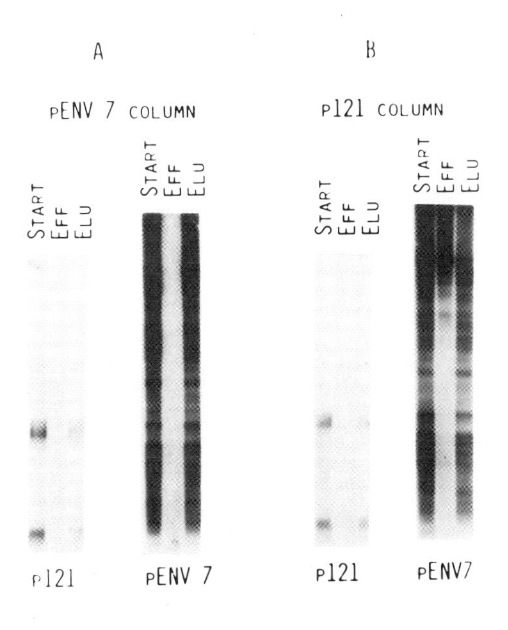

Fig. 3. Western blot analysis of p121 and pENV 7 proteins separated by gel electrophoresis then reacted with AIDS serum fractionated on recombinant protein columns. A) AIDS serum fractionated on pENV 7column. B) AIDS serum fractionated on p121 column. Start - unfractionated AIDS serum, Eff - Effluent, Elu - Eluate.

Recombinant proteins were separated by 12% polyacrylamide gel electrophoresis and electrophoretically transferred to nitrocellulose. Strips were cut from the nitrocellulose and incubated with serum or serum fractions at a dilution equivalent to 1/100 of untreated serum. The reactivities of serum fractionated on the pENV 7 column are shown in Fig. 3A. The unfractionated serum reacted with both p121 and pENV 7 proteins. The effluent reacted with neither p121 nor pENV 7 and the eluant reacts with both proteins. This would indicate that pENV 7 contains most or all of the antigenic determinants found in p121. Figure 3B shows the reactivities of serum fractionated on the p121 column. The unfractionated serum reacted with both p121 and pENV 7 proteins as did the eluate. However, the effluent from the p121 column did not react with the p121 protein but did react with the pENV 7 protein. These data demonstrate that the AIDS serum contained antibodies to determinants found in pENV 7 but not in p121. When the p121 effluent was reacted with viral proteins in a western blot, no reactivity to gp41 protein was seen (data not shown). This would indicate that the p121 effluent was devoid of gp41 reactivity. Since the p121 effluent did react with the pENV 7 proteins, apparently the AIDS serum used in these experiments contained an antibody fraction reactive with the protein portion of the viral gp120. Because the gp120 protein is heavily glycosylated it is of interest that AIDS serum contains antibodies to the protein portion.

Thus by using recombinant antigens representing different domains of viral proteins as affinity reagents, we have identified antibody populations in AIDS serum reactive with independent domains of the viral proteins.

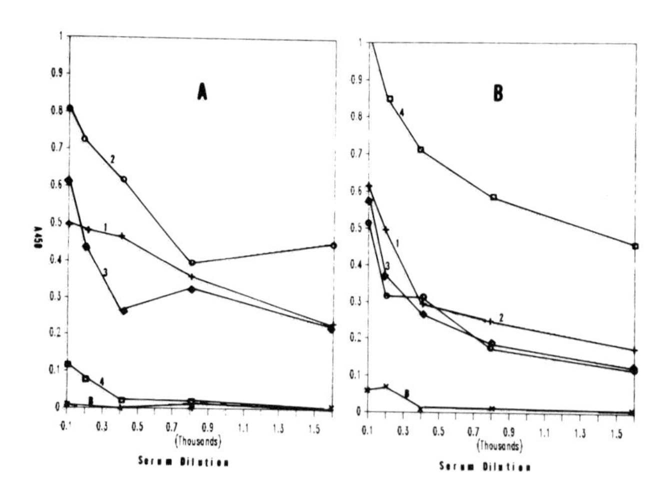

Fig. 4. ELISA of 4 separate AIDS sera reacted with pGAG 1 or pENV 9 protein, as described in text. (1)+, (2)○, (3)■, (4)□ - 4 AIDS sera; **x**-average of normal human sera. **A**-pGAG 1; **B**-pENV 9.

Recombinant *gag* and *env* Proteins as Diagnostic Tools.

Recombinant proteins were purified from *E. coli* cultures containing vectors pGAG 1 and pENV 9. The proteins were suspended in 60mM carbonate buffer pH9.5. Immulon II (Dynatech) ELISA plates were incubated overnight at 4°C with pGAG 1 or pENV 9 at concentrations of 100 ng/well and 20 ng/well respectively. The plates were washed, blocked and then reacted with 4 AIDS patients sera and 4 normal human serum controls (Fig. 4). Serum 4 reacted only weakly with the *gag* protein, as opposed to sera 1, 2, and 3, which reacted strongly with both the *env* and *gag* specific proteins. These ELISA data directly parallel western blot analysis of the four patients sera against gel separated viral proteins (data not shown). These results demonstrate the potential of recombinant proteins as diagnostic tools.

DISCUSSION

We have engineered HTLV-III *gag* and *env* open reading frame fragments into expression vectors and have demonstrated that *E. coli* harbouring these vectors produce recombinant proteins sharing immunological characteristics with viral proteins.

The recombinant antigens described here (pGAG 1, pENV 7, pENV 9) reacted specifically with AIDS sera but not with sera from normal controls. Monoclonal and polyclonal anti-p24 antibodies raised to either purified p24 protein or whole virus, reacted with the pGAG 1 recombinant protein. Monoclonal antibody to viral gp41 reacted with the pENV 9 protein which established the location of at least part of the coding sequence for gp41 within the *env* open reading frame.

By western blot analysis, rabbit anti-pENV 7 serum reacted with all known mature and precursor forms of HTLV-III envelope protein. It also reacted with pENV 9 protein as expected since the clones contain overlapping *env* sequences. Because proteins synthesized in *E. coli* are *not* glycosylated, the antisera was expected to be directed toward protein determinants.

We have also demonstrated that AIDS serum could be fractionated using recombinant antigens in immunoadsorbtion experiments. Using this method we have identified a population of antibodies in AIDS serum that reacted with pENV 7 (containing gp120 and gp41 sequences) but not with p121 antigen (containing only gp41 sequences). This indicates that pENV 7 contains antigenic determinants for gp120 and that patients serum contains antibodies to the protein portion of gp120. We infer from these two experiments that there are antibodies in AIDS serum directed against the protein backbone of gp120 not effected by post translational modification.

Purified pGAG 1 and pENV 9 antigens when used as the immunoabsorbent in ELISA were able to discriminate reactivity to gag and env antibodies in AIDS sera. We believe that these recombinant proteins will be of value in investigating the serology of the dynamics of antibody response in individuals exposed to HTLV-III.

We conclude from these data that recombinant HTLV-III proteins produced by E. coli share structural features with viral proteins that make them useful immunological tools for the study of host virus interactions and the development of diagnostics and vaccine candidates for AIDS. Recombinant proteins able to induce antibodies reactive with env proteins are currently under investigation as vaccine candidates.

ACKNOWLEDGMENTS

We thank J. Neumann, D. McCabe, E. Bullitt, H. Sundemeyer, C. Greenamoyer, D. Cox and M. Lemon for technical assistance; Drs. R.C. Gallo and R. Ting for monoclonal antibodies and HTLV-III clones, and S. Nowell for preparation of the manuscript.

REFERENCES

1) Wong-Staal F & Gallo RC (1985). Human T-lymphotropic retroviruses. Nature 317:395.
2) Sarngadharan MG, Popovic M, Bruch L, Schupbach J, and Gallo RC (1984). Antibodies reactive with human T-lymphotropic retroviruses (HTLVIII) in the serum of patients with AIDS. Science 224:506.
3) Schupbach J, Popovic M, Gilden RV, Gonda MA, Sarngadharan MG, and Gallo RC (1984). Serological analysis of a subgroup of human T-lymphotropic viruses (HTLVIII) associated with AIDS. Science 224:503
4) Chang TW, Kato I, McKinney S, Pranab C, Barone AD, Wong-Staal F, Gallo RC, and Chang NT (1985). Detection of Antibodies to Human T-Cell Lymphotropic virus-III

(HTLVIII) with an Immunoassay Employing a Recombinant Escherichia coli-Derived Viral Antigen Peptide. Bio/Technology 3:905.
5) Weiss RA, Clapham R, Cheingsong-Popov R, Dalgleish AG, Carne IV and Weller D (1985). Neutralization of human T-lymphotrophic virus type III by sera of AIDS and AIDS-risk patients. Nature 316:72
6) Robert-Guroff M, Brown M, and Gallo RC (1985). HTLV-III neutralizing antibodies in patients with AIDS and AIDS related complex. Nature 316:72.
7) Sarngadharan MG, Bruch L, Popovic M, and Gallo RC (1985). Immunological properties of the gag protein p24 of the acquired immunodeficiency syndrome retrovirus (human T-cell leukemia virus type III). Proc. Natl. Acad. Sci. USA 82:3481.
8) Kalyanraman VS, Cabradilla,CD, Getchell JP, Narayanan R, Braff EH, Chermann J-C, Barre-Sinoussi F, Montagnier L, Spira TJ, Kaplan J, Fishbein D, Jaffe HW, Curran JW, Francis DP (1984). Antibodies to the core protein of lymphadenopathy-associated virus (LAV) in patients with AIDS. Science 225:321.
9) Barin F, McLane MF, Allan JS, Lee TH, Groopman JE, Essex M (1985). Virus Envelope of HTLVIII represents Major Target for Antibodies in AIDS Patients. Science 228:1094.
10) Robey WG, Safai B, Oroszlan S, Arthur LO, Gonda MA, Gallo RC, Fischinger PJ (1985). Characterization of Envelope and Core Structural Gene products of HTLVIII with Sera from AIDS Patients. Science 228:593.
11) Ratner L, Haseltine W, Patarca R, Livak K, Starcich B, Josephs SF, Doran ER, Rafalski JA, Whitehorn EA, Baumeister K, Ivanoff L, Petteway Jr. SR, Pearson ML, Lautenberger JA, Papas TS, Ghrayeb J, Chang NT, Gallo RC, and Wong-Staal F (1985) Complete nucleotide sequence of the AIDS virus, HTLV-III. Nature 313:277.
12) Wain-Hobson S, Sonigo P, Danos O, Cole S, and Alizon M (1985). Nucleotide sequence on the AIDS virus, LAV. Cell 40:9
13) Allan JS, Coligan JE, Barin F, McLane MF, Sodroski JG, Rosen CA, Haseltine WA, Lee TH, Essex M (1985). Major Glycoprotein Antigens That Induce Antibodies in AIDS Patients Are Encoded by HTLVIII. Science 228:1091.
14) Guarente L, Roberts G, Ptashne M (1980). Improved method for maximizing expression of a cloned gene: A bacterium that synthesizes β-globin. Cell 20:543.
15) Towbin H, Gordon J, and Staehelin T. (1979). Electrophoretic transfer of proteins from polyacrylamide gels to nitrocellulose sheets, procedure and some

applications. Proc. Natl. Acad. Sci. USA 27:4350.

16) Laemmli UK, (1970). Cleavage of Structure Proteins during the assembly of the head of Bacteriophage T4. Nature 227:680.
17) Ivanoff LA, Towataru T, Korant BD, and Petteway, Jr. SR (Submitted for publication). Expression and site-specific mutagenesis of the poliovirus 3C protease in Escherichia coli.
18) Porath J, and Korstiansen T, (1975). Biospecific Affinity Chromatography and Related Methods. In The Proteins, Vol. 1. Edited by H. Neurath and R.L. Hill, Academic Press, NY p 95.

Viruses and Human Cancer, pages 29–41

SPECIFICITIES OF MONOCLONAL ANTIBODIES GENERATED AGAINST A GENETICALLY ENGINEERED ENVELOPE GENE PRODUCT OF THE AIDS RETROVIRUS

James C. Stephans, Diana Jin Lee,
Elizabeth T. Miller, Gary A. Van Nest,
Jay A. Levy, Carlos George-Nascimento,
Debbie Parkes, Philip J. Barr, Chuck Staben,
Dino Dina, Paul A. Luciw and Kathelyn S. Steimer

Chiron Research Laboratories, Chiron Corporation
4560 Horton St., Emeryville, California 94608

ABSTRACT Murine monoclonal antibodies have been produced to a genetically engineered polypeptide expressed in yeast representing the amino acid sequence of the envelope glycoprotein, gp120, of the ARV-2 isolate of the AIDS retrovirus. Of the 89 monoclonal antibodies obtained, 76 (84%) reacted with epitopes in the amino-terminal half of the polypeptide while only 13 (16%) reacted with epitopes in the carboxyl-terminal half. Eight of the monoclonal antibodies reacting with the recombinant polypeptide also reacted with fully glycosylated gp120 in immunoblots of ARV-2 virus. The monoclonal antibodies reactive with viral gp120 were tested in immunoblot assays with two additional virus strains, ARV-3 and ARV-19. All 8 monoclonals reacted with gp120 from ARV-19. However, only 6 reacted with ARV-3 gp120. The two monoclonals that failed to react with gp120 of ARV-3 may be useful reagents for studying antigenic variation within the envelope glycoprotein of different virus strains.

INTRODUCTION

A novel human retrovirus, referred to as lymphadenopathy associated virus (LAV, 1), human T lymphotropic retrovirus, type III (HTLV-III, 2) and AIDS-associated retrovirus (ARV, 3), has been identified as the etiological agent of acquired immune deficiency syndrome (AIDS). Two glycoproteins are associated with the membrane of this AIDS retrovirus. They are derived during virus maturation by cleavage of a common precursor polypeptide, gp160. The larger subunit, gp120, originates from the amino terminal portion of gp160 (4). Analogy with other retroviruses suggests that this glycoprotein is the external subunit of the envelope glycoprotein complex. The smaller envelope glycoprotein subunit, gp41, is derived from the carboxyl terminal domain of gp160 (5). This protein is thought to span the virus membrane and serve as an anchor for the viral envelope glycoprotein complex.

Monoclonal antibodies to viral envelope glycoproteins have been useful tools in other virus systems in identifying epitopes that elicit neutralizing antibodies (6) and for studying strain variation (7). To provide immunological reagents for beginning to address similar issues in the AIDS retrovirus system we have produced mouse monoclonal antibodies to the viral glycoprotein, gp120. Several attempts to obtain such monoclonals by immunizing mice with virus purified on sucrose gradients did not yield hybridomas producing antibodies against either of the viral envelope glycoproteins. The glycoproteins of retroviruses represent only a small proportion of the total protein of virions (8). Thus, the probability of obtaining anti-envelope monoclonal antibodies by immunizing with virus would be expected to be quite low. As an alternative strategy, we have immunized mice with purified recombinant polypeptides produced in yeast from the envelope gene of the virus. We report here the isolation of monoclonal antibodies to the viral envelope glycoprotein gp120 using this approach. Presented is a preliminary characterization of these antibodies that includes a demonstration of the utility of these antibodies for evaluating sequence variation in independent virus isolates.

METHODS

Purification of Genetically Engineered gp120 From Yeast Extracts

Yeast expressing the product of the gp120 coding region of the ARV-2 envelope gene (referred to as env-2) were harvested and the cells disrupted with glass beads. Details of the purification procedure will be described elsewhere

(K. Steimer, et al., in preparation). Briefly, env-2 was precipitated with 30% ammonium sulfate and the precipitate was solubilized by boiling for 10 min in 2.3% SDS/67.5 mM Tris-HCl/5% β-mercaptoethanol (pH 7.0). The resuspended protein was fractionated by gel filtration on an ACA-34 column (LKB) equilibrated with phosphate-buffered saline containing 0.1% SDS. Fractions containing env-2 were pooled and concentrated by ultrafiltration on an Amicon YM5 membrane. This procedure yielded env-2 that was approximately 95% pure as judged by Coomassie staining of SDS polyacrylamide gels.

Immunization of Mice

Balb/c mice were immunized intraperitoneally (IP) with 100 μg of purified env-2 in complete Freund's adjuvant (CFA) and boosted three times over three week intervals with 100 μg of env-2 in incomplete Freund's adjuvant (IFA). Two weeks after the final injection of antigen in IFA, the mice received 20 μg of env-2 without adjuvant in 0.1 ml of saline; 10 μg injected IP and 10 μg injected intravenously (IV). Three days later, the mice were sacrificed, and their spleens were removed for fusion.

Cell Fusion and Tissue culture

Fusion of spleen cells with the murine cell line P3x63Ag8.653, hybrid selection, and cell culture methods were as described previously (9).

Enzyme Linked Immunosorbent Assay (ELISA) for Env-2 Antibodies

Microtiter plates (Immulon I, Dynatech) were coated with 5 μg of purified env-2 per ml. Culture supernatants were assayed for specific antibodies as described previously (10,11).

Virus Preparation

HUT-78 cells chronically infected with the ARV-2, ARV-3 or ARV-19 virus isolates were diluted into fresh culture medium (RPMI 1640, 10% fetal calf serum) to a density of approximately 1 x 10^5 cells/ml. After 5d, the culture supernatants were collected and the virus concentrated by high speed centrifugation. The viral pellet from 400 ml of culture medium was resuspended in 1 ml of electrophoresis sample buffer, boiled, and 200 µl electrophoresed on SDS polyacrylamide gels for electroblotting.

Immunoblot Procedure

Electrophoresis was on standard Laemmli discontinuous SDS polyacrylamide gels (12) using a Bio-Rad mini gel apparatus. A 0.5 cm well for prestained molecular weight standards and an 8 cm well for the sample were prepared in the stacking gel. The sample (100-200 µl), containing either approximately 50 µg of recombinant polypeptide or concentrated virus (see above), and standards (5 µl) were loaded and the gels electrophoresed. Following electrophoresis, proteins were electroblotted onto nitrocellulose filters (13). Strips of the blotted samples (2 mm wide) were cut and each strip processed separately. Carnation nonfat milk was used in the diluents as described previously (14). Human sera were tested diluted 1/50 and monoclonal antibody supernatants were tested undiluted. When human sera were being tested, the conjugate was horseradish peroxidase conjugated goat antiserum to human immunoglobulin G (Cappel Laboratories). The conjugate for blots with monoclonal antibodies was horseradish peroxidase conjugated goat antiserum to mouse immunoglobulin (Tago). The chromogen was 4-chloro-1-Napthol (Bio-Rad).

Immunofluorescence

HUT-78 cells chronically infected with the ARV-2 isolate of AIDS-associated retrovirus were fixed with 1% paraformaldehyde and spotted onto microscope slides. The slides were air-dried, fixed for 10 min with ice cold acetone, and the acetone was allowed to evaporate before adding the hybridoma supernatants. Slides were incubated for 1 h at 37°C with culture supernatants and 1 h at 37°C with FITC conjugated goat anti-mouse immunoglobulin (Becton Dickinson).

RESULTS

Yeast have been genetically engineered to produce several polypeptides from the gp120 coding region of the envelope (env) gene of the ARV-2 isolate of the AIDS retrovirus. Env-2 consists of amino acids 26-490 from the env open reading frame and corresponds to the major portion of the gp120 polypeptide. Env-1, consisting of amino acids 26-276 from the env open reading frame, corresponds to approximately the amino-terminal half of the gp120 polypeptide. Figure 1 shows examples of western blots with a selected AIDS patient's serum with these two polypeptides. Immunoreactive species at 53,000 and 25,000 daltons (as expected for these proteins) were seen in immunoblots of env-2 and env-1, respectively. No visible reactive species were seen in western blots with normal human serum. The yeast vector system for expressing these polypeptides will be described elsewhere (P. Barr *et al* manuscript in preparation).

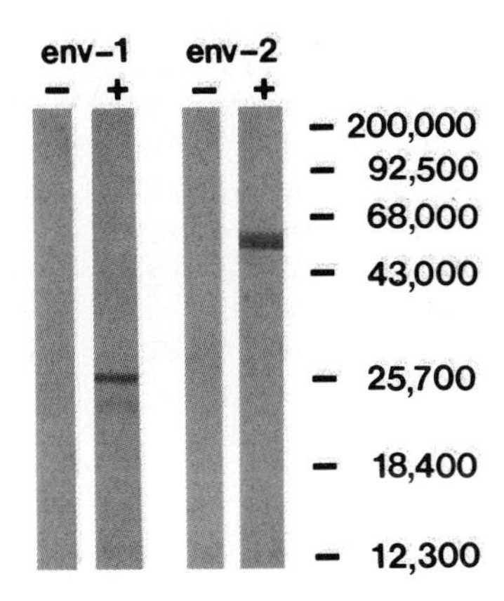

FIGURE 1. Immunoblots of env-1 and env-2 with serum from a normal donor (-) and an AIDS patient (+). Molecular weight markers are indicated in the right margin. Sera were tested diluted 1/50.

When we attempted to purify env-2 from yeast, it was discovered that this protein was present in the insoluble fraction complexed with yeast proteins. This is often observed with heterologous proteins expressed in yeast (D. Parkes, unpublished). To purifiy env-2, it was necessary to solubilize it in detergents under denaturing conditions. As a consequence, the env-2 used for immunizing mice and for assaying for env-2 antibodies was denatured.

From the fusion of spleen cells from mice immunized with purified env-2, 912 proliferating clones were obtained. Two hundred and fourteen of these clones were producing antibodies detected in an ELISA for env-2 antibodies. The antibody-positive clones were expanded, the cells were frozen and a sufficient quantity of each culture supernatant was collected for a preliminary characterization of these hybridomas. Ninety-three of the original clones testing positive were still producing env-2 antibodies at the time that the cells were frozen.

Supernatants from the hybridomas that were positive for anti-env-2 antibodies by ELISA were tested for reactivity with env-2 in immunoblot assays. Most of these (89) clearly reacted with the env-2 protein band. When the supernatants from monoclonals reactive in env-2 immunoblots were tested in immunoblot assays with env-1, 76 (84%) were positive, indicating that they recognized epitopes present in the amino-terminal half of the gp120 polypeptide. The remaining 13 (16%) recognized only env-2 and must therefore have been specific for epitopes in the carboxyl-terminal half. Figure 2 shows examples of the patterns of reactivity observed with monoclonals reactive with the carboxyl- or amino-terminal portions of the gp120 polypeptide.

We next tested the env-2 hybridoma supernatants in virus immunoblot assays to identify those that reacted with viral gp120. Ten hybridomas produced antibodies that reacted with viral gp120. Supernatants from the remaining 79 did not react with any viral proteins. The hybridomas that were positive for viral gp120 were all subcloned. Of the original 10 isolates, 8 maintained stable antibody production. Figure 3 shows the reactivity with ARV-2 virus of supernatants from these clones. One of these hybridomas (B3) was from the group producing antibodies that reacted with epitopes in the carboxyl terminal half of gp120. The remaining seven reacted with the amino terminal half of the polypeptide.

All of the viral gp120 immunoblot positive monoclonal antibodies were tested for binding to gp120 on infected cells by indirect immunofluorescence.

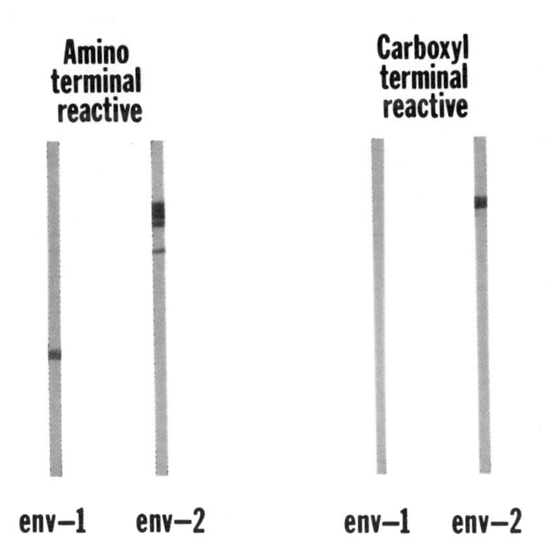

FIGURE 2. Examples of immunoblots of monoclonal specific for the carboxyl terminal (env-1$^-$, env-2$^+$) and amino-terminal (env-1$^+$, env-2$^+$) halves of the polypeptide backbone of gp120. Strips from immunoblots of env-1 or env-2 were reacted with undiluted supernatants.

We could not detect any specific immunofluorescence of infected cells. Under identical conditions, we were able to readily detect cytoplasmic immunofluorescence of infected cells with culture supernatants from several hybridomas producing monoclonal antibodies to the viral core antigen $p25^{gag}$.

Sequence comparisons of different isolates of the AIDS retrovirus have revealed that there is considerable variation within the envelope gene of these viruses (15,16). Furthermore, this variation is predominantly located in the amino terminal portion in the gp120 coding region. The monoclonals that we have generated were raised against a recombinant polypeptide from the ARV-2 isolate of the AIDS retrovirus. It might be expected that monoclonal antibodies against epitopes coded by the conserved regions as well as the variable region of the gp120 gene would be obtained. The former would be useful reagents for detecting gp120 from

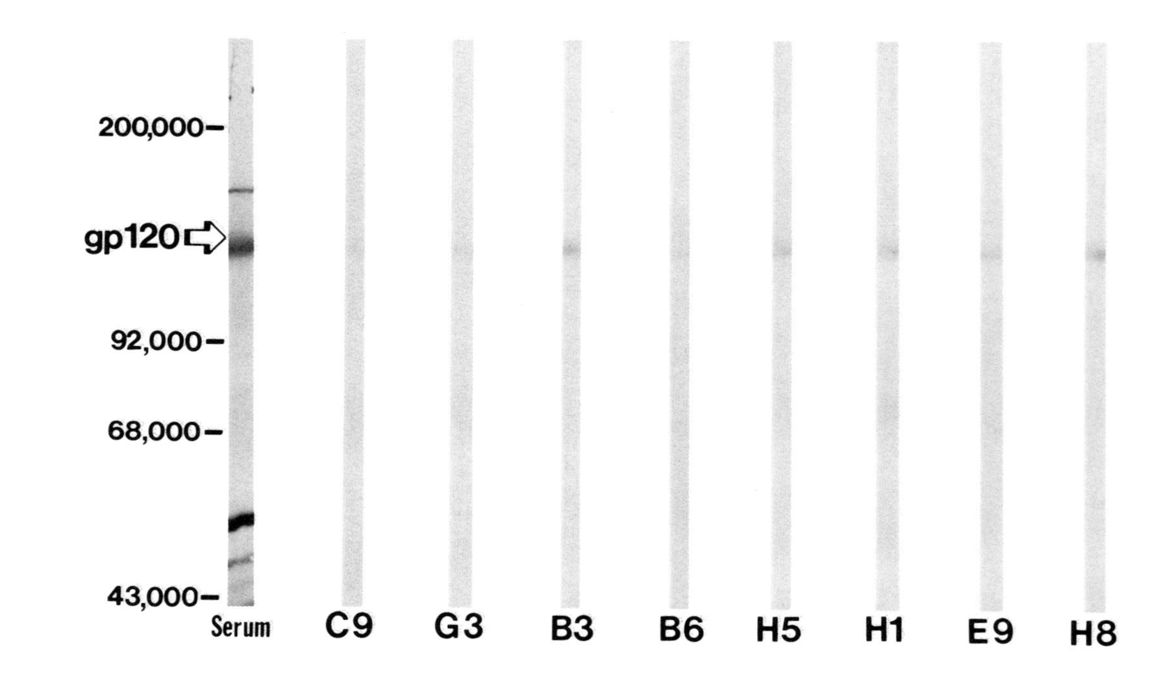

FIGURE 3. Immunoblots with culture supernatants from hybridomas producing antibodies that react with ARV-2 virus gp120. For reference, an immunoblot with serum from an AIDS patient (serum) is included.

all virus strains whereas the latter would be useful for studying strain variation. We have tested these monoclonals for reactivity with gp120 of two additional virus isolates; ARV-19 and ARV-3. The ARV-19 isolate was chosen because we know from studies with oligonucleotide probes prepared to regions of the ARV-2 env gene that it is quite closely related to ARV-2. Similar studies with ARV-3

TABLE 1
SPECIFICITIES OF gp120 MONOCLONAL ANTIBODIES FOR THE GLYCOPROTEIN OF VARIOUS VIRUS ISOLATES[a]

Hybridoma	Amino (N) or Carboxyl (C) Reactive	Reaction with gp120 of: ARV-2	ARV-3	ARV-19
B6	N	+	+	+
C9	N	+	-	+
G3	N	+	+	+
B3	C	+	+	+
H5	N	+	+	+
H1	N	+	-	+
E9	N	+	+	+
H8	N	+	+	+

[a]Supernatant from each hybridoma was reacted with immunoblots of each virus. Blots showing a visible band in the gp120 region were scored as "+".

revealed substantial differences in the variable regions of the ARV-3 gp120 coding sequences (17). These results are summarized in Table 1.

All of these monoclonals reacted in immunoblots with the gp120 species of both ARV-2 and ARV-19. However, only 6 reacted with gp120 in immunoblots of ARV-3 virus. The two monoclonals failing to react with ARV-3 gp120 (C9 and H1) both mapped to the amino-terminal half of the gp120 polypeptide.

DISCUSSION

Monoclonal antibodies reacting with fully glycosylated gp120 of the ARV-2 isolate of AIDS-associated retrovirus were produced by immunizing mice with a recombinant polypeptide (env-2) representing the predicted amino acid sequence of this glycoprotein. Since virtually unlimited quantities of purified env-2 were available, it was feasible to hyperimmunize mice with this antigen to generate numerous env-2-reactive monoclonal antibodies. Of the 89 monoclonal antibodies characterized, 10 (11%) were found to react in virus immunoblot assays with viral gp120. The monoclonals that did not react with viral gp120 may have failed to do so because they may recognize epitopes that are masked by glycosylation of the viral glycoprotein. Alternatively, they may be of such low affinity or titer that their reactivity with viral gp120 under the conditions that we have used is not visualized.

The majority (84%) of env-2-reactive monoclonal antibodies also reacted with env-1, a recombinant polypeptide corresponding to approximately the amino-terminal half of env-2. This suggested that mice immunized with env-2 were more responsive to epitopes in this region than to epitopes in the carboxyl-terminal half of the polypeptide. We have also observed a dominant polyclonal antibody response to the amino-terminal half of the polypeptide when rabbits were immunized with env-2 (data not shown).

The env-2 monoclonal antibodies that reacted with viral gp120, were tested in immunofluorescence assays for binding to gp120 associated with the surface of infected cells. None of the eight monoclonals tested reacted with cell-associated gp120. It is possible that these monoclonals all recognize sequential epitopes that are not accessible in native gp120 or that the density of gp120 at the cell surface is not high enough for immunofluorescence with monoclonal antibodies to be visible. Alternatively, these antibodies may be specific for epitopes that are masked by the association of viral gp120 with the infected cell membrane and/or with gp41, the viral envelope glycoprotein that presumably anchors gp120 to the surface of infected cells and virions. Radioimmunoprecipitation studies of labelled infected cells solubilized with non-ionic detergents are in progress to distinguish between these possibilities.

There is considerable strain variation in the sequence of the envelope gene of different AIDS retrovirus isolates, particularly in the region of the envelope gene coding for the amino-terminal half of the gp120 polypeptide (15,16,17). These differences tend to be clustered in two regions each of approximately 30 nucleotides. Since 7 of the viral gp120 positive monoclonals reacted with epitopes in this amino-terminal half, we considered that one or more might react with epitopes in variable regions. Results of immunoblots of these monoclonals with two additional strains, ARV-19 which is closely related to ARV-2 in regions where this variation is clustered, and ARV-3, which is quite different from ARV-2 in these regions, indicated that two monoclonals possibly react with epitopes localized in variable regions. Both monoclonals reacted with ARV-2 and ARV-19 but not ARV-3. Clearly, epitope mapping studies and studies with additional virus strains are required before we will be able to come to any firm conclusions regarding the utility of these monoclonals for studying strain variation.

In conclusion, we report here the isolation of the first monoclonal antibodies specific for gp120, the external glycoprotein of the AIDS retrovirus. These monoclonals have been used to begin addressing critical issues in the AIDS retrovirus system such as antigenic variation in this envelope glycoprotein and the effects of carbohydrates on immunogenicity and antigenicity of this glycoprotein.

REFERENCES

1. Barré-Sinoussi F, Chermann JC, Rey F, Nugeyre MT, Chamaret S, Gruest J, Dauguet C, Axler-Blin C, Brun Vézinet F, Rouzioux C, Rozenbaum W, Montagnier L (1983). Isolation of a T-lymphotropic retrovirus from a patient at risk for acquired immunodeficiency syndrome (AIDS). Science 220:868.
2. Popovic M, Sarngadharan MG, Read E, Gallo RC (1984). Detection, isolation and continuous production of cytopathic retrovirus (HTLV-III) from patients with AIDS and pre-AIDS. Science 224:497.
3. Levy JA, Hoffman AD, Kramer SM, Landis JA, Shimabukuro JM, Oshiro LS (1984). Isolation of lymphocytopathic retroviruses from San Francisco patients with AIDS. Science 225:840.
4. Allan JS, Coligan JE, Barin F, McLane MF, Sodroski JG, Rosen CA, Haseltine WA, Lee TH, Essex M (1985). Major glycoprotein antigens that induce antibodies in AIDS patients are encoded by HTLV-III. Science 228:1091.

5. Veronese F, DeVico AL, Copeland TD, Oroszlan S, Gallo RC, Sarngadharan MG (1985). Characterization of gp41 as the transmembrane protein coded by the HTLV-III/LAV envelope gene. Science 229:1402.
6. Sherry B, Mosser AG, Colonno RJ, Rueckert RR (1986). Use of monoclonal antibodies to identify four neutralization immunogens on a common cold picornavirus, human rhinovirus 14. J Virol 57:246.
7. Laver WG, Gerhard W, Webster RG, Frankel ME, Air GM (1979) Antigenic drift in type A influenza virus: Peptide mapping and antigenic analysis of A/PR/8/34 (HON1) variants selected with monoclonal antibodies. Proc Natl Acad Sci USA 76:1425.
8. Weiss R, Teich N, Varmus H, Coffin J (1982). "Molecular Biology of Tumor Viruses." Cold Spring Harbor: Cold Spring Harbor Laboratories.
9. Oi VT, Herzenberg LA (1980). Immunoglobulin-producing hybrid cell lines. In Mishell B, Shiigi S (eds): "Selected Methods in Cellular Immunology," San Francisco: W.H. Freeman and Co, p 351.
10. Voller A, Bidwell DE, Bartlett A (1979). "The Enzyme Linked Immunosorbent Assay." Rue de Pre, Gurnsey, U.K.: Borough House.
11. McHugh Y, Walthall B, Steimer K (1983). Serum-free growth of murine and human lymphoid and hybridoma cell lines. Biotechniques June/July:72.
12. Laemmli UK (1970). Cleavage of structural proteins during the assembly of head of bacteriophage T4. Nature 227:680.
13. Towbin HT, Staehelin T, Gordon J (1979). Electrophoretic transfer of proteins from polyacrylamide gels to nitrocellulose sheets: Procedure and some applications. Proc Natl Acad Sci USA 76:4350.
14. Johnson DA, Gautsch JW, Sportman JR, Elder JR (1984). Improved technique utilizing nonfat dry milk for analysis of proteins and nucleic acids transferred to nitrocellulose. Gene Anal Techn 1:3.
15. Rabson AB, Martin MA (1985). Molecular organization of the AIDS retrovirus. Cell 40:478.
16. Hahn BH, Gonda MA, Shaw GM, Popovic M, Hoxie JA, Gallo RC, Wong-Staal F (1985). Genomic diversity of the acquired immune deficiency syndrome virus HTLV-III: Different viruses exhibit greatest divergence in their envelope genes. Proc Natl Acad Sci USA 82:4813.

17. Staben C, Shimabukuro JM, Stephans JC, Barr PJ, Sabin EA, Parkes D, Luciw PA, Steimer KS, Levy JA, Dino D (in press). The nature and significance of sequence variation among independent AIDS retrovirus isolates. In: Vaccines '86. Cold Spring Harbor Laboratories, New York.

Viruses and Human Cancer, pages 43–57

CROSSREACTING ANTIBODIES TO GAG PROTEINS OF HTLV-I AND HTLV-III IN PATIENTS WITH MYCOSIS FUNGOIDES OR ITS PRODROME LARGE-PLAQUE PARAPSORIASIS[1]

Annamari Ranki[2], and Kai Krohn[2]

Department of Dermatology, Helsinki University 00170 Helsinki, Institute of Biomedical Sci., Univ. of Tampere, 33101 Tampere, Finland, and Laboratory of Tumor Cell Biology, National Cancer Institute, NIH, Bethesda, Maryland 20892

ABSTRACT Serological evidence for retrovirus infection in mycosis fungoides (MF) and in its prodrome, large-plaque parapsoriasis (PPs) was looked for using ELISA and Western blot methods and partially purified HTLV-I and HTLV-III preparations. In ELISA test, 6 of 15 MF and 4 of 12 PPs patients had positive or borderline reactions against HTLV-I. The corresponding figure for HTLV-III was 1 in both groups. Western blots revealed that both virus preparations, but especially HTLV-I, contained proteins of probable cellular origin which reacted nonspecifically with patient sera but also with many of the control sera consisting of other dermatological patients and healthy persons. These nonspecific reactions could be abolished when immunoglobulins, purified with ammonium sulphate precipitation, were used. By so doing, 15 MF or PPs sera reacted against the p19 or p24 antigens of HTLV-I and 7 sera against corresponding HTLV-III gag proteins. The results are taken to suggest that an as yet undefined retrovirus, antigenically somewhat related to HTLV-I and III, is involved in the pathogenesis of MF.

[1]This work was supported by grants from The Yamagiwa-Yoshida Memorial International Foundation, and by H. Rosenberg Foundation (AR). Kai Krohn has a Fogarty International Visiting Scientist Award.
[2]Present address: LTCB, NCI, NIH, Bethesda, MD 20892

INTRODUCTION

Mycosis fungoides (MF) is a cutaneous lymphoma with a slow, chronic clinical course. The majority of the cells in the skin infiltrates of MF express surface antigens of mature T lymphocytes and thus, MF belongs to the group of cutaneous T cell lymphomas (1). The onset of MF is usually insidious, often presenting with chronic dermatitis not responding to steroid therapy, called large-plaque parapsoriasis (PPs, 2). On the other hand, MF may ultimately result in a visceral lymphoma with a more or less heterogeneic histological picture (3). The etiology of MF is unknown but a bulk of indirect evidence suggest a role for retroviruses in MF. In 1979, C-type virus-like particles were first observed in Langerhans cells in the skin and in dendritic cells in the lymph nodes of MF patients (4). This finding has been recently confirmed by Slater _et al_. (5). In 1980 a retrovirus, named HTLV-I was isolated from two patients with skin lesions resembling MF (6). It was, however, later found that these patients actually had a leukemic disease, adult T cell leukemia/lymphoma (ATL), endemic in Japan and Caribbean islands (7,8). Using an ELISA test, it has been demonstrated that the Japanese and Caribbean ATL patients almost invariably have antibodies against the structural proteins of HTLV-I (9-11). Healthy controls from these endemic areas for ATL show anti-HTLV-I antibodies in up to 16% (8). On the contrary, Caucasian CTCL patients have previously failed to show HTLV-I antibodies (less than 1% positive, 10). The incidence of HTLV-I antibodies in healthy individuals in U.S. and Europe is below 1%, too (12).

When a more sensitive indirect ELISA test became available, Saxinger, et al. (13,14) reported HTLV-I antibodies in 15% of Danish MF patients, while healthy individuals from Scandinavia were antibody negative (14, 15). Using a monoclonal anti-HTLV-I antibody (16), HTLV-I related p19 antigen has been demonstrated immunohistochemically in lymph node and skin lesions from a MF patient (17) and, also, in a long-term culture of skin- and lymph node-derived cells of MF patients (18).

As ELISA tests for viral antibodies in patient material, however, is subject to pitfalls due to the presence of cellular antigens in the viral preparations used and to circulating immune complexes in the sera (Saxinger et al., in preparation), we subjected our material of 26 Finnish patients with histologically verified MF or parapsoriasis en plaque to immunoblotting (Western blot). We here provide evidence suggesting that most of the antibody response against HTLV-I and III seen in such patients is directed against the gag proteins of these viruses. Our findings thus suggest that MF is associated with a retrovirus but the low titer antibodies restricted against the core proteins of the above viruses, indicates that this virus is partially related but not identical to the presently known human retroviruses.

MATERIALS AND METHODS

Patients. Fifteen patients with histologically verified mycosis fungoides (MF;19) and 12 patients with parapsoriasis en plaque were studied. The subjects represented consecutive cases seen in the Department of Dermatology, University of Helsinki during the years 1982-85. Seven of the MF and 7 of the parapsoriasis patients were males and none of the patients belonged to any AIDS risk group or had otherwise been exposed to HTLV-I or HTLV-III.

As controls, 42 consecutive out-patients with diverse dermatological disorders, 33 healthy blood donors and 6 patients with various immune disorders like SLE, polyendocrinopathy and chronic mucocutaneous candidiasis, were studied.

Serum samples were drawn from each of the patients on several occasions, from 3 to 24 months apart. Additional plasma samples were obtained from heparinized blood, collected for immunological *in vitro* studies, and the immunoglobulins were separated with ammonium-sulfate precipitation and dialyzed.

HTLV-I and HTLV-III Antibody Assays

Viral antigen: The virus preparations used in the ELISA or Western blot assays were prepared as follows:

HTLV-I was grown in C10/MJ cell line (20) and HTLV-IIIb in H9 cells (21). Both viruses were purified from the culture supernatants by sucrose gradient centrifugation. Both preparations represented 5000 x concentration of the original supernatant. Protein concentrations were 5.5 mg/ml for HTLV-I and 9.4 mg/ml for HTLV-III.

ELISA. Antibodies for HTLV-I and HTLV-III were initially screened with the enzyme linked solid phase immunoassay method (ELISA), as described earlier (18).

Western blot. Western blot analysis of the sera were performed as previously described (22,23), with minor modifications in order to increase the sensitivity of the assay. Virus preparations (5 μg/gel) were boiled in sample buffer containing 1.0% SDS, and 1.0% 2-mercapto-ethanol and run for 18-24 hrs, using a constant current of 4.5 mA/gel. After 60 min equilibration in Tris-HCl-glycine with 20% methanol and 0.1% SDS, the protein transfer onto 0.45 μ pore size nitrocellulos paper (Schleicher & Schuell, Keene, NH) was performed by cross electrophoresis using constant voltage (30 volts, 110-150 mA) for 24-48 hrs. After the transfer, the filter was blocked in the blotting buffer (Tris-HCl 0.1M, NaCl 0.05 M, pH 7.4, containing 1% v/v normal goat serum and 5% w/v defatted dried milk powder for 60 min at 37°C. Patient or control sera were then added in 1:100 dilution, and the strips were incubated on a rocking platform, for 24 hrs at 4°C. Purified immunoglobulin was used respectively, in a concentration corresponding to serum dilution 1:100. After washing with TBS-Tween 20, biotinylated goat anti-human IgG (diluted 1:1000, BRL, Bethesda, MD), was added and incubated for 30 min at 37°C. Thereafter, avidin - peroxidase (Tago, Inc., Burlingame, CA) in 1:1000 dilution was added, and the strips were incubated for 30 min at 37°C, washed and stained with 4-chloro-1-naphthol (Sigma, St. Louis, MS; 0.5 mg/ml in TBS with 20% methanol and 0.01% H_2O_2.

RESULTS

ELISA

Initial screening for the presence of HTLV-I/HTLV-III antibodies in the patient materials was performed with ELISA. As presented in Figure 1, 6 of the 15 MF and 4 of the 12 PPs patients had antibodies reacting with the HTLV-I antigen. Reactions with HTLV-III

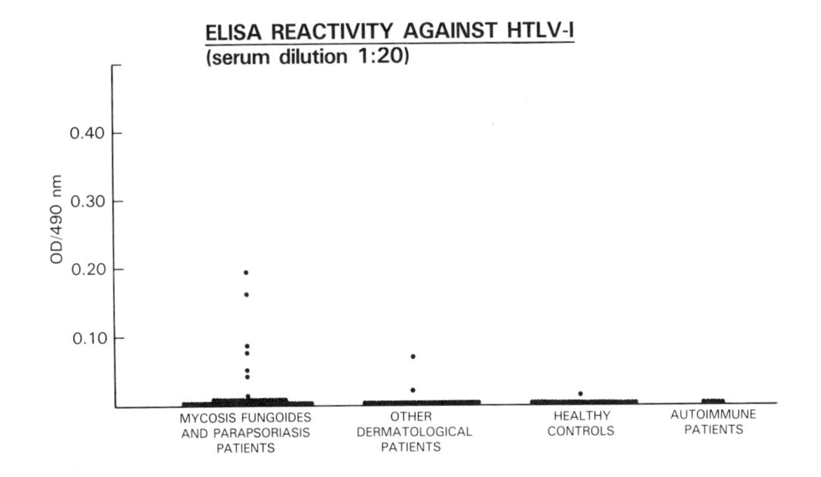

FIGURE 1. Antibodies against HTLV-I as recognized by ELISA assay.

were less frequent, seen in 1 of the MF and 1 of the PPs patients (Figure 2). These antibody reactions were characteristically of low titer (1:100-1:200). The dermatological control sera gave borderline (within 2-3 SD) or negative results in HTLV-I ELISA and negative results in HTLV-III ELISA except for one case which later showed reactivity in the Western blot too. Healthy control sera and the 6 autoimmune/immune disorder sera always gave negative results.

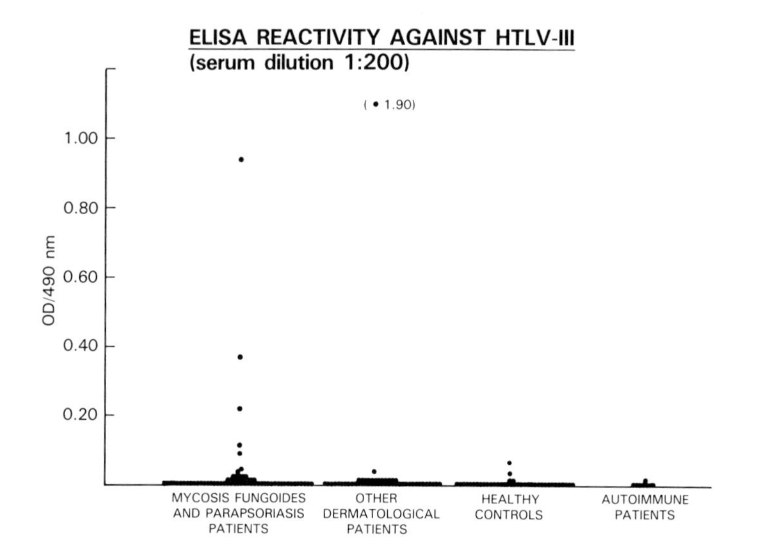

FIGURE 2. Antibodies against HTLV-III as recognized by ELISA assay.

Western Blot

In immunoblotting, 9 of the 15 MF patients and six of the 12 PPs patients recognized low molecular weight antigens in HTLV-I preparations and, respectively, 4 of the MF and 3 of the PPs patients recognized similar antigens in HTLV-III preparations (Table 1). The bands in Western blots were not very strong but became clearly demonstrable when purified immunoglobulins were used Representative experiment is shown in Fig. 3.
In several cases of all patient groups, purified immunoglobulins gave clear bands in Western blots despite negative ELISA results. As can be seen, most sera recognized the gag (core) proteins p19 and p24 in the HTLV-I preparation. With HTLV-III, most reactions were against gag protein p24, too, None of the sera from healthy control persons showed similar bands.

TABLE 1
ANTIBODIES RECOGNIZING HTLV-I AND HTLV-III VIRAL ANTIGENS AS DEMONSTRATED BY WESTERN BLOT ANALYSIS IN MF AND LARGE-PLAQUE PARAPSORIASIS PATIENTS

Diagnosis/ patient No.	Viral proteins recognized HTLV-I	HTLV-III
MF/1	p24	-
MF/2	-	p24
MF/3	p19,31	-
MF/4	p24,31	-
MF/5	p24	-
MF/6	24	-
MF/7	nd	-
MF/8	-	nd
MF/9	p24,43	-
MF/10	p24	+
MF/11	-	-
MF/12	-	p24
MF/13	nd	-
MF/14		p34,53
MF/15	p19	nd
PPs/16	nd	p15
PPs/17	-	p24
PPs/18	p15,24,31	-
PPs/19	p19	-
PPs/20	p24	p24
PPs/21	-	-
PPs/22	-	-
PPs/23	nd	-
PPs/24	p24	-
PPs/25	nd	-
PPs/26	nd	-
PPs/27	-	-

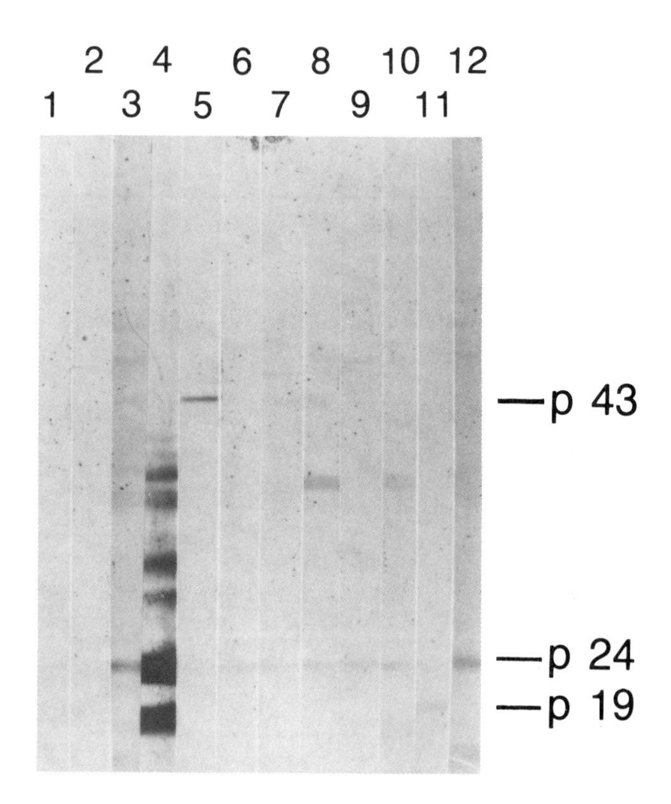

FIGURE 3. Demonstration of antibodies with Western blot against HTLV-I gag-proteins in MF patients. Lanes 1 and 2: healthy controls, lane 4: ATL patient, lanes 3, 5 to 12: Mycosis fungoides and large-plaque parapsoriasis patients

In almost all sera, including the dermatological and healthy control sera, weak reactions against high molecular weight proteins were seen. These bands were not recognized by the positive control serum and are presumably of cellular origin. However, in altogether 4 dermatological patient sera, antibodies against HTLV-I p24 and HTLV-III p24 or p41 were seen. These patients were diagnosed as having non T-cell skin lymphoma, Epidermodysplasia verruciformis, Lichen corneus obtusus or Erythema fixum, respectively. Two were females and two were heterosexual males without any known risk to having been exposed to HTLV-III.

DISCUSSION

The association of HTLV-I and ATL has been well documented (24). However, the exact mechanism of how this virus causes the leukemic process is still not clear. ATL and mycosis fungoides patients show similar skin lesions, even histologically (25), but MF patients do not become leukemic and do not show hypercalcemia as do patients with ATL (26). Most serological studies including MF patients have been done on U.S. or Japanese patients and as ATL and HTLV-I have now been shown to be endemic for Japan, the Caribbean Islands and Africa (24), the findings of these studies cannot be directly applied to European populations. In fact, all HTLV-I positive U.S. CTCL patients have been of Caribbean or African origin except for a few cases of Sezary's syndrome. Also, there has been discrepancy in defining the diagnostic criteria of MF.

Our Finnish patient material has been studied repeatedly for the histological findings allowing the diagnosis of MF and only cases repeatedly fullfilling the criteria have been included. The large-plaque parasoriasis patients have been under careful clinical follow-up and other reasons for the eczematous lesions have been excluded and the lesions have been confirmed histologically. The patients included in this study had normal findings in bone marrow biopsy, normal immunological findings including lymphocyte subsets and their function, no signs of hypercalcemia or any infections of opportunistic character (data not shown).

In the present study, we could demonstrate antibodies towards HTLV-I and HTLV-III with ELISA. Our results thus confirm those obtained by Saxinger et al. (14), but we found antibodies in a higher frequency (30% vs 15% in the Danish material). The antibodies in the present material have, however, been negative for HTLV-I in preliminary competitive ELISA assays (Saxinger, unpublished observation). This suggests the presence of not HTLV-I but a related virus.

The ELISA test for viral antibodies is subject to pitfalls due to nonspecific cross-reactivity. The HTLV virus preparations, derived from lymphoid cell cultures, contain cellular antigens and some of the antibody reactions observed in ELISA may be directed towards these non-viral structures. In HTLV-III studies, it has been observed that sera containing circulating immune complexes frequently give rise to nonspecific ELISA reactions. This could, in part, explain the positive reactions found in some of our dermatological control sera. It would therefore, be of importance to more extensively characterize the antigens recognized by the MF sera and to determine whether they are of viral or cellular origin.

In the more sensitive and specific Western blot analysis, we could demonstrate that most MF and other dermatological patients recognized several high molecular weight proteins, not recognized by the positive control sera (ATL/AIDS). As weak reactions against these bands were seen even in healthy control sera, it was concluded that these reactions were nonspecific. In addition, the MF and PPs sera recognized antigens belonging to the gag proteins of HTLV-I or III (p19 and p24 for HTLV-I and p15 and p24 for HTLV-III). Even these reactions did not sem to be totally specific for MF and PPs, as 4 of the 42 dermatological control patients showed positive reactions, although less strong. However, using purified immunoglobulin from the patient and control sera, these bands became more clear in MF and Pps patients. Our results thus demonstrate low titer antibodies in patients with either MF or PPs against the gag proteins of HTLV-I and III. Although these findings suggest an association of the disease with infection with a retrovirus, it must be taken into consideration that these reactions might be of nonspecific nature, as a result of crossreactivity between some tissue antigens and the viral antigens. The fact that some of the dermatological controls also showed antibodies, favors this interpretation. Another possibility to explain some of the reactions would be a crossreaction between some cytokeratin epitope and the p19 antigen of HTLV-I (B. Haynes, personal communication). Such an interpretation is, however, unlikely, as the cross-reactivity has been demonstrated only with p19, not with p24, which was the main target antigen in our sera. A more plausible interpretation would then be that the observed reactions suggest a retrovirus infection in our patient material. This virus could be either HTLV-I or more likely a hitherto undescribed human T-cell tropic retrovirus with close homology to the presently known human retroviruses HTLV-I and III.

Whatever the case, the expression of the viral proteins *in vivo* is probably abortive so that only part of the genome, starting from the 5' end of it, would be transcribed. In our material of HTLV-III exposed homosexual men, we noticed that in every occasion of seroconversion, the first antibodies to appear were directed against the gag proteins p15 or p24 (Krohn et al., in preparation). First, after a period of several months, an antibody response towards all other viral proteins appeared. It is assumed that the infection may first remain latent and that cofactors are necessary for the full expression of the viral genome. A similar situation might prevail in MF, leading to a chronic/localized disease, rather than to a leucemic disease like ATL.

The cells initially infected by the proposed virus might be few in number which is the case in HTLV-III infection (27). These cells could be T lymphocytes or even Langerhans cells. In HTLV-III infection it has recently been shown that macrophages are able to actively produce the virus (28,29), and we also know that in visna virus infection, the macrophages are the primary target for the virus (30). As it has recently been demonstrated, a clonal proliferation of T cells takes place in MF (31). We can hypothesize that this occurs in response to a latent/chronic viral infection giving rise to a continuous antigen stimulation. The skin lesions of MF with a dense lymphoid cell infiltrates, but only a few atypical cells would fit into this hypothesis. Virus isolation studies, currently underway from the patient material presented herein, will ultimately prove the justification of this hypothesis. In our preliminary studies with long term cultures of peripheral blood lymphocytes from MF patients, cultured in the presence of mitogen and IL-2, reverse transcriptase activity, specific for retroviruses, was seen in 2 of 6 cases (own unpublished observation).

ACKNOWLEDGMENTS

The histopathological diagnoses were confirmed by Dr. Kirsti-Maria Niemi, University of Helsinki, Finland, who is gratefully acknowledged. We also wish to thank Dr. Carl Saxinger, LTCB, NCI, NIH, Bethesda, MD for performing some preliminary ELISA assays and Professor Vaino Havu, University of Turku, Finland and Dr. Raimo Suhonen, Mikkeli Central Hospital, Finland, for providing us with some of the patient material.

REFERENCES

1. Lever W, Schaumburg-Lever G (1975). Histopathology of the skin. Philadelphia: Lippincott.
2. Samman PD (1972). The natural history of parapsoriasis en plaque and pre-reticulotic poikiloderma. Br J Dermatol 87:405.
3. Sheen SR, Banks PM, Winkelmann RK (1984). Morphologic heterogeneity of malignant lymphomas developing in mycosis fungoides. Mayo Clin Proc 59:95.
4. Van der Loo EM, Van Muijen GNP, Van Vloten WA, Peens W, Scheffer E, Meijer CJL (1979). C-type virus-like particles specially localized in Langerhans cells and related cells of skin and lymph nodes of patients with mycossis fungoides and Sezary's syndrome. Virchow Archives B Cell Path 31:193.
5. Slater DN, Rooney NS, Bleehen S, Hamad A (1985). The lymph node in mycosis fungoides: a light and electron microscopy and immunohistological study supporting the Langerhans' cell - retrovirus hypothesis. Histopathology 9:587.
6. Poiesz BJ, Ruscetti FW, Reitz M, et al. (1981). Isolation of a new type-C retrovirus (HTLV) in primary uncultured cells of a patient with Sezary T-cell leukemia. Nature 294:268.
7. Blattner WA, Kalyanaraman VS, Robert-Guroff M, et al. (1982). The human Type-C retrovirus, HTLV, in blacks from the Caribbean region, and relationship to adult T-cell leukemia/lymphoma. Int J Cancer 30:257.

8. Hinuma Y, Komoda H, Chosa T, Kondo T, Kohakura M, et al. (1982). Antibodies to adult T-cell leukemia-virus-associated antigen (ATLA) in sera from patients with ATL and controls in Japan: a nationwide seroepidemiologic study. Int J Cancer 29:631.
9. Takatsuki K, Uchiyama T, Ueshima Y, Hattori T (1979). Adult T-cell leukemia: further clinical observations and cytogenetic and functional studies of leucemic cells. Jap J Clin Oncol 3:317.
10. Sarngadharan MG, Schupbach J, Kalyanaraman VS, Robert-Guroff M, Oroslan S, Gallo RC (1983). Immunological characterization of the natural antibodies to human T-cell leukemia virus in human sera. Hematology and Blood Trans 28:498.
11. Markham PD, Salahuddin SZ, Popovic M, Sarngadharan MG, Gallo RC (1985). Retrovirus infections in man: human T-cell leukemia (lymphotropic) viruses. In "Viral Mechanisms of Immunosuppression", p. 49.
12. Gallo RC, Kalyanaraman VS, Sarngadharan MG, Sliski A, Vonderheid EC et al. (1983). Association of the human Type C retrovirus with a subset of adult T cell cancers. Cancer Res 43:3892.
13. Saxinger C, Gallo RC (1983). Application of the indirect enzyme-linked immunosorbent assay microtest to the detection and surveillance of human T cell leukemia-lymphoma virus. Lab Invest 49:371.
14. Saxinger C, Lange Wantzin G, Thomsen K, Hoh M, Gallo RC (1985). Occurrence of HTLV-I antibodies in Danish patients with cutaneous T-cell lymphoma. Scand J Haematol 34:455.
15. Blomberg J, Faldt R (1985). Antibodies to human adult T cell leukemia virus type I associated antigens in Swedish leukemia patients and blood donors. Brit J Haematology 60:555.
16. Robert-Guroff M, Ruscetti FW, Posner LE, Poiesz BJ, Gallo RC (1981). Detection of the human T-cell lymphoma virus p19 in cells of some patients with cutaneous T-cell lympohoma and leukemia using a monoclonal antibody. J Exp Med 154:1957.

17. Turbitt M, Mackie R (1985). p19 antigen in skin and lymph nodes of patients with advanced mycosis fungoides. Lance ii: 945.
18. Kaltoft K, Thesrup-Pedersen K, Jensen JR, Bisballe S, Zachariae H (1984). Establishment of T and B cell lines from patients with mycosis fungoides. Brit J Dermatol 111:303.
19. Lutzner M, Edelson R, Schein P, et al (1975). Cutaneous T-cell lymphomas: The Sezary syndrome, mycosis fungoides and related disorders. Ann Intern Med 83:534.
20. Popovic M, Sarin PS, Robert-Guroff M, et al. (1983). Isolation and transmission of human retrovirus (human T-cell leukemia virus). Science 219:856.
21. Popovic M, Sarngadharan MG, Read E, Gallo RC (1984). Detection, isolation, and continuous production of cytopathic retroviruses (HTLV-III) from patients with AIDS and pre-AIDS. Science 224:497.
22. Towbin H, Stahelin T, Gordon J (1979). Electrophoretic transfer of proteins from polyacrylamide gels to nitrocellulose sheets: procedure and some applications. Proc Natl Acad Sci 76:4350.
23. Burnett WN (1981). "Western Blotting". Electrophoretic transfer of proteins from SDS-polyacrylamide gels to unmodified nitrocellulose and radiographic detection with antibody and radioiodinated protein A. Anal Biochem 112:195.
24. Blattner WA, Blayney DW, Robert-Guroff M, Sarngadharan MG, Kalyanaraman VS, et al (1983). Epidemiology of human T-cell leukemia/lymphoma virus (HTLV). J Infect Dis 47:406.
25. Jaffe ES, Cossman J, Blattner WA, Robert-Guroff M, Blayney DW, et al (1984). The pathologic spectrum of adult T-cell leukemia/lymphoma in the United States. Am J Surg Pathol 8:263.
26. Uchiyama T, Yodoi J, Sagawa K, Takatsuki K, Uchino H (1977). Adult T-cell leukemia: clinical and hematologic features of 16 cases. Blood 50:481.
27. Harper M, Marselle LM, Gallo RC, Wong-Staal, F (1986). Detection of lymphocytes expressing human T-lymphotropic virus type III in lymph nodes and peripheral blood from infected individuals by in situ hybridization. Proc Natl Acad Sci, in press.

28. Salahuddin SZ, Rose RM, Groopman JE, Markham PD, Gallo RC (1986). Human T-lymphotropic virus type III (HTLV-III) infection of human alveolar macrophages. Blood, in press.
29. Gartner S, Markovits P, Markovitz D, Betts RF, Popovic M. Virus isolation from and identification of HTLV-III/LAV-producing cells in brain tissue from an AIDS patient, submitted.
30. Narayan O, Wolinsky JS, Clemens JE, et al. (1982). Slow virus replication: the role of macrophages in the persistence and expression of visna viruses of sheep and goats. J Gen Virol 59:345.
31. Weiss LM, Hu E, Wood GS, Moulds C, Cleary ML, Warnke R, Sklar J (1985). Clonal rearrangements of T-cell receptor genes in mycosis fungoides and dermatopathic lymphadenopathy. N Engl J Med 313:539.

Viruses and Human Cancer, pages 59–79

IMMUNOPATHOGENESIS OF HTLV-III/LAV INFECTION

Joseph B. Margolick,[1] H. Clifford Lane, and
Anthony S. Fauci

Laboratory of Immunoregulation, National Institute
of Allergy and Infectious Diseases, National
Institutes of Health, Bethesda, Maryland 20892

ABSTRACT The acquired immunodeficiency syndrome (AIDS) is caused by the human retrovirus termed human T lymphotropic virus (HTLV)-III or lymphadenopathy associated virus (LAV). This virus has the extraordinary property of selective tropism for a specific subset of T lymphocytes termed helper or inducer T cells defined by the T4 or Leu 3 phenotypic marker. Infection of the $T4^+$ subset results in a cytopathic effect on the target cell. Since this subset is responsible for the induction of a wide range of immunologic functions, a broad range of immunologic abnormalities ensues. The immunologic defect is manifested by both quantitative and qualitative abnormalities in $T4^+$ cell functions. Of note is the fact that there is a selective and early defect in the subset of T4 cells which is responsible for antigen recognition. Furthermore, abnormalities in antigen-presenting cells (monocyte/macrophages) compound the defect in response to specific antigens. The presence of additional abnormalities of monocytes likely reflects the infection of these cells with the HTLV-III/LAV. B cells are also abnormal in that they are polyclonally activated and defective in their response to de novo antigens. Abnormalities of natural killer cells and cytotoxic T cells likely

[1]Present address: Department of Environmental Sciences, Johns Hopkins University School of Hygiene and Public Health, Baltimore, Maryland 21205

reflect a defect in inductive signals delivered by the $T4^+$ cell. Certain of the monocyte and cytotoxic cell functional abnormalities can be corrected in vitro by T4 cell-derived soluble mediators such as interleukin-2 and interferon-γ, further implicating the lack of an inductive T4 cell signal as the cause of these abnormalities.

Following infection of suspensions of T4 cells with the AIDS retrovirus, a small proportion of cells survive and latently harbor virus in the absence of spontaneous virus secretion. Of note is the fact that these cells can be induced to secrete virus under appropriate conditions, representing important implications for the study of latent infection in the host. Studies on the differential effects of the retrovirus on various cloned populations of T4 cells as well as the direct stimulatory effect of the virus on subpopulations of B cells will be discussed.

INTRODUCTION

The acquired immunodeficiency syndrome (AIDS) was first recognized as a new clinical entity in 1981 when the Centers for Disease Control reported outbreaks of Pneumocystis carinii pneumonia (1) and Kaposi's sarcoma (2) in previously healthy homosexual men. The striking occurrence of these and other illnesses previously associated almost exclusively with immunocompromised hosts quickly led to the realization that the basic lesion in AIDS is a profound impairment of cellular immunity (3-5). Although humoral immunity was initially thought to be spared, it was subsequently learned that B cell function in patients with AIDS is also abnormal. However, the bacterial diseases associated with humoral immunodeficiency do not occur with increased frequency in adult patients with AIDS, and a profound impairment in cellular immunity has remained the hallmark of AIDS. More recently, with the discovery that AIDS is caused by a human retrovirus (6), a clearer understanding of the mechanisms underlying the development of this selective cellular immunodeficiency has emerged. This paper will summarize current knowledge concerning the cellular basis of immune abnormalities in patients with AIDS and will describe how the properties of the AIDS retrovirus may account for these abnormalities. For reasons to be explained below, it is believed that most, if not all, of these abnormalities derive from the defective functioning of the $T4^+$ (helper-inducer) subset of T lymphocytes.

IMMUNOLOGIC ABNORMALITIES

Quantitative Abnormalities of T Cells

Early descriptions of patients with AIDS noted that most of the patients had decreased numbers of circulating lymphocytes (2-5). AIDS was recognized shortly after the development and characterization of monoclonal antibodies, such as OKT4 and OKT8, that defined the helper-inducer ($T4^+$ or $CD4^+$) and suppressor-cytotoxic ($T8^+$ or $CD8^+$) subsets of T lymphocytes, respectively. With the use of these antibodies it became apparent that almost all of the lymphopenia was due to decreased numbers of helper-inducer T cells (3, 5, 7, 8). This decrease, along with the presence of normal or increased numbers of suppressor-cytotoxic T cells in most patients, led to reversal of the T4/T8 ratio, i.e., from greater than 1.0 to 2.0 in normal individuals to less than 1.0 in most patients with AIDS (8). Thus, the reversed T4/T8 ratio in AIDS could be clearly distinguished from the reversed ratios seen in immunologically normal individuals with certain viral infections, such as cytomegalovirus or Epstein-Barr virus (EBV), who have normal numbers of $T4^+$ cells but elevated numbers of $T8^+$ cells (9).

Functional Abnormalities of T Cells

Multiple functional T cell abnormalities have been reported in AIDS, including in vivo defects such as decreased or absent delayed hypersensitivity responses (3, 4, 7, 10), in vitro defects such as decreased mitogen- and antigen-induced blastogenesis (8, 11-13), decreased lymphokine production (11, 14), decreased cytotoxicity (15), decreased allogeneic (4, 16, 17) and autologous (12) mixed lymphocyte reactions, decreased T cell help for B cell immunoglobulin (Ig) synthesis (18, 19), and decreased expression of interleukin-2 (IL-2) receptors (20) and generation of IL-2 (11, 21, 22). These responses are normally initiated by the $T4^+$ helper cell (8, 23), the cell, as discussed above, that is most consistently decreased in number in patients with AIDS. Therefore, an important question is whether these abnormalities reflect simply the depletion of $T4^+$ helper cells or whether an additional defect in T cell function exists.

To study this question, purified $T4^+$ cells from patients with AIDS were compared to equal numbers of normal $T4^+$ cells. The responses of AIDS $T4^+$ cells to pokeweed mitogen, as indicated by proliferation, production of IL-2 and interferon (IFN)-γ, and expression of IL-2 receptors, were all normal (13), indicating that qualitative abnormalities in T4 cell functions were not present in addition to quantitative deficiencies of cell numbers. In contrast, purified T4 cells from patients with AIDS exhibited variable help for Ig synthesis, while $T8^+$ cells from patients with AIDS suppressed Ig synthesis normally in the presence of normal T4 cells (19). Most importantly, purified AIDS T4 cells consistently demonstrated the same greatly reduced proliferative responses to soluble antigen that were seen in bulk lymphocyte populations (13). This result suggests that patients with AIDS have an intrinsic defect in antigen-induced responses which may be fundamental to the immunopathogenesis of AIDS. In support of this concept is the fact that the lack of antigen-induced responses was pronounced even in patients with only mildly reduced numbers of T4 cells (13). In addition, a recent prospective study of patients with AIDS-related complex found that development of AIDS was more highly correlated with lack of antigen-induced lymphokine release than with numbers of T4 cells (24). The most likely explanation for this selective loss of antigen-specific T cell function is the relative or absolute depletion of the antigen-reactive T4 cell population. This population of T4 cells has been reported to bear the 4B4 antigen (25), while the complementary suppressor-inducer subset of T4 cells bears the 2H4 antigen (26). Thus far, however, it has not been shown in any prospective study that there is a preferential depletion of $T4^+/4B4^+$ cells in individuals infected with the AIDS retrovirus. Studies to answer this question are currently underway.

A further possibility to explain the lack of antigen-induced T4 cell responses in patients with AIDS, not mutually exclusive with depletion of antigen-reactive cells, is that antigen-reactive T4 cells are viable but functionally impaired. This possibility is consistent with the variable help provided by purified AIDS T4 cells to allogeneic B cells, as mentioned above. Further support for this possibility comes from experiments which evaluated the ability of purified viable T4 and T8 cells from patients with AIDS to be clonally expanded under appropriate culture conditions (i.e., limiting dilution in the presence of phytohemagglutinin [PHA], irradiated feeder cells, and IL-2). Both T4 and T8 cells from patients with AIDS were demonstrated to have cloning efficiencies which were approximately 85% less than purified T4 and T8 cells from healthy hetero- or homosexual donors (27). These results suggest an intrinsic defect affecting individual T4 and T8 cells in patients with AIDS approximately equally, but the basis for this defect remains to be established. Of note, the T cell clones that were ultimately obtained from patients with AIDS proliferated normally in response to PHA and provided normal help or suppression ($T4^+$ or $T8^+$ clones, respectively) for B cell Ig synthesis, suggesting that some T4 and T8 cells in patients with AIDS are not functionally impaired.

Functional and phenotypic abnormalities of suppressor-cytotoxic ($T8^+$) cell function in patients with AIDS have also been described. The proportion of such cells expressing the HLA-DR and T10 markers has been reported to be increased (3, 10); this was originally interpreted as suggesting an increased number of activated cells, although more recently evidence has been presented that the cells with these markers represent immature rather than activated cells, whose presence in the circulation correlates with the degree of depletion of $CD4^+$ cells (28). In keeping with this concept, $T8^+$ cell function is decreased rather than increased in patients with AIDS. For example, specific killing of cytomegalovirus-infected target cells by AIDS $T8^+$ cells was markedly reduced compared to similar cells from healthy donors (15). Of great significance is the finding that this functional defect was partially reversed by the addition of IL-2 to the cultures, showing that cytotoxic functional potential was present in the AIDS cells but required the presence of an inductive signal normally provided by T4 cells. Similar results were also obtained with natural killer (NK) cells (15).

Functional Abnormalities of B Cells

Serum Ig are characteristically elevated in patients with AIDS, who frequently have circulating immune complexes and autoantibodies (29-32). These findings reflect an intense polyclonal activation of B cells, similar to that seen in systemic lupus erythematosus (19). Although this observation is now well established, the mechanisms for such an intense state of B cell activation have been elusive. Many possible mechanisms have been considered, including 1) induction or transformation of B cells by a B-lymphotropic virus such as EBV, an infectious agent which is present in virtually all patients with AIDS (33) and which can induce hypersecretion of Ig by B cells in vitro; 2) a direct effect of the AIDS retrovirus, which can infect B cells under certain conditions, such as when they have already been transformed with EBV (34); 3) absence of normal regulation by functional T cells; 4) excessive secretion of B cell stimulating factors by virally transformed B cells, as can occur with human T lymphotropic virus, type I (HTLV-I)-transformed cells in vitro (35); or some combination of these mechanisms. In this regard, Pahwa et al. (36) have reported that a noninfectious component of the AIDS retrovirus induced as much as an 8-fold increase in the number of Ig-secreting cells present in cultures of peripheral blood mononuclear cells. This study did not address the question of a direct effect of the virus on B cells. However, very recent studies by Schnittman et al. (37) have demonstrated that highly enriched B cell populations cultured in the presence of AIDS retrovirus exhibited marked increases in both proliferation and Ig production within 48 hours after initiation of the cultures. This effect occurred in a much shorter period of time than is required for productive infection of T cells by the virus. Moreover, the effect was transient and did not lead to transformation of the B cells. Thus, it is unlikely that infection of the B cells was involved in these effects. Further studies are underway to confirm these findings and to define the properties of the AIDS retrovirus that are responsible for them.

Patients with AIDS fail to respond with production of specific antibody after exposure to antigens, either new antigens such as keyhole limpet hemocyanin or recall antigens such as tetanus toxoid (19, 38). Possible explanations for this finding include a lack of the antigen-specific T4 cell function needed to trigger an appropriate immune response, and a lack of resting B cells capable of being triggered. In any case, one implication of this finding is that the use of the usual serologic criteria for the diagnosis of infections is unreliable in patients with AIDS (39, 40).

Functional Abnormalities of Mononuclear Phagocytes

Although patients with AIDS generally have normal numbers of circulating monocytes, these cells exhibit a number of in vitro functional abnormalities, many of which appear to be due to a lack of IFN-γ, i.e., they can be reversed by the addition of this lymphokine to the cultures. Thus, defective extracellular killing of the parasitic organism Toxoplasma gondii in vitro by monocytes from patients with AIDS was enhanced by the addition of IFN-γ to the cultures (14). AIDS monocytes express lower levels of class II HLA antigens than normal monocytes (42-44), and this can also be reversed by exogenous IFN-γ (43, 44). Finally, monocytes from patients with AIDS increase H_2O_2 release and cytotoxicity normally in response to IFN-γ (14). These data all suggest that abnormalities of function of mononuclear phagocytes in AIDS are a result at least in part of lack of inductive signals from helper-inducer lymphocytes. Whether this accounts for other reported in vitro and in vivo functional abnormalities of these cells, such as spontaneous secretion of IL-1 and prostaglandin E_2 (45), diminished response to inducers of IL-1 release, diminished chemotactic responses (45), and impaired Fc receptor-mediated clearance (46) remains to be determined. Other possible explanations include a lack of regulatory signals other than IFN-γ from T cells, direct infection of mononuclear phagocytes by the AIDS retrovirus, and secondary effects arising from the presence of multiple infections in patients with AIDS.

Serologic Abnormalities and Circulating Suppressor Factors

A number of immune-related serologic abnormalities have been reported in patients with AIDS, and the importance of these is not currently known. Antilymphocyte antibodies directed against both $T4^+$ and $T8^+$ lymphocytes have been reported (30, 31), as have elevated circulating levels of β-2-microglobulin (47). In addition, elevations of α-1-thymosin (48) and an acid-labile IFN-α (49) as well as low levels of thymulin (50) have been reported. Sera from patients with AIDS have also been reported to contain factors that suppress immune functions of normal mononuclear cells in vitro. Among the responses inhibited are IFN-α-induced NK activity, mitogen- and alloantigen-induced proliferation (17, 51) and IL-2 production (51), and B cell responses to mitogens (52). The identity and significance of these factors in the immunopathogenesis of AIDS remain unknown.

THE AIDS RETROVIRUS AND IMMUNOPATHOGENESIS OF AIDS

It is now well established that the causative organism of AIDS is a newly discovered human retrovirus. Although several names for this virus have been used, including lymphadenopathy-associated virus (LAV) (53) and HTLV-III (54), we will refer to it simply as the AIDS retrovirus. This virus has several extraordinary properties that appear to be crucial to its ability to cause AIDS. First, the virus has a remarkable selective tropism for $T4^+$ lymphocytes (55). It is extremely cytopathic for these cells in vitro (55, 56); infected cells release infectious virus into the cultures before dying 10-14 days after infection. Several studies have in fact provided evidence that the T4 (CD4) molecule itself is necessary for viral infection of these cells, since antibodies directed against this molecule have been shown to inhibit infection (56-58) and only cells expressing the CD4 antigen were able to be infected with the AIDS retrovirus (57, 59). Lymphocytes which do not express detectable T4 antigen as detected by immunofluorescence using the monoclonal antibody OKT4 can also be infected, but only if these cells express other epitopes of the CD4 antigen (60). These data suggest that either the CD4 molecule or another molecule closely associated with it on the lymphocyte surface serves as the receptor for the virus.

Second, the virus, or viral products, may interfere directly with the function of the T4 lymphocyte. For example, infected lymphocytes lose the CD4 antigen even before they die. Moreover, our preliminary experiments indicate that very low levels of virus can significantly inhibit antigen- and mitogen-induced lymphocyte proliferative responses in vitro. In these experiments peripheral blood mononuclear cells from a tetanus toxoid immunized donor were stimulated with tetanus toxoid on day 0, exposed to purified virus for 1 hour on day 3, and then restimulated with tetanus toxoid or PHA on day 21.

Proliferative responses to both stimuli were inhibited in cells that had been exposed to virus. When the cells which had been exposed to virus were not initially stimulated with tetanus toxoid, subsequent antigen- but not mitogen-induced responses were inhibited (Table 1). These effects of the virus were noted despite the absence of sufficient virus replication in either experiment to produce detectable

TABLE 1
INHIBITION OF PROLIFERATIVE RESPONSES OF PERIPHERAL BLOOD MONONUCLEAR CELLS BY PRIOR EXPOSURE TO AIDS RETROVIRUS

Stimulating antigen (Day 0)	Virus dilution (Day 3)	Restimulation[a]		
		PHA[b]	TT[b]	KLH[b]
KLH	0	51,182[c]	2,973	29,349
	10^{-4}	8,767	888	392
	10^{-5}	27,979	2,066	14,804
TT	0	52,449	19,126	2,786
	10^{-4}	6,598	341	0
	10^{-5}	24,101	10,959	1,161
0	0	23,447	12,700	6,712
	10^{-4}	7,815	97	0
	10^{-5}	30,409	2,863	2,436

[a]Restimulation on day 21.
[b]PHA = phytohemagglutinin; TT = tetanus toxoid; KLH = keyhole limpet hemocyanin.
[c]Data expressed as Δcpm compared to cells restimulated with medium (including irradiated autologous peripheral blood mononuclear cells) alone.

amounts of reverse transcriptase. Moreover, cell viability in the cultures exposed to virus was comparable to that in the unexposed cultures. These data suggest a functional inhibition of antigen- and mitogen-induced T cell responses that cannot be accounted for by direct destruction of T4 cells by retroviral infection. In support of this concept, Pahwa et al. (36) have recently reported inhibition of antigen- and mitogen-induced lymphoproliferative responses in vitro by purified, noninfectious preparations of AIDS retrovirus particles. Additionally, in vitro inhibition of lymphocyte proliferative responses by HTLV-I and -II encoded envelope proteins, which have some homology to the corresponding proteins encoded by the AIDS retrovirus, has also been reported (61). The potential importance of these mechanisms of virus-related immunosuppression is emphasized by the finding that the proportion of susceptible lymphocytes actually infected with the AIDS retrovirus at a given time appears to be relatively small, at least with current techniques of detection (62). Thus, it is likely that mechanisms other than direct cell destruction by the virus contribute to immunosuppression in individuals infected with the AIDS retrovirus.

In this connection, persistence of viral infection in a small population of cells may be a critical consideration in both the pathogenesis and treatment of AIDS-related immunodeficiency. Among the types of cells in which the virus could potentially persist in vivo are brain cells (63, 64), B lymphocytes infected with EBV (34), and T4 lymphocytes which are latently rather than productively infected with the virus but may release infectious virus under suitable conditions. An experimental model for the latter has recently been provided by a $T4^+$ cell line, A3.01 (65). These cells undergo lytic infection with the AIDS retrovirus, but a small proportion of infected cells are not killed. These surviving cells harbor the retroviral genome, expression of which can be induced by exposure of the cells to IUdR and detected by measuring reverse transcriptase activity in the culture supernatants (Fig. 1). Of note, the latently infected cells lose expression of the CD4 molecule. Thus, the AIDS retrovirus appears to be capable of latent as well as productive infection of certain human cells. In this regard, Hoxie et al. (66) reported persistent infection of normal human lymphocytes in vitro over a period of 3-4 months. However, the possibility of a sustained low level of lymphocyte death due to retroviral infection was not conclusively excluded.

As mentioned above, loss of antigen-specific T cell function may be one of the fundamental immunologic lesions of infection with the AIDS retrovirus. This has led to the postulate that antigen-specific or antigen-induced T4 cells may be more susceptible than other T4 cells to depletion by the virus. However, there is as yet no direct experimental evidence to substantiate this hypothesis. In one study, no

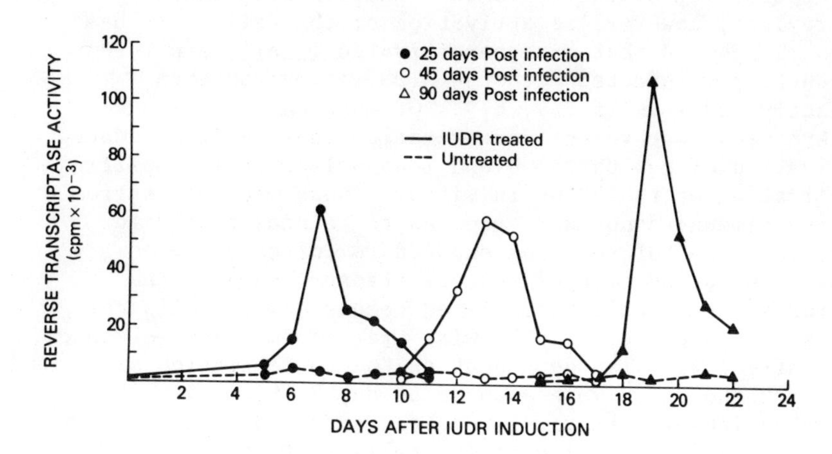

FIGURE 1. Inducibility of AIDS retrovirus after treatment with IUdR. Cells surviving AIDS retrovirus infection were maintained in culture for 25, 45, or 90 days. Aliquots (1 x 10^6) were treated with IUdR (100 μg/ml) for 24 hours, washed 3 times in medium, and cocultured with 1 x 10^6 A3.01 cells. Supernatants were tested daily for reverse transcriptase activity. Untreated survivor cells were also cocultured with A3.01 cells and monitored for reverse transcriptase activity. Taken from reference 65.

predilection of the virus for a particular T4 subset was noted, although only one phenotypic subset marker, Leu 8, was investigated (56). Of particular interest in this regard will be the results of pending studies to determine if cells expressing the phenotypic marker associated with the antigen-specific subset of $T4^+$ cells, i.e., the 4B4 antigen (25), exhibit an increase in susceptibility to infection with the virus, or in relative rate of depletion in vivo in individuals who are infected with the AIDS retrovirus.

One factor that appears to have a distinct influence on susceptibility of T4$^+$ cells to infection with the AIDS retrovirus, however, is activation of the cells. It has been recognized that mitogen-activated T cells are more productively infected with the AIDS retrovirus than nonactivated T cells (56, 67). Of more physiologic relevance is our recent demonstration that antigen-induced activation of peripheral blood mononuclear cells amplifies replication of the virus in vitro. Thus, when cells from a tetanus-immune donor were exposed to tetanus toxoid for 3 days in vitro before being exposed to infectious virus, subsequent production of reverse transcriptase in the cultures was greatly potentiated compared with cells not exposed to antigen but otherwise treated in the same manner; this effect was also dependent on the concentration of virus to which the cells were exposed on day 3 (Fig. 2). This potentiation of retroviral replication could lead either to the more rapid depletion of T4$^+$ cells or to the release of large amounts of free virus that could interfere with T4 cell function as discussed above. Thus, this finding may provide a mechanism for the frequently recognized correlation between development of AIDS and history of multiple infections, especially parasitic and viral infections that provide a chronic stimulation of the host immune system, and may explain, at least in part, why some individuals infected with the AIDS retrovirus eventually develop AIDS while others do not.

HOST DEFENSE MECHANISMS AGAINST THE AIDS RETROVIRUS

Most individuals infected with the AIDS retrovirus produce antibodies against several viral proteins, although exceptions have been noted (68). However, it has not been established which, if any, of these antibodies confer a protective effect on the host. Some studies have reported a neutralizing effect of antibodies on viral infectivity (69), but this has not been a consistent finding (70). Recent data by Rook et al. (manuscript submitted) have demonstrated that sera from many individuals infected with the AIDS retrovirus contain antibodies that are capable of mediating antibody-dependent cellular cytotoxicity (ADCC) against a target cell line infected with the AIDS retrovirus. The finding that levels of ADCC were higher in asymptomatic

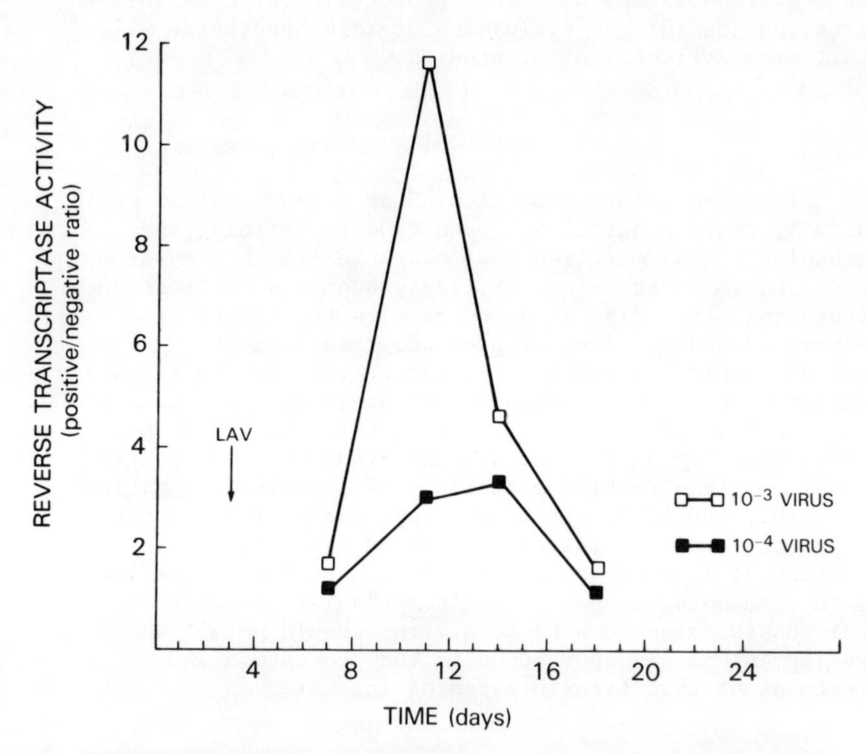

FIGURE 2. Reverse transcriptase production by tetanus toxoid (TT)-stimulated peripheral blood mononuclear cells (PBMC). PBMC from a TT-immune donor were cultured at 2×10^6/ml in 24-well culture plates in RPMI 1640 with 10% fetal calf serum in the presence or absence of 10 ug/ml TT. On day 3, 8×10^6 cells were resuspended in 1 ml of medium and cocultured with a 10^{-3} (□—□) or 10^{-4} (■—■) dilution of stock virus (NY3) for 1-3 hours at 37°C. The cells were then washed three times in fresh medium and incubated at 2×10^6/ml as above but without antigen. Culture medium was changed twice weekly and supernatants were stored at -70°C. Mg^{++} dependent reverse transcriptase activity of supernatants was assayed in duplicate, with all supernatants from a given experiment run in the same assay. Blastogenic response of 10^5 PBMC exposed to TT was 26,380 ($^\Delta$cpm) with 18 hour pulse on day 4-5.

infected individuals than in those with AIDS suggests that these antibodies may have some protective value against the virus, but definitive evaluation of this hypothesis will await prospective clinical studies.

CONCLUSION

Since the recognition of AIDS as a new clinical entity in 1981, rapid progress has been made in defining the immunologic abnormalities that characterize this syndrome. These abnormalities are most likely due to a combination of mechanisms, including 1) a decrease in the number of helper-induced ($T4^+$) cells present, resulting in commensurately reduced helper-inducer function to stimulate the function of other immunocompetent cells; 2) intrinsic functional abnormalities of T cells that may or may not depend directly on decreased helper-inducer function, such as the impaired cloning efficiency of T cells from patients with AIDS, and 3) direct immunosuppressive effects of the AIDS retrovirus, or molecules derived form it, on the function of T, and possibly B, cells. Further clarification of the immunologic and virologic mechanisms at work in individuals infected with this virus should permit the development of new approaches to the prevention and treatment of this life-threatening infection.

REFERENCES

1. Centers for Disease Control (1981). Pneumocystis pneumonia - Los Angeles. Morbid Mortal Weekly Rep 20:250.
2. Centers for Disease Control (1981). Kaposi's sarcoma and Pneumocystis pneumonia among homosexual men - New York City and California. Morbid Mortal Weekly Rep 25:305.
3. Gottlieb MS, Schroff R, Schander HM, Weisman DO, Fan PT, Wolf RA, Saxon A (1981). Pneumocystis carinii pneumonia and mucosal candidiasis in previously healthy homosexual men: evidence for a new acquired cellular immunodeficiency. N Engl J Med 305:1421.
4. Masur H, Michelis MA, Greene JB, Onorato I, Vande Stouwe RA, Holzman RS, Wormser G, Brettman L, Lange M, Murray HW, Cunningham-Rundles S (1981). An outbreak of community-acquired P. carinii pneumonia: initial

manifestation of cellular immune dysfunction. N Engl J Med 305:1431.

5. Siegal FP, Lopez E, Hammer GS, Brown AE, Kornfeld SJ, Gold J, Haset J, Hirschman SZ, Cunningham-Rundles C, Adelsberg BR, Parham DM, Siegal M, Cunningham-Rundles S, Armstrong D (1981). Severe acquired immunodeficiency in male homosexuals, manifested by chronic perianal ulcerative herpes simplex lesions. N Engl J Med 305:1439.
6. Fauci AS, Masur H, Markham PD, Hahn BH, Lane HC (1985). The acquired immunodeficiency syndrome: an update. Ann Intern Med 102:800.
7. Mildvan D, Mathur U, Enlow RW, Roman PL, Winchester RJ, Cop C, Singman H, Adelsberg BR, Springland I (1982). Opportunistic infections and immunodeficiency in homosexual men. Ann Intern Med 96:700.
8. Fauci AS, Macher AM, Longo Dl, Lane HC, Rook AH, Masur H, Gelmann EP (1984). Acquired immunodeficiency syndrome: epidemiologic, clinical, immunologic, and therapeutic considerations. Ann Intern Med 100:92.
9. Carney WP, Rubin RH, Hoffman RA, Hansen WP, Healey K, Hirsch MS (1981). Analysis of T lymphocytes subsets in cytomegalovirus mononucleosis. J Immunol 126:2114.
10. Schroff RW, Gottlieb MS, Prince HE, Chai LL, Fahey JL (1983). Immunological studies of homosexual men with immunideficiency and Kaposi's sarcoma. Clin Immunol Immunopathol 27:300.
11. Ciobanu N, Welk K, Kruger G, Venuta S, Gold J, Feldman S, Wang CY, Koziner B, Moore MAS, Safai B, Mertelsmann R (1983). Defective T-cell responses to PHA and mitogenic monoclonal antibodies in male homosexuals with acquired immune deficiency syndrome and its in vitro correction by interleukin-2. J Clin Immunol 3:332.
12. Gupta S, Safai B (1983). Deficient autologous mixed lymphocyte reaction in Kaposi's sarcoma associated with deficiency of Leu-3 positive responder cells. J Clin Invest 71:296.
13. Lane HC, Depper Jm, Greene WS, Whalen G, Waldmann TA, Fauci AS (1985). Qualitative analysis of immune function in patients with the acquired immunodeficiency syndrome: evidence for a selective defect in soluble antigen recognition. N Engl J Med 313:79.
14. Murray HW, Rubin BY, Masur H, Roberts RB (1984). Impaired production of lymphokines and immune (gamma)

interferon in the acquired immunodeficiency syndrome. N Engl J Med 310:883.

15. Rook AH, Masur H, Lane HC, Frederick W, Kasahara T, Macher AM, Djeu JY, Manischewitz JF, Jackson L, Fauci AS, Quinnan GV Jr (1983). Interleukin-2 enhances the drpressed natural killer and cytomegalovirus-specific cytotoxic activities of lymphocytes from patients with the acquired immune deficiency syndrome. J Clin Invest 72:398.
16. Amman AJ, Abrams D, Conant M, Dhudwin D, Cowan M, Volberding P, Lewis B, Casavant C (1983). Acquired immune dysfunction in homosexual men: immunologic profiles. Clin Immunol Immunopathol 27:315.
17. Cunningham-Rundles S, Michelis MA, Masur H (1983). Serum suppression of lymphocyte activation in vitro in acquired immunodeficiency disease. J Clin Immunol 3:156.
18. Benveniste E, Schroff R, Stevens RH, Gottlieb MS (1983). Immunoregulatory T cells in men with a new acquired immunodeficiency syndrome. J Clin Immunol 3:359.
19. Lane HC, Masur H, Edgar LC, Whalen G, Rook AH, Fauci AS (1983). Abnormalities of B lymphocyte activation and immunoregulation in patients with the acquired immunodeficiency syndrome. N Engl J Med 309:453.
20. Munn CG, Reuben JM, Hersh EM (1984). Interleukin-2 receptor expression on lymphocytes of patients with AIDS during in vitro mitogen stimulation. Lymphokine Res 3:97.
21. Kirkpatrick CH, Davis KC, Horsburgh CR Jr, Cohn DL, Penley K, Judson FN (1985). Interleukin-2 production by persons with the generalized lymphadenopathy syndrome or the acquired immune deficiency syndrome. J Clin Immunol 5:31.
22. Murray JL, Hersh EM, Reuben JM, Munn CG, Mansell PWM (1985). Abnormal lymphocyte response to exogenous interleukin-2 in homosexuals with acquired immune deficiency syndrome (AIDS) and AIDS related complex (ARC). Clin Exp Immunol 60:25.
23. Reinherz EL, Morimoto C, Fitzgerald KA, Hussey RE, Daley FJ, Schlossman SF (1982). Heterogeneity of human T4 inducer T cells defined by a monoclonal antibody that delineates two functional subpopulations. J Immunol 128:463.
24. Murray HW, Hillman JK, Rubin BY, Kelly CD, Jacobs JL, Tyler LW, Donelly DM, Carriero SM, Godbold JH, Roberts

RB (1985). Patients at risk for AIDS-related opportunistic infections: clinical manifestations and impaired gamma interferon production. N Engl J Med 313:1504.

25. Morimoto C, Letvin NL, Boyd AW, Hagan M, Brown HM, Kornacki MM, Schlossman SF (1985). The isolation and characterization of the human helper inducer T cell subset. J Immunol 134:3762
26. Morimoto C, Letvin NL, Distaso JA, Aldrich WA, Schlossman SF (1985). The isolation and characterization of the human suppressor inducer T cell subset. J Immunol 134:1508.
27. Margolick JB, Volkman DJ, Lane HC, Fauci AS (1985). Clonal analysis of T lymphocytes in the acquired immunodeficiency syndrome: evidence for an abnormality affecting individual helper and suppressor T cells. J Clin Invest 76:709.
28. Salazar-Gonzalez JF, Moody DJ, Giorgi JV, Martinez-Maza O, Mitsuyasu RT, Fahey JL (1985). Reduced ecto-5'-nucleotidase activity and enhanced OKT10 and HLA-DR expression on CD8 (T suppressor/cytotoxic) lymphocytes in the acquired immune deficiency syndrome: evidence of CD8 cell immaturity. J Immunol 135:1778.
29. McDougal JS, Hubbard M, Nicholson JKA, Jones BM, Holman RC, Roberts J, Fishbein CB, Jaffe HW, Kaplan JE, Spira TJ, Evatt BL (1985). Immune complexes in the acquired immunodeficiency syndrome (AIDS): relationship to disease manifestation, risk group, and immunologic deficit. J Clin Immunol 5:130.
30. Williams RC, Masur H, Spira TJ (1984). Lymphocyte-reactive antibodies in acquired immune deficiency syndrome. J Clin Immunol 4:118.
31. Tomar RH, John PA, Hennig AK, Kloster B (1985). Cellular targets or antilymphocyte antibodies in AIDS and LAS. Clin Immunol Immunopathol 37:37.
32. Walsh CM, Nardi MA, Karpatkin S (1984). On the mechanism of thrombocytopenia in sexually active homosexual men. N Engl J Med 311:635.
33. Quinnan GV, Masur H, Rook AH, Armstrong G, Frederick WR, Epstein J, Manischewitz JF, Macher AM, Jackson L, Ames J, Strauss SE (1984). Herpesvirus infections in the acquired immune deficiency syndrome. JAMA 252:72.
34. Montagnier L, Gruest J, Chamaret S, Dauguet C, Axler C, Guétard D, Nugeyre MT, Barré-Sinoussi F, Chermann J-C, Brunet JB, Klatzmann D, Gluckman JC (1984). Adaptation of lymphadenopathy associated virus (LAV) to

replication in EBV-transformed B lymphoblastoid cell lines. Science 225:63.
35. Salahuddin SZ, Markham PD, Lindner SG, Gootenberg J, Popovic M, Hemmi H, Sarin PW, Gallo RC (1984). Lymphokine production by cultured human T cells transformed by human T cell leukemia lymphoma virus I. Science 223:703.
36. Pahwa S, Pahwa R, Saxinger C, Gallo RC, Good RA (1985). Influence of the human T-lymphotropic virus/lymphadenopathy-associated virus on functions of human lymphocytes: evidence for immunosuppressive effects and polyclonal B cell activation by banded viral preparations. Proc Natl Acad Sci USA 82:8198.
37. Schnittman S, Lane HC, Higgins SE, Folks T, Fauci AS. Direct polyclonal activation of human B lymphocytes by the acquired immunodeficiency syndrome virus. Science, in press.
38. Ammann AJ, Schiffman G, Abrams D, Volberding P, Ziegler J, Conant M (1984). B-cell immunodeficiency in acquired immune deficiency syndrome. JAMA 251:1447.
39. Dylewski J, Chou S, Merigan TC (1983). Absence of detectable IgM antibody during cytomegalovirus disease in patients with AIDS. N Engl J Med 209:493.
40. Luft DJ, Conley J, Remington JS, Laverdiere M, Levine JF (1983). Outbreak of central-nervous system toxoplasmosis in Western Europe and North America. Lancet 1:781.
41. Roberts CJ (1984). Coccidioidomycosis in acquired immune deficiency syndrome. Depressed humoral as well as cellular immunity. Am J Med 76:734.
42. Belisto CV, Sanchez MR, Baer RL, Valentine R, Thorbecke GJ (1984). Reduced Langerhans' cell Ia antigen and ATPase activity in patients with the acquired immunodeficiency syndrome. N Engl J Med 310:1279.
43. Heagy W, Kelley VE, Strom TB, Mayer K, Shapiro HM, Mandel R, Finberg R (1984). Decreased expression of human class II antigens on monocytes from patients with acquired immune deficiency syndrome: increased expression with interferon-gamma. J Clin Invest 74:2089.
44. Murray HW, Gellene RA, Libby DM, Rothermel CD, Rubin BY (1985). Activation of tissue macrophages from AIDS patients: in vitro response of AIDS alveolar macrophages to lymphokines and interferon-gamma. J Immunol 135:2374.

45. Smith PS, Ohura K, Masur H, Lane HC, Fauci AS, Wahl SM (1984). Monocyte function in the acquired immune deficiency syndrome: defective chemotaxis. J Clin Invest 71:2121.
46. Bender BS, Frank MM, Lawley TJ, Smith WJ, Brickman CM, Quinn TC (1985). Defective reticuloendothelial system Fc-receptor function in patients with acquired immunodeficiency syndrome. J Infect Dis 152:409.
47. Bhalla RB (1983). Abnormally high concentrations of beta 2 microglobulin in acquired immunodeficiency syndrome (AIDS) patients. Clin Chem 19:1560.
48. Hersh EM, Reuben JM, Rios A, Mansell PW, Newell GR, McClur JE, Goldstein AL (1983). Elevated serum thymosin alpha-1 levels associated with evidence of immune dysregulation in male homosexuals with a history of infectious diseases or Kaposi's sarcoma. N Engl J Med 308:45.
49. DeStefano E, Friedman RM, Friedman-Kien AE, Godert JJ, Henriksen D, Preble OT, Sonnabend JA, Vilcek J (1982). Acid-labile human leukocyte interferon in homosexual men with Kaposi's sarcoma and lymphadenopathy. J Infect Dis 145:451.
50. Dardenne M, Bach JF, Safai B (1983). Low serum thymic hormone levels in patients with acquired immunodeficiency syndrome. N Engl J Med 309:48.
51. Siegal JP, Djeu JY, Stocks NI, Masur H, Gelmann EP, Quinnan GV Jr (1985). Sera from patients with the acquired immunodeficiency syndrome inhibit production of interleukin 2 by normal lymphocytes. J Clin Invest 75:1975.
52. Laurence J, Mayer L (1984). Immunoregulatory lymphokines of T hybridomas from AIDS patients: constitutive and inducible suppressor factors. Science 225:66.
53. Barré-Sinoussi F, Chermann J-C, Rey F, Nugeyre MT, Chamaret S, Gruest J, Dauguet C, Axler-Blin C, Brun-Vezinet F, Rouzioux C, Rozenbaum W, Montagnier L (1983). Isolation of a T lymphotropic retrovirus from a patient at risk for acquired immune deficiency syndrome (AIDS). Science 220:868.
54. Gallo RC, Salahuddin SZ, Popovic M, Shearer GM, Kaplan M, Haynes BF, Palker TF, Redfield R, Oleske J, Safai B, White G, Foster P, Markham PD (1984). Frequent detection and isolation of cytopathic retroviruses (HTLV-III) from patients with AIDS and at risk for AIDS. Science 224:500.

55. Klatzmann D, Barré-Sinoussi F, Nugeyre MT, Dauguet C, Vilmer E, Griscelli C, Brun-Vezinet F, Rouzioux C, Gluckman JC, Chermann J-C, Montagnier L (1984). Selective tropism of lymphadenopathy associated virus (LAV) for helper-inducer T lymphocytes. Science 225:59.
56. McDougal JS, Mawle A, Cort SP, Nicholson JKA, Cross GD, Scheppler-Campbell JA, Hicks D, Sligh J (1985). Cellular tropism of the human retrovirus HTLV-III/LAV. I. Role of T cell activation and expression of the T4 antigen. J Immunol 135:3151.
57. Dalgleish AG, Beverley PCL, Clapham PR, Crawford DH, Greaves MF, Weiss RA (1984). The CD4 (T4) antigen is an essential component of the receptor for the acquired immunodeficiency syndrome retrovirus. Nature 312:763.
58. Klatzmann D, Champagne E, Chamanet E, Gruest J, Guétard D, Hercend T, Gluckman JC, Montagnier L (1984). The T4 molecule behaves as the receptor for human retrovirus LAV. Nature 312:767.
59. Popovic M, Read-Connole E, Gallo RC (1984). T4 positive human neoplastic cell lines susceptible to and permissive for HTLV-III. Lancet 2:1472.
60. Hoxie JA, Flaherty LE, Haggarty BS, Rackowski JL (1986). Infection of T4 lymphocytes by HTLV-III does not require expression of the OKT4 epitope. J Immunol 136:361.
61. Cianciolo GJ, Copeland TD, Oroszlan S, Snyder R (1985). Inhibition of lymphocyte proliferation by a synthetic peptide homologous to retroviral envelope proteins. Science 230:453.
62. Shaw GM, Hahn BH, Arya SK, Groopman JE, Gallo RC, Wong-Staal F (1984). Molecular characterization of human T cell leukemia (lymphotropic) virus Type III in the acquired immune deficiency syndrome. Science 226:1165.
63. Ho DD, Rota TR, Schooley RT, Kaplan JC, Allan JD, Groopman JE, Resnick L, Felsenstein D, Andrews CA, Hirsch MS (1985). Isolation of HTLV-III from cerebrospinal fluid and neural tissues of patients with neurologic syndromes related to the acquired immunodeficiency syndrome. N Engl J Med 313:1493.
64. Resnick L, DiMarzo-Veronese F, Schupbach J, Tourtellotte WW, Ho DD, Muller F, Shapshak P, Vogt M, Groopman JE, Markham PD, Gallo RC (1985). Intra-blood-brain-barrier synthesis of HTLV-III-specific IgG in patients with neurologic

symptoms associated with AIDS or AIDS-related complex. N Engl J Med 313:1498.

65. Folks T, Powell DM, Lightfoote MM, Benn S, Martin MA, Fauci AS (1986). Induction of HTLV-III/LAV expression from a nonvirus-producing T cell line: implications for latent infection in man. Science 231:600.
66. Hoxie JA, Haggarty BS, Rackowski JL, Pillsbury N, Levy JA (1985). Persistent noncytopathic infection of normal human T lymphocytes with AIDS-associated retrovirus. Science 229:1400.
67. Montagnier L, Chermann JC, Barré-Sinoussi F, Chamaret S, Gruest J, Nugeyre MT, Rey F, Dauguet C, Axler-Blin C, Vezinet-Brun F, Rouzioux C, Saimot G-A, Rozenbaum W, Gluckman JC, Klatzmann D, Volmer E, Griscelli C, Foyer-Gazengel C, Brunet JB (1984). A new human T-lymphotropic virus: characterization and possible role of lymphadenopathy and acquired immune deficiency syndromes. In Gallo RC, Essex ME, Gross L (eds): "Human T Leukemia Lymphoma Viruses," Cold Spring Harbor, NY: Cold Spring Harbor Laboratories, p. 363.
68. Salahuddin SZ, Markham PD, Redfield RR, Essex M, Groopman JE, Sarngadharan MG, McLane MF, Sliski A, Gallo RC (1984). HTLV-III in symptom-free seronegative persons. Lancet 2:1418.
69. Robert-Guroff M, Brown M, Gallo RC (1985). HTLV-III-neutralizing antibodies in patients with AIDS and AIDS-related complex. Nature 316:72.
70. Weiss RA, Clapham PR, Cheingsong-Popov R, Dalgleish AG, Carne CA, Weller IVD, Tedder RS (1985). Neutralization of human T-lymphotropic virus type III by sera of AIDS and AIDS-risk patients. Nature 316:69.

Viruses and Human Cancer, pages 81–91

PREVALENCE OF ANTIBODIES TO HUMAN T-LYMPHOTROPIC VIRUS TYPES I AND III (LAV) IN HEALTHY VENEZUELAN POPULATIONS

F. Merino[1], L. Rodriguez[1],
S. Dewhurst, F. Sinangil, D.J. Volsky*

Molecular Biology Laboratory
Department of Pathology and Microbiology
University of Nebraska Medical Center
Omaha, Nebraska 68105

and

[1]Departamento Medicina Experimental,
Instituto Venezolano de Investigaciones
Cientificas Apdo, 1827 Caracas 1010A, Venezuela

ABSTRACT. Serum samples from blood donors from major cities in Venezuela and from healthy subjects residing in various regions of the country were tested for antibodies to HTLV-III/LAV virus. The antibody prevalence varied from 0.6 to 6.0% when tested by an indirect immunofluorescence assay (IF) using a HTLV-III/LAV-producer cell line. Titers ranged from 1:40 to 1:320. 2.3% of the sera tested were confirmed positive for specific anti-HTLV-III/LAV antibodies by Western blot and radioimmunoprecipitation assays. These observations indicate that HTLV-III/LAV or an antigenically related virus is indigenous and endemic to certain South American populations. Comparison of the HTLV-I and HTLV-III/LAV antibody distributions in Venezuela showed a different epidemiological pattern. Our data suggest that environmental factors can play role in retrovirus transmission in tropical populations.

This work was supported by anAmerican Foundation for AIDS Research and by a grant CA37465 from the National Institute of Health (to DJV). *Corresponding author.

INTRODUCTION

Human retroviruses are implicated in the etiology of adult T-cell leukemia (HTLV type I) and acquired immunodeficiency syndrome, AIDS (HTLV-III/LAV) (1). The HTLV-I virus has been shown to be endemic in Japan (2), the Caribbean basin including mainland parts of South America (3,4), and in Africa (5). HTLV-III/LAV infection is highly prevalent in Central Africa (6-9) and in some of the Caribbean islands (9,10). We have recently reported the presence of antibodies to HTLV-III/LAV or a closely related virus among aboriginal Indians living in the Amazonas territory of Venezuela (12) and in some Venezuelan patients with malaria (13). The results of a wide-scale testing for anti-HTLV-III/LAV antibodies in the general Venezuelan population, presented here, suggest that HTLV-III/LAV, like HTLV-I (3), might be indigenous to tropical Latin American populations without prior history of AIDS.

RESULTS

We obtained 169 serum samples from blood bank donors from the main hospitals in Caracas, Maracaibo, Valencia, Maracay, Cumana, and Barcelona, and 464 sera from healthy subjects residing in rural or more developed communities throughout Venezuela. In addition, we tested sera from 99 patients with leukemia, 36 patients with hemophilia and 50 patients with Chagas' disease. The results of HTLV-III/LAV seroepidemiological surveys among Amazonian Indians and patients with malaria have been reported (12,13). None of the individuals tested had apparent AIDS, or AIDS-related disease. Viral antibodies were detected by a sensitive and specific indirect immunofluorescence assay, IF, using an HTLV-III/LAV-positive cell line CEM/LAV-N1T (Ref. 14). Presence of antibodies to specific HTLV-III/LAV antigens was confirmed by Western blotting (WB) and radioimmunoprecipitation (RIP).

The overall frequency of samples positive in the IF assay in the general population was 6.8% (Table I and Figure 1). Antibody titers ranged from 1:40 to 1:320. Thirty-two sera (3.0%) were positive for HTLV-III/LAV antibodies by all three assays used in the survey (IF,

TABLE 1. PREVALENCE OF ANTIBODIES TO HTLV-III/LAV IN VARIOUS VENEZUELAN POPULATION GROUPS

Population group tested	Number of sera tested			Number of sera positive for HTLV-III/LAV by IF				Number of sera positive for HTLV-III/LAV by IF, WB & RIP			
	Total	M	F	Total	% of tested	M	F	Total	% of tested	M	F
Blood bank donors[a]	169	149	20	1		1	0	0			
Other blood donors[b]	464	167	297	28	0.6	15	13	11	2.3	6	5
Indians	224	108	116	22	9.8	12	10	9	4.0	6	3
Leukemia patients	99	62	37	0				0			
Hemophiliacs	36	36	0	2	5.5	2		2	5.5	2	
Malaria patients	24	19	5	12	50.0	10	2	8	33.3	7	1
Chagas' patients	50	25	25	8	16.0	5	3	2	4.0	0	2
TOTAL	1066	573	500	73	6.8	50	28	32	3.0	20	11

[a]Obtained via blood banks in major metropolitan hospitals; [b]Obtained from healthy volunteers residing in rural communities and small towns throughout Venezuela (c.f. also map in Fig. 1). IF: indirect immunofluorescence assay performed using acetone-fixed smears of CEM/LAV-N1 cells (12,14); WB and RIP: Western blot and radioimmunoprecipitation assays, respectively, performed as described previously (14); M: Male; F: Female.

WB, RIP). None of the randomly chosen serum samples from 169 healthy blood donors in 6 major Venezuelan cities was positive.

Geometric mean antibody titers for all groups of seropositives ranged between 76 and 160. Most of the low titer positive sera contained antibodies to the 42 kd polypeptide. In addition, each serum contained antibodies recognizing at least one other HTLV-III/LAV-specific polypeptide, i.e., p32, p68, gp120 and gp160. The presence of antibodies to the core proteins of 18 and 25 kd was verified by the more sensitive RIP assay.

DISCUSSION

Prevalence and geographic distribution of antibodies to HTLV-III/LAV-like retrovirus in Venezuela

Several investigators have suggested that AIDS and AIDS retrovirus originated in central Africa (1,6-9). It has also been suggested that the HTLV-III-like virus of African green monkeys may have been transmitted to man coincident with the recognition of AIDS on that continent (15). However, the first report of AIDS in Africa appeared only in 1983 (Ref. 16), a high frequency of virus-infected healthy subjects and absence of AIDS has been described in several African countries (6,8), and contradictory results regarding HTLV-III/LAV antibody prevalence in these populations have also been reported (17,18). The HTLV-III/LAV seroepidemiological data presented here and published elsewhere (12,13) clearly indicate that HTLV-III/LAV or an antigenically-related virus might be indigenous to certain populations in Venezuela and perhaps in other tropical areas of South America.

The results of testing for HTLV-III/LAV antibodies in Venezuela showed marked variations in antibody prevalence by population group, geographical region or associated tropical diseases (Figure 1).

The low antibody prevalence, 0.6%, among blood donors from seven different cities in the country, geographically distant from each other, was similar to that observed in USA or Europe. A significantly higher frequency was observed among apparently healthy subjects residing in different rural communities throughout the country. This discrepancy is not due to differences in

Fig. 1. Map of Venezuela showing the frequency of HTLV-III/LAV antibodies as tested by an IF assay in the various populations/communities. Panare, Makaritare, Pemon and Yanoami represent Amazonian Indian tribes.

the socioeconomical conditions of the studied communities since all were of a similar degree of development.

No seropositive individuals were found in 3 different communities in the State of Falcon; namely, Adicora (a small rural town at the seashore in the arid northern part of the state), Chruguara (a small city in the southern mountain areas), and Coro, the state capital. Similar results were obtained in the neighboring State of Zulia. Seropositive individuals were found in the village of Bobures but not in the nearby communities of Caja Seca and La Conquista. High antibody prevalence was found in Torondoy, a settlement in the Andes.

In the State of Aragua, several rural communities with more or less similar degrees of development, but differing in their geographical and environmental conditions, were studied. Choroni and Ocumare are small towns located on the seashore. Uraca, Los Tanques and Cuyagua are located at different altitudes in the nearby mountains. Magdaleno, San Francisco, Villa de Cura and La Guacamaya are located on the southern plains. Whether the observed differences are due to the different environmental conditions is speculative. It should be noted that in areas with the higher antibody prevalence, such as La Guacamaya, arthropode transmitted tropical diseases such as leishmaniasis and Chagas' disease are frequently observed.

In addition to the above described ethnic groups and localities (see also the map in Fig. 1.), antibodies to HTLV-III/LAV have been detected among the aboriginal indians inhabiting the Amazonas Territory in Venezuela (12). In this geographical area malaria as well as other tropical diseases are highly prevalent. Perhaps not surprisingly, antibodies to HTLV-III/LAV or a related virus have been detected in a high proportion (29.2%) of patients acutely infected with malaria _P. vivax_ or _P. falciparum_ (13). A high prevalence of anti-HTLV-III/LAV antibodies has also been reported in African patients with acute malaria or other parasitic infections (8). The significance of these findings is unclear. Immunoregulatory disturbances, such as the a depletion of T4- positive peripheral blood lymphocytes with a relative increase of the T8-positive subset, polyclonal hypergammaglobulinemia, circulating immunocomplexes and auto-antibodies, have been associated with acute malaria. The possible association, if any, between these abnormalities and HTLV-III/LAV expression has to be determined. Even more intriguing is the question of the possible mechanism by virtue of which malaria patients acquire the viral infection. Two possibilities can be suggested. One is that the virus might be transmitted by an arthropod. Anti-HTLV-III/LAV antibodies were detected in areas where arthropode-transmitted diseases are common i.e., the States of Aragua, Zulia, Amazonas (Fig. 1). Similar observations were made in Africa (8). The other possibility is that a preexisting viral genome in T cells is activated via stimulation of these cells by parasite antigens. Support for this hypothesis can be found in a recent report by Zagury et al (19) showing that immunologic activation of HTLV-III/LAV-infected cells leads to increased expression of the virus and cell death.

The explanation for the presence of anti-HTLV-III/LAV antibodies in normal, healthy Venezuelan subjects is not clear. AIDS is uncommon in Venezuela. Only 8 cases have been reported to the World Health Organization by the National AIDS Commission of the Health Department (20). For several of the populations tested (e.g., La Guacamaya) immunological studies did not reveal any major immune functions. Most probably, in tropical negroid and non-negroid populations HTLV-III/LAV-like virus constitutes a natural infection that does not cause major disease. Alternatively, some cofactors required for the pathogenicity of this virus might be lacking. It is also possible that the virus against which antibodies have been detected in healthy (in terms of clinical AIDS) Venezuelan (Table 1, Fig. 1) and African (7-9) populations might be a non-pathogenic ancestor of the AIDS-retrovirus, or a new member of the family of human retroviruses. Isolation and characterization of the South American retrovirus will allow us to distinguish between these possibilities.

Seroepidemiology of HTLV-III/LAV and HTLV-I in Venezuela

It has been previously reported that HTLV-I antibodies are highly prevalent in Venezuela (3,4). Notable geographical variation in antibody prevalence has also been described (3). Specific antibody frequency varied from less than 1% in Caracas to 13.7% in the Amazonas region and the State of Zulia. Transmission by environmental factors among tropical populations was suggested (3). Since most of the populations tested for anti-HTLV-III/LAV antibodies in the present study were previously assayed for HTLV-I, we could compare the prevalence of antibodies to these two viruses among various Venezuelan populations and geographical locations. As shown in Table II, specific antibodies to HTLV-I and to HTLV-III/LAV showed a different epidemiological pattern. HTLV-I antibodies were frequently observed in the western part of the country where HTLV-III/LAV was not detected (except for Torondoy in the State of Merida, where the situation was reversed. HTLV-III/LAV and HTLV-I had a similar distribution in the central part of the country. They were not detected in blood donors from Caracas, and they

TABLE 2. ANTIBODIES TO HTLV-I AND HTLV-III/LAV IN VENEZUELAN POPULATIONS.

Location of serum donors	Frequency of specific antibodies to: (in % of positive sera) HTLV-I*	HTLV-III/LAV**
Aragua:		
La Guacamaya	9.3	8.3
Falcon:		
Coro	3.9	<1%
Churuguara	3.9	<1%
Tachira:		
San Cristobal	5.9	<1%
Zulia:		
Caja Seca	9.8	<1%
Maracaibo	7.9	<1%
Paraguaipoa	13.7	<1%
Amazonas Teritory:		
Yanoama indians	13.7	3.3%
Patients with Chagas' disease	13.7	29.2%

*Compiled from data presented in Ref. 3. Antibodies were detected by ELISA assay and p24 specificity determined in competition experiments. **Positive by IF, WB, RIP in parallel (c.f. Ref. 12-14).

were frequently observed in the rural communities, such as La Guacamaya in the State of Aragua, as well in Chagas' disease patients residing in rural areas of the nearby States Aragua and Guarico. A similar frequency was observed in the Yanoama indians in the Amazonas region.

Our studies thus indicate that both HTLV-I and HTLV-III/LAV are not only indigenous to but highly endemic in South America and suggest, as postulated for Africa (6,8), a dependence on environmental transmission factors. The association of retrovirus infection with parasitic diseases, i.e., Leishmania, malaria, Chagas' disease requires further evaluation. Likewise, the possible role of these viruses in regional variants of cancer and immunodeficiency-related diseases remains to be determined.

ACKNOWLEDGMENTS

We thank Dr. Gregorio Godoy, Universidad de Oriente, Dr. Norma Bosh, Banco Municipal de Sangre, Caracas, Dr. Anabel Merino, Blood Bank, Hospital Universitario, Caracas, Dr. Luigi De Salvo, Institute Hematologica de Occidente, Maracaibo, and Dr. Tebaida Tapia, Hematology Service Hospital Central, Maracay for kindly providing some of the sera studied. We also thank M. Hedenskog and B. Ward for excellent technical assistance and G. Barton for typing the manuscript.

REFERENCES

1. Wong-Staal F, Gallo RC. (1985). Human T-lymphotropic retrovirus. Nature 317:395.

2. Hinuma Y, Komoda H, Chosa T, Kondo T, Kohakura M, Takenaka T, Kikuchi M, Ichimaru M, Yunoki K, Sato I, Matsuo R, Takiuchi Y, Uchino H, and Hanaoka M. (1982) Antibodies to adult T-cell leukemia-virus associated antigen (ATLA) in sera from patients with ATL and controls in Japan: a nationwide sero-epidemiologic study. Int J Cancer 29:631.

3. Merino F, Robert-Guroff M, Clark J, Biondo-Bracho M, Blattner WA, Gallo RC. (1984) Natural antibodies to human T-cell leukemia/lymphoma virus in healthy Venezuelan populations. Int J Cancer 24: 501.

4. Robert-Guroff M, Shupbach J, Blayney DW, Kalyanaraman VS, Merino F, Sarngadharan MG, Clark J, Saxinger WC, Blattner WA, Gallo RC (1984) Seroepidemiologic studies on human T-cell leukemia/lymphoma virus type I In Gallo RC, Essex ME, Gross L (eds): "Human T-cell leukemia/lymphoma virus," Cold Spring Harbor Laboratory, New York p 285.

5. Saxinger W, Blattner WA, Levine PH, Clark J, Biggar R, Hoh M, Moghissi J, Jacobs P, Wilson L, Jacobson P, Crookes R, Strong M, Ansari AA, Dean AG, Nkrumah FK, Mourali N, Gallo RC. (1984) Human T-cell leukemia virus (HTLV-I) antibodies in Africa. Science 225: 1473.

6. Saxinger, WC, Levine, PH, Dean, AG, de The G. (1985) Evidence for exposure to HTLV-III in Uganda before 1973. Science 227:1036.

7. Biggar RJ, Melbye M, Kestens L, de Peyter M, Saxinger C, Bodner AJ, Paluko L, Blattner WA, Gigase PL. (1985) Seroepidemiology of HTLV-III antibodies in a remote population of eastern Zaire Br Med J 290:808.

8. Biggar RJ, Johnson BK, Oster C, Sarin PS, Ocheng D, Tukei P, Nsanze H, Alexander S, Bodner AJ, Siongok TA, Gallo RC, Blattner RC. (1985) Regional variation in prevalence of antibody against human T-lymphotropic virus types I and III in Kenya, East Africa. Int J Cancer 35:763.

9. Gazzolo L, Robert-Guroff M, Jennings A, Duc Dodon M, Najberg G, De-The G, (1985) Type-I and Type-III antibodies in hospitalized and out patients Zairans. Int J Cancer 36:373.

10. Gazzolo L, Gessain A, Robin Y, Robert-Guroff M, De-The G. (1984) Antibodies to HTLV-III in Haitian immigrants in French Guiana. N Engl J Med 311:1252.

11. Pitchenick AE, Spira TJ, Elie R, Fischl MA, Getchell JP, Arnoux E, Pierre GD, LaRoche AC, Guerin JM, Malebranche R. (1985) Prevalence of HTLV-III/LAV antibodies among Haitians. N Engl J Med 313:1705.

12. Rodriguez L, Dewhurst S, Sinangil F, Merino F, Godoy G, Volsky DJ. (1985) Natural antibodies to HTLV-III/LAV among aboriginal amazonian indians in Venezuela. Lancet ii:1098.

13. Volsky DJ, Merino F, Rodriguez L, Wu YT, Stevenson M, Dewhurst S, Sinangil F, Godoy G. Antibodies to HTLV-III/LAV in Venezuelan patients with acute malaria infectious (P. falciparum and P. vivax) New Engl J Med, in press.

14. Casareale D, Dewhurst S, Sonnabend J, Sinangil F, Purtilo DT, Volsky D. (1984) Prevalence of AIDS-associated retrovirus and antibodies among male homosexuals at risk for AIDS in Greenwich Village. Aids Res 1:407.

15. Kanki PJ, Alroy J, Essex M. (1985) Isolation of T-Lymphotropic Retrovirus Related to HTLV-III/LAV from Wild-Caught African Green Monkeys. Science 230:951.

16. Clumeck N, Mascart-Lemone F, de Maubegue J, Brenez D, Marcelis L. (1983) Acquired immune deficiency syndrome in black Africans. Lancet ii:642.

17. Hunsmann G, Schneider J, Wendler I, Freming AF. (1985) HTLV-III positivity in Africans. Lancet ii:952.

18. Montagnier L. (1985) Lymphadenopathy-associated virus: from molecular biology to pathogenicity. Ann Int Med 103:689.

19. Zagury D, Bernard J, Leonard R, Cheynier R, Feldman M, Sarin PS, Gallo RC. (1986) Long-term cultures of HTLV-III-infected T cells: A model of cytopathology of T-cell depletion of AIDS. Science 231:850-853.

20. Brunet JB, Ancelle RA. (1985) The international occurrence of the acquired immunodeficiency syndrome. Ann Int Med 103:670.

Viruses and Human Cancer, pages 93–103
Published 1987 by Alan R. Liss, Inc.

ISSUES IN THE SEROEPIDEMIOLOGY OF HUMAN RETROVIRUSES

Paul H. Levine[1], William A. Blattner[1],
Robert J. Biggar[1], Jeffrey Clark[1], Stanley Weiss[1],
Marjorie Robert-Guroff[2], and W. Carl Saxinger[2]

Environmental Epidemiology Branch, Division of Cancer Etiology and Laboratory of Tumor Cell Biology, Division of Cancer Therapy National Cancer Institute, National Institutes of Health, Bethesda, Maryland 20892

ABSTRACT The development of sensitive serologic tests and the availability of purified antigens have improved the capacity of the laboratory to dissect the immune response to human retroviruses with great precision. The diversity of immunologic responses to infection with HTLV-I and HTLV-III has, however, thus far limited the application of the more specific assays in sero-epidemiologic studies. While the presence of acute T-cell leukemia/lymphoma (ATLL) and the acquired immuno-deficiency syndrome (AIDS) highlight areas where HTLV-I and HTLV-III respectively, can be isolated, it is difficult to interpret seroepidemiologic data where viral antibodies are detected with no apparent sentinel disease. In this respect, we review our approaches for clarifying the seroepidemiology of HTLV-I and HTLV-III and suggest approaches for clarifying discrepant data from different laboratories.

[1]Environmental Epidemiology Branch, 3C25 Landow Building, National Cancer Institute, National Institutes of Health, Bethesda, Maryland 20892

[2]Laboratory of Tumor Cell Biology, National Cancer Institute, National Institutes of Health, Bethesda, Maryland, 20892

INTRODUCTION

Since the isolation of human T-cell leukemia virus (HTLV)-I and the identification of its strong association with adult T-cell leukemia/lymphoma (ATLL) (1, 2), there has been an explosion of information on human T-lymphotropic retroviruses. Rapid advances in the laboratory have resulted in a series of serologic assays which have been applied to a variety of populations. The discoveries of additional lymphotropic retroviruses with varying degrees of homology (3-6), the realization that additional human retroviruses are likely to be identified, and the knowledge that a variety of antibodies can be produced in individuals infected by the same virus have led us to re-evaluate data in the literature in order to clarify the patterns of infection and disease associated with HTLV-I and -III. In this review, we summarize some of the methodologic controversies and the relevant data that have led to our current approach to retrovirus seroepidemiology.

We have used two distinct approaches to investigate the geographic distribution of HTLV-I: 1) searching for clinically diagnosed ATLL as a sentinel disease, and 2) the screening of sera from well-defined populations for HTLV-I antibody distribution.

The rationale for using ATLL as a marker for HTLV-I endemicity is based on the well-documented association of HTLV-I and ATLL in southwest Japan (7-9), where the unusual geographic restriction of both the virus and the tumor remains unexplained. The data from Japan suggest that HTLV-I is not highly infectious, with transmission mediated primarily by infected cells. The Caribbean basin, a second endemic area for HTLV-I and ATLL (10, 11), has been shown to resemble Japan with regard to both the age and sex distribution of the virus and to the incidence of ATLL in the general population. The importance of place of birth in determining risk of developing ATLL was demonstrated in Japan (9), and has also been noted in Caribbean-born migrants to Great Britain (11, 12) and Holland (13).

Although it is not possible to define the precise extent of HTLV-I infection outside of the highly endemic areas noted above, it is reasonable to assume that the presence of HTLV-I associated malignancies in a region serves as confirmation of serologic studies identifying viral antibodies in the general population. Therefore, the identification of such cases in natives of Nigeria (14), Taiwan (15), Israel (16), Italy (17), Alaska (18), Ecuador (16), Brazil (18), and the Southeast United States (19), attest to the widespread distribution of the virus. Based on our experience with more than 42,000 cases, and relying on specific competition assays (8, 20) as well as Western blot, we have tried to develop a map that approximates the worldwide distribution of HTLV-I infection (Fig. 1). It should be emphasized that the pictorial presentation of virus distribution is handicapped by the focal nature of infection but, although the precise percentages could be affected by viruses or other factors immunologically related to HTLV-I, we have adopted the most conservative approach which is to reserve defining an area as endemic only if there is both seropositivity in the normal population (8, 10, 21, 22) and HTLV-I positive ATLL patients. Because of the logistic difficulties of documenting the presence of virus in tumor cells obtained from geographic locales far from the virology laboratories, seropositivity, in conjunction with classic ATLL features, has been sufficient for the designation of HTLV-I-related ATLL in current studies.

Serosurveys in Africa have led to widely varying estimates of HTLV-I prevalence (21, 23-27). Since no assay has yet proven to be a "gold standard" comparable to the viral capsid antibody test used worldwide to detect infection with the Epstein-Barr virus, we would interpret the conflicting data in the literature as follows: 1) Our findings of seropositivity in surveys of sub-Saharan Africa, which used reliable competition assays, are supported by the identification of HTLV-I associated ATLL in native West Africans (14, 28) and, therefore, the presence of HTLV-I in this area seems likely. Our estimate of seroprevalence in West Africa (Fig. 1) is further supported by the studies of Hunsmann et al (26), who observed viral antibodies in sera from 6 of 22 patients with chronic lymphocytic leukemia and non-Hodgkin's lymphoma, as well as 6 (3.7%) of 161 healthy blood donors. Thus, in Nigeria, HTLV-I antibody was detected in two laboratories using different assays, the first using a competition assay with specific heterologous antisera (18) and the second using both an immunofluorescent assay and immunoprecipitation (26). 2) The ELISA test, which has proven to be a reliable predictor of specific HTLV-I antibody in most geographic locales, has been less reliable for African sera (29, 30, submitted for publication). Thus African serologic data based solely on the HTLV-I ELISA (23) resulted in an overestimate of virus prevalence. We emphasize that the correlation between malaria and ELISA reactivity to retrovirus tests (29) has not been found in any laboratory using more specific assays.

The discrepancies between laboratories can only be resolved by exchanging serum, an effort that is currently under way. At the same time, we are also searching for

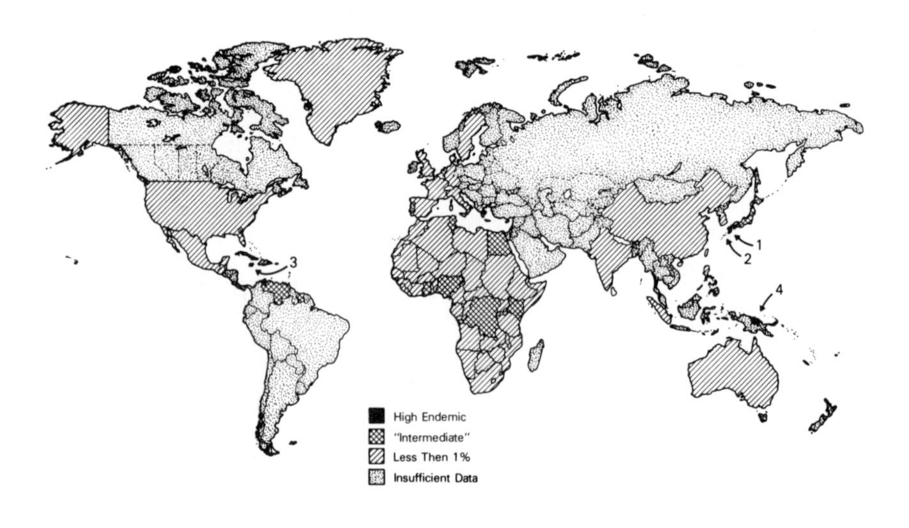

FIGURE 1. Pattern of HTLV-I infection derived from our laboratory data. Most current studies indicate highly endemic regions are in Southwest Japan [1], including Okinawa [2], and the Caribbean basin [3], all of these areas having a high proportion of ATLL. Another possible endemic area, as yet unconfirmed, is New Guinea [4]. Intermediate areas include portions of Africa, Latin America, Alaska, Lapland, Japan, and the Southeast United States.

cases of ATLL and attempting viral isolation in African cases to confirm the presence of HTLV-I related disease. Identification of additional HTLV-I-associated cases will allow us to investigate more fully those factors distinguishing the virus-susceptible, disease-prone individual from the asymptomatic carrier. The much greater impact of other infectious diseases and the decreased life span in some of the populations with significant seropositivity may prevent the appearance of some of the ATLL which one might expect in a heavily infected HTLV-I positive area, since ATLL is primarily a disease of adults (31). Viral isolation is an important component of this effort since the growing library of new retroviruses, some apparently pathogenic and some not (6), could perhaps explain in part the discrepancies in African serosurveys.

Still problematic are the areas where virus antibody has been reported but HTLV-I-positive cases of lymphoma have not yet been documented, such as New Guinea (32), the Amazon Valley (33), and the Florida Everglades (30). Among the more intriguing reports is that by Kazura and his colleagues who found HTLV-I antibody in more than 18% of sera obtained from donors in Papua, New Guinea. Although this population is also in a region endemic for malaria, Japanese soldiers were present in that island during World War II, making it an important area for subsequent epidemiologic studies.

Other regions where intensive investigations are needed include Latin America and the southern United States. We have observed significant percentages of antibody-positive healthy donors in Indian tribes in southern Florida (30) and Reeves et al also have found Indians in Panama to be seropositive (34), the latter being particularly significant because of the detection of an ATLL case positive for HTLV-I in a Panamanian Mestizo (unpublished data). The findings of Merino et al (33) and Robert-Guroff et al (22) suggest that HTLV-I could be endemic among Indians and Eskimos in the Western hemisphere and offer an opportunity for useful seroepidemiologic studies supported by virus isolation and characterization. Urban areas also appear to be potentially important foci of infection. Among black parenteral drug abusers in New York City, a high seroprevalence for HTLV-I has been found (35, 36) and a cluster of recent ATLL cases among blacks in New York City has also been reported (37). A review of lymphoproliferative malignancies in that region in the previous two years did not reveal any instances of unsuspected ATLL (Dosik unpublished data), suggesting that

HTLV-I may have been introduced to this region in the relatively recent past.

Although many areas of the world need to be surveyed, it is apparent that HTLV-I is not a major problem in most temperate regions. Serosurveys have been completed in those areas indicated in Figure 1; these data indicate that HTLV-I infection is a rare event in most parts of the world. We find the ELISA test is a sensitive, economical screening tool and, therefore, most of our attention now is devoted to determining the role of various assays in confirming the specificity of the reactivity. Because of the likelihood of other retroviruses sharing antigens detected by current methods, it is important to isolate viruses in areas where antibodies are detected but disease is not seen in order to confirm the specificity of antibody.

While many of the issues relevant to HTLV-III have already been discussed above, such as the need to clarify the relative sensitivity and specificity of the various serologic assays, there are some additional aspects to HTLV-III/LAV that require discussion because of the apparent threat to the general population. This has led to a more widespread testing of the population for HTLV-III antibodies and, as a result, a chance for danger of misinterpretation of the data.

The ELISA test has become the standard for screening large populations (38), and the need for using confirmatory tests such as radioimmunoprecipitation or Western blot is well recognized. It is important that the limitations of the assay be realized and that the interpretations of the test results be made in the context of the purpose for testing. It has been reported that a virus-positive antibody-negative state exists (39) and certainly, in theory, one must have viremia prior to the development of antibodies. With this in mind, it is unlikely that the current screening test in use will be able to safeguard completely the blood supply.

Unlike the situation with HTLV-I, it is far more difficult to conduct a dispassionate seroepidemiologic study for HTLV-III because of the political and psychological implications of the results to both individuals and entire populations. While there are still many detailed studies and interlaboratory comparisons that are necessary to interpret a number of reports in the literature, recent observations (6, Essex this volume) put some of the earlier seroepidemiologic studies of HTLV-III in perspective. It is quite clear that HTLV-III or a closely related virus was in

Central Africa in the 1970's, as we have previously reported (40, 41), but it is also quite possible that our assays, which were developed and calibrated on sera from United States AIDS patients, were detecting additional antibodies from more recently isolated retroviruses (6) which share antigens with HTLV-III/LAV.

In our original study of Ugandan children, there was no illness suggestive of AIDS in the region at that time, and among the possible alternatives, we noted that "it is conceivable that the observed antibody reactivities against HTLV-I and HTLV-III result from unusual cross-reactivity of antibodies against an undescribed variant of HTLV, that is, a Type 4 (40)." While sera from that particular group of individuals is no longer available, sera from other Ugandans collected during the same time period are being evaluated by a number of investigators.

In summary, because many of the problems faced in the laboratory are similar for HTLV-I and HTLV-III, our approaches to their solution are also similar. It is apparent that confirmation of antibody specificity requires an experienced laboratory. Comparative studies in a number of laboratories are critical to be able to understand the spectrum of the immune response to infection with these viruses and, until more experience is gained with a wide variety of virus-associated diseases and with more normal populations, there will continue to be controversies over the interpretation of data where the full pattern of antibody reactivity on the Western blot is not seen. As part of our collaborative serologic studies, we are attempting to develop panels of antisera that can be tested in a number of laboratories to develop reference reagents for the detection of specific antibodies and, as we are able to obtain specific reactivity in a variety of clinical situations, it should be possible to determine more precisely the spectrum of antibodies as they appear in the course of infection. We are continuing to attempt to isolate virus from those areas where there is serologic reactivity to antigens of HTLV-I and -III in order to determine whether there may be viruses of varying pathogenicity. Finally, we are attempting to investigate the effect of host reactivity since, as with EBV, it is quite possible that the same virus can be responsible for a variety of effects which could, in part, be modified by host genetics.

REFERENCES

1. Poiesz BJ, Ruscetti FW, Gazdar AF, et al (1980). Detection and isolation of Type C retrovirus particles from fresh and cultured lymphocytes of a patient with cutaneous T-cell lymphoma. Proc Natl Acad Sci, USA 77:7415.
2. Blattner WA, Clark JW, Gibbs WN, et al (1984). HTLV: Epidemiology and relationship to human malignancy. In Gallo RC, Essex M, Gross L. (eds): "Cancer Cells, Volume 3: Human T-Cell Leukemia/Lymphoma Viruses." New York: Cold Spring Harbor Press, p 267.
3. Kalyanaraman VS, Sarngadharan MG, Robert-Guroff M, et al (1982). A new subtype of human T cell leukemia virus (HTLV-II) associated with a T cell variant of hairy cell leukemia. Science 218:517.
4. Barre-Sinousi F, Cherman JC, Rey F, et al (1983). Isolation of a T-lymphotropic retrovirus from a patient at risk for acquired immunodeficiency syndrome (AIDS). Science 220:868.
5. Gallo RC, Salahuddin SZ, Popovic M, et al (1984). Frequent detection and isolation of cytopathic retroviruses (HTLV-III) from patients with AIDS and at risk for AIDS. Science 224:500.
6. Kanki PJ, Barin F, M´Boup S, et al (1986). New human T-lymphotropic retrovirus related to simian T-lymphotropic virus Type III ($STLV-III_{AGM}$). Science 232:238.
7. Takatsuki K, Uchiyama J, Sagawa K, et al (1977). Adult T-cell leukemia in Japan. In Seno S, Takaku K, Roino S (eds): "Topics in Hematology." Amsterdam: Excerpta Medica, p 73.
8. Robert-Guroff M, Nakao Y, Notake K, et al (1982). Natural antibodies to human retrovirus HTLV in a cluster of Japanese patients with adult T-cell leukemia. Science 215:975.
9. Hinuma Y, Komoda H, Chosa T, et al (1982). Antibodies to adult T-cell leukemia-virus-associated antigen (ATLA) in sera from patients with ATL and controls in Japan: A nationwide sero-epidemiologic study. Int J Cancer 29:631.
10. Blattner WA, Kalyanaraman VS, Robert-Guroff M, et al (1982). The human Type-C retrovirus, HTLV, in blacks from the Caribbean region, and relationship to adult T-cell leukemia/lymphoma. Int J Cancer 30:257.

11. Catovsky D, Greaves MF, Rose M, et al (1982). Adult T-cell lymphoma-leukemia in blacks from the West Indies. Lancet 1:639.
12. Greaves MF, Verbi W, Tilley R, et al (1984). Human T-cell leukemia virus (HTLV) in the United Kingdom. Int J Cancer 33:795.
13. Vyth-Dreese FA, De Vries JE (1982). Human T-cell leukaemia virus in lymphocytes from T-cell leukaemia patient originating from Surinam. Lancet 2:993.
14. Williams OK, Saxinger C, Junaid A, et al (1984). HTLV-associated lymphoproliferative disease: A report of two cases in Nigeria. Br Med J 288:1495.
15. Kuo T-T, Chan H-L, Su I-J, et al (1985). Serological survey of antibodies to the adult T-cell leukemia virus-associated antigen (HTLV-A) in Taiwan. Int J Cancer 36:345.
16. Blayney DW, Jaffe ES, Fisher RI, et al (1983). The human T-cell leukemia/lymphoma virus, lymphoma, lytic bone lesions, and hypercalcemia. Ann Int Med 98:144.
17. Pandolfi F, Manzari V, De Rossi G, et al (1985). T-helper phenotype chronic lymphocytic leukaemia and "adult T-cell leukaemia" in Italy: Endemic HTLV-I-related T-cell leukemias in Southern Europe. Lancet 2:633.
18. Gallo RC, Blattner WA (1984). Human T-cell leukemia/lymphoma viruses: ATL and AIDS. In Devita VT, Hellman S, Rosenberg SA (eds): "Important Advances in Oncology." New York: JB Lipincott Co, p 104.
19. Blayney DW, Blattner WA, Robert-Guroff M, et al (1983). The human T-cell leukemia-lymphoma virus in the Southeastern United States. JAMA 250:1048.
20. Saxinger C, Gallo RC (1983). Application of the indirect enzyme linked immunosorbent assay micro test to the detection and surveillance of human T-cell leukemia/lymphoma virus. Lab Invest 49:371.
21. Saxinger W, Blattner WA, Levine PH, et al (1984). Human T-cell leukemia virus (HTLV-I) antibodies in Africa. Science 225:1473.
22. Robert-Guroff M, Clark J, Lanier AP, et al (1985). Prevalence of HTLV-I in Arctic regions. Int J Cancer 36:651.
23. Biggar RJ, Johnson BK, Oster C, et al (1985). Regional variation and prevalence of antibody against human T lymphotropic virus Types I and III in Kenya, East Africa. Int J Cancer 35:763.

24. Gazzolo L, Robert-Guroff M, Jennings A, et al (1985). Type I and Type III HTLV antibodies in hospitalized and outpatient Zairians. Int J Cancer 36:373.
25. Biggar RJ, Saxinger C, Gardiner C, et al (1984). Type I HTLV antibody in urban and rural Ghana, West Africa. Int J Cancer 34:215.
26. Hunsmann G, Schneider J, Schmitt J, et al (1983). Detection of serum antibodies to adult T-cell leukemia virus in non-human primates and in people from Africa. Int J Cancer 32:329.
27. Weiss RA, Cheingsong-Popov R, Clayden S, et al (1986). Lack of HTLV-I antibodies in Africans. Nature 319:794.
28. Stewart JSW, Matutes E, Lampert JA, et al (1984). HTLV-I positive T-cell lymphoma/leukemia in an African resident in the United Kingdom. Lancet 2:984.
29. Biggar RJ, Gigase PL, Melbye M, et al (1985). ELISA HTLV retrovirus antibody reactivity associated with malaria and immune complexes in healthy Africans. Lancet 1:520.
30. Levine PH, Saxinger WC, Clark J, et al (1986). HTLV-I geographic distribution and identification of new high-risk populations. J Cell Biochem (Suppl 10A):192.
31. Blattner WA, Blayney DW, Robert-Guroff M, et al (1983). Epidemiology of human T-cell leukemia/lymphoma virus (HTLV). J Inf Dis 147:406.
32. Kazura J, Forsythe K, Lederman M, et al (1985). Household cluster of human T-cell leukemia/lymphoma virus Type I antibodies in Papua, New Guinea. Clin Res 33:4548.
33. Merino F, Robert-Guroff M, Clark J, et al (1984). Natural antibodies to human T-cell leukemia/lymphoma virus in healthy Venezuelan populations. Int J Cancer 34:501.
34. Reeves WC, Saxinger C, Brenes MM, et al (1986). HTLV-I seroepidemiology and risk factors in metropolitan Panama. Submitted.
35. Weiss SH, Blattner WA, Ginzburg HM, et al (1985). Retroviral antibodies in parenteral drug abusers from an AIDS epidemic region. American Society for Microbiology, 25th Interscience Conference on Antimicrobial Agent and Chemotherapy 25:228.
36. Robert-Guroff M, Weiss SH, Giron JA, et al (1986). Prevalence of antibodies to HTLV-I, -II, and -III in intravenous drug abusers from an AIDS endemic region. JAMA, in press.

37. Dosik H, Anandakrishnan A, Denic S, et al (1985). Adult T cell leukemia/lymphoma. A cluster in Brooklyn, a new endemic area? Blood 66 (Suppl 1):187A.
38. Weiss SH, Goedert JJ, Sarngadhuran MG, et al (1984). Screening test for HTLV-III (AIDS agent) antibodies: Specificity, sensitivity and applications. JAMA 253:29.
39. Groopman JE, Hartzband PI, Shulman L, et al (1985). Antibody seronegative human T-lymphotropic virus type III (HTLV-III) infected patients with acquired immunodeficiency syndrome or related disorders. Blood 66:742.
40. Saxinger WC, Levine PH, Dean AG, et al (1985). Evidence for exposure to HTLV-III in Uganda before 1973. Science 227:1036.
41. Saxinger C, Levine PH, Dean A, et al (1985). Unique pattern of HTLV-III (AIDS-related) antigen recognition by sera from African children in Uganda (1972). Cancer Res 45:4624s.

Viruses and Human Cancer, pages 105–107
Published 1987 by Alan R. Liss, Inc.

WORKSHOP SUMMARY: THE EPIDEMIOLOGY OF AIDS AND RELATED RETROVIRUSES

Paul H. Levine

Environmental Epidemiology Branch, National Cancer Institute, National Institutes of Health, 3C25 Landow Building, Bethesda, Maryland 20892

The epidemiology of AIDS has been under intensive investigation by a number of research groups. All agree that AIDS is now widespread and the number of cases is increasing throughout the world. In order to be able to understand the disease and its etiology, this session was organized with an emphasis on the laboratory problems that have hampered clearer understanding of the seroepidemiology of the causative agent, HTLV-III/LAV, and other human retroviruses discussed elsewhere in the symposium.

Seroepidemiology laboratory issues were reviewed by Dr. Levine, who presented the increasing incidence of AIDS in groups with and without known AIDS risk factors. AIDS in persons with no known risk factors is now third in frequency behind male homosexuals and intravenous drug users, and ahead of transfusion recipients. The rate of increase in this group was as great as that of the other groups. Over the past six years, AIDS has clearly become of worldwide concern, cases now having been reported throughout Europe, in Central Africa, and more recently in the Soviet Union. Prospective cohort studies are becoming particularly important in clarifying the attack rate of the causative agent, HTLV-III/LAV. Among a cohort of HTLV-III/LAV antibody-positive Manhattan homosexuals, 32% developed AIDS within three years and, in a cohort of hemophilia patients, the six year attack rate following seroconversion was reaching 15% (Goedert et al, 1986).

The increasing awareness of a prolonged incubation period for this disease was marked by another recent paper from the Centers for Disease Control (Lui et al, 1986), which used maximum likelihood techniques to provide a mean latent period of 4.5 years for AIDS following HTLV-III/LAV infection due to transfusion exposure with a 90% confidence interval ranging from 2.6 to 14.2 years. Dr. Gallo cautioned that his past experience with AIDS has shown that reliance on current data is safer and less likely to lead to unfounded speculation. These data formed the basis for discussion, which was to approach the following questions:

- Can laboratory assays be better utilized to evaluate the epidemiology of AIDS?
- How can we interpret the seroepidemiology of HTLV-III/LAV in areas where clinical AIDS is not apparent?

Discussion of the laboratory problems relevant to AIDS epidemiology centered on the report of Groopman et al (1985), who stated that they now have observed nine cases of individuals excreting HTLV-III/LAV in the absence of detectable antibody. Although the frequency and duration of a virus-positive antibody-negative state is clearly low, the data point out the dangers of any statement that a laboratory test can identify more than 98% of virus-positive blood donors.

Turning to retrovirus serology in areas where AIDS is not apparent, Dr. Essex reviewed the available assays and some of their strengths and weaknesses. Regardless of the assay used, Dr. Essex emphasized the importance of using purified antigens for most reliability. Relying on the radio-immunoprecipitation assay for his own seroepidemiologic studies, his slides demonstrated the need for a pattern of antibodies to specific viral antigens rather than antibody reactivity to any single virus- associated peptide. It was noted, however, that virus has been isolated from individuals with apparent antibodies to only one viral protein, and Dr. Goudsmit speculated that the testing of serial samples, particularly by Western blot, was of potential value since antibodies to different proteins occur sequentially.

In regard to the epidemiology of retroviruses in various populations, the data continue to suggest the existence of HTLV-III/LAV or a related virus in Central Africa prior to the awareness of clinical AIDS in this or any other region. First reported by Saxinger et al (1985) in a series of Ugandan children whose sera were collected in the early 1970s, Dr. Essex confirmed the presence of antibody reacting with HTLV-III antigen when he reported a positive serum sample (using RIP) from a man from Kinshasa, Zaire (then Leopoldville, Belgian Congo) collected in 1959. Although the sera tested by Essex, Saxinger, and their colleagues had reactivity to several viral antigens now well-associated with HTLV-III/LAV and demonstrated the pattern seen in well-documented HTLV-III/LAV infection, the discussion tended to favor the hypothesis that a related retrovirus could have been identified, particularly with Dr. Kanki et al (1986) suggesting the presence of another retrovirus (HTLV-IV) related to STLV-III in West Africa. Antibody to this virus is not being associated with clinical illness. The need to isolate retroviruses related to HTLV-III/LAV, should such other retroviruses exist, is of obvious importance for a number of reasons, one of them being the ability to interpret the seropositivity in various populations where AIDS has not been identified. One such population reported at this symposium was an isolated tribe living in the Amazon region of South America (Volsky et al, 1986). It is apparent that advances in our understanding of the epidemiology of retrovirus related to AIDS will be greatly facilitated by exchange of sera between laboratories, allowing further definition of the immune response to various retrovirus antigens.

REFERENCES

Groopman JE, Hartzband PI, Shulman L, et al (1985). Antibody seronegative human T-lymphotropic virus type III (HTLV-III)-infected patients with acquired immunodeficiency syndrome or related disorders. Blood 66:742-747.

Goedert JJ, Biggar RJ, Weiss SH, et al (1986). Three-year incidence of AIDS in five cohorts of HTLV-III-infected risk group members. Science 231:992-995.

Kanki PJ, Barin F, M´Boup S, et al (1986). New human T-lymphotropic retrovirus related to simian T-lymphotropic virus Type III ($STLV-III_{AGM}$). Science 232:238-243.

Lui K-J, Lawrence DN, Morgan WM, et al (1986). A model-based approach for estimating the mean incubation period of transfusion-associated acquired immunodeficiency syndrome. Proc Natl Acad Sci, USA 83:3051-3055.

Saxinger WC, Levine PH, Dean AG, et al (1985). Evidence for exposure to HTLV-III in Uganda before 1973. Science 227: 1036-1038.

Volsky DJ, Rodriguez L, Dewhurst S, et al (1986). Anti-HTLV-III/LAV antibodies among native populations in Venezuela. J Cell Biochem (Suppl 10A):206.

Viruses and Human Cancer, pages 109–129

BIOLOGY AND LEUKEMOGENICITY OF HUMAN T-CELL LEUKEMIA VIRUSES[1]

Mordechai Aboud[2], David W. Golde, William Wachsman,
Joseph D. Rosenblatt, Alan J. Cann,
Richard B. Gaynor, Dennis J. Slamon, Irvin S.Y. Chen

Division of Hematology-Oncology, Department of Medicine,
UCLA School of Medicine, Los Angeles, CA 90024

INTRODUCTION

Endogenous as well as exogenous retroviruses have long been identified in a wide variety of animal species, and their etiological implication with various animal malignancies has been well established (1). However, no clear evidence for such retroviral involvement in human malignancies has been available until the last few years. Many animal retroviruses share, to a variable extent, antigenic and sequence homologies which have facilitated the detection of new animal retroviruses by using immunologic or nucleic acid probes derived from previously isolated viruses. However, no retroviral markers could be detected in primary human tumors with any of the available animal retroviral probes. Furthermore, retrovirally induced leukemias/lymphomas in animals usually involve abundant viral replication and chronic viremia which allow the demonstration of virus budding in the primary tumor cells, as well as free virions. Although data suggesting the existence of human retroviruses have been reported

[1] This work was supported by NCI grants CA30388, CA32737, CA38597, CA09279, and CA16042; grants PF-2182 and JFRA-99 from the American Cancer Society; and grants from the California Institute for Cancer Research and California University-wide Task Force on AIDS.
[2] Present address: Department of Microbiology and Immunology, Faculty of Health Sciences, Ben gurion University of the Negev, Beer Sheva, Israel.

from time to time (2-6), budding or free virions of such viruses have never been unequivocally demonstrated in primary human tumors.

A retrovirus with properties distinct from other animal retroviruses is the bovine leukemia virus (BLV), which has been implicated as the etiological agent of lymphoid neoplasms in cattle (7). Unlike the other retroviruses, expression of BLV could be demonstrated only in cultured cell lines derived from the leukemic animals, but not in their primary tumor cells (8,9). Moreover, BLV does not share antigenic or sequence homology with other animal retroviruses (10). Therefore, before its isolation in cell culture, no evidence could be obtained with previously available retroviral probes for the existence of this virus. A similar approach has been used in subsequent attempts to identify human retroviruses. Three different types of such viruses have been isolated thus far, currently referred to as human T-cell leukemia virus type I (HTLV-I), type II (HTLV-II), and type III (HTLV-III) (11).

HTLV-associated T-cell Disorders

HTLV-I has been firmly established by seroepidemiological and molecular studies as the etiological agent of a subtype of adult T-cell leukemia (ATL), which is endemic to Southwestern Japan, the Caribbean basin, South Africa, and certain parts of the southeastern United States (see ref 12 for a review). Some cases of this disease have also been found in other places such as Israel, Venezuela, Brazil, Guatamala, Ecuador, and Alaska (13). It is a clinically aggressive disease with a rapid course, usually appearing in human adults, probably due to a long latency period, often involving hepatosplenomegaly, lymphoadenopathy, hypercalcemia, and cutaneous lesions. The malignant cells are a subset of mature T cells with OKT4 and Leu 3 surface markers characteristic of helper T cells. However, these malignant cells are frequently found to suppress *in vitro* immunoglobulin synthesis (14), suggesting a functional alteration of these cells. Moreover, this malignancy is frequently accompanied by various severe opportunistic infections such as Pneumocystis carinii pneumonia, further suggesting that the infection of T cells with HTLV-I may result in immunological deficiency due to abrogation of their specific function. HTLV-I has also been isolated from some AIDS patients (15-17), but these isolates are believed to represent opportunistic infections rather than to play an etiological role.

HTLV-II has been found to be associated with two cases of T-cell variants of hairy-cell leukemia (18-21). Unlike ATL, this rare type of leukemia is a clinically benign disease (see review in ref. 22). The first patient (Mo), diagnosed nine years ago, is still alive and well. We have recently diagnosed a second patient (NRA, ref. 21). While the first case was not characterized in detail at the molecular level, we have obtained molecular and immunological indications for an etiological role of HTLV-II in the second case (21). The HTLV-II provirus was found integrated in the genome of the leukemic cells of the second patient in an oligoclonal pattern. However, despite the presence of this provirus, viral RNA expression could not be detected in these cells. These two features of the disease are also unique to the HTLV-I-associated ATL (23) and the BLV-associated bovine leukosis (10). They would be highly unexpected in HTLV-II-associated hairy-cell leukemia if the patient had this virus merely as an opportunistic infection. If this was the case, a random polyclonal integration of the provirus would be more likely. In addition, the patient's serum contained antibodies against HTLV-II antigens, which confirm the _in vivo_ association of the isolated virus with the patient and exclude a trivial argument that this virus merely reflected a laboratory contaminant. Sohn et al. (24) have recently described two cases of a new variant of chronic T-cell lymphocytic leukemia that clinically and pathologically resembles hairy-cell leukemia, in which the malignant cells were of T-suppressor antigenic phenotype. However, although these two cases seem to be clinically similar, one patient had antibodies against HTLV-I, and the other against HTLV-II. Therefore, the significance of these findings as indicating a possible HTLV etiology for this malignancy is unclear, as no molecular evidence for clonal provirus integration was obtained. HTLV-II has also been isolated from one hemophiliac and one AIDS patient, but without evidence of T-cell malignancy (25,26).

A common feature of HTLV-I- and HTLV-II-associated diseases that distinguishes them from leukemias induced by most other replication-competent retroviruses is the lack of chronic viremia. Similar lack of viremia is also observed with BLV, the etiological agent of lymphoproliferative diseases in cattle; including lymphoid leukemia lymphosarcoma and a condition known as persistent lymphocytosis (10). The latter is a benign disease involving a polyclonal persistent lymphoid proliferation that sometimes develops into a monoclonal lymphosarcoma.

HTLV-III has been repeatedly isolated from AIDS patients, and accumulating evidence strongly associates it etiologically with this syndrome (27-30). AIDS is prevalent in homosexual communities, but is also frequently found in hemophiliac patients receiving blood transfusions or blood products, intravenous drug users, Haitian and African blacks, and in heterosexual partners and children of high-risk individuals (31). HTLV-III exhibits an $OKT4^{+}$ and Leu 3^{+} helper T cell tropism (27), but is rather cytotoxic, killing the infected lymphocytes *in vitro* (27), and has not been associated with T-cell transformation.

HTLV Replication and Infection in Culture

A common feature of HTLV-I and HTLV-II is that, although the infected individuals contain specific antibodies against antigens of these viruses, and although their proviral genome is readily detected in the DNA of the primary leukemic cells, these cells do not express viral antigens, nor do they produce viral RNA or virus particles unless they are cultured *in vitro*. Furthermore, these viruses are poorly infectious (12). Efficient *in vitro* infection with these viruses can usually be demonstrated only by co-cultivation of the target cells with X-ray irradiated or mitomycin-C treated virus-producing cell lines (12). Successful infection with cell-free virions of HTLV-I and HTLV-II has been reportedly obtained only with highly concentrated virus stocks (12). This low infectivity suggests that transmission of these viruses among humans may require prolonged intimate contact, which may account for the limitation of HTLV-I-associated ATL to specific endemic areas and for the low frequency of HTLV-II-associated T-cell variant hairy-cell leukemia.

Both HTLV-I and HTLV-II exhibit helper T-cell tropism. From early reports, it appeared that B cells or other T-cell subsets of patients with HTLV-associated diseases were seemingly uninfected (see review in ref. 12). In addition, *in vitro* infection of HTLV-I and HTLV-II by co-cultivation of X-ray irradiated or mitomycin-C treated infected cell lines with cells from umbilical cord blood, bone marrow, or peripheral blood results in a predominant transformation of mature $OKT4^+$ Leu 3^+ T cells (12,21,32), further indicating the tropism of these viruses to helper T cells. Nevertheless, Epstein-Barr virus (EBV)-transformed B-cell lines harboring HTLV-I were subsequently established from some ATL patients (33,34). We established a similar EBV-transformed B-cell line from the first patient with the T-cell variant hairy-cell leukemia (Mo), and found it to harbor HTLV-II (32). Furthermore, both HTLV-I and HTLV-II were later shown to also infect less mature T cells derived from fetal thymus *in vitro* (12), as well as additional hematopoietic cell types such as normal suppressor/cytotoxic ($OKT8^+$) T cells and non-T bone marrow cells (35-38). We were able to introduce an infectious cloned HTLV-II genome into B cells by DNA transfection, which resulted in a productive infection (39). In addition, HTLV-I was found capable of infecting non-hematopoietic human cells such as fibroblasts, endothelial cells, osteosarcoma cells (40-43), and non-human cells such as monkey (44) and rabbit (45) lymphocytes, and a cat cell line, 8C (46). HTLV-II cannot infect most fibroblasts (20,40,46), but it does infect non-human lymphocytes such as marmoset T cells (40).

In vitro Cell Transformation by HTLV-I and HTLV-II

Co-cultivation of peripheral and cord blood T cells with either HTLV-I- or HTLV-II-producing cell lines results in transformation, usually of helper T cells with OKT4 and Leu 3 markers (12,48). In comparative studies, we have shown that although these two viruses are involved in two distinct malignancies, they have a similar in vitro transforming efficiency. In addition, we could not detect any apparent differences in the properties of the cells transformed by either of them (48). Normal T lymphocytes require mitogen or antigenic activation in order to express IL-2 receptors, which allow them to grow in culture in the presence of IL-2. Moreover, even under such conditions, they usually have a limited life span. By contrast, HTLV-I- and HTLV-II-transformed T cells develop into immortalized cell lines expressing IL-2 receptors independently of mitogen activation. They also constitutively produce IL-2, thus requiring no additional exogenous IL-2 for their growth (12,21,48). In addition, such transformed cells produce several other lymphokines such as macrophage migration inhibitory factor, leukocyte inhibitory factor, migration enhancing factor, macrophage activating factor, interleukin 3, fibroblast activating factor, B-cell growth factor, erythroid potentiating factor, and γ-interferon (12,21,48). They also express HLA-Dr, HLA-A and HLA-B antigens that are not found in normal non-activated T lymphocytes (12). In addition, rearrangement of T-cell receptor genes has also been found in some of these transformed cells (48). Of particular interest are the findings that in vitro infection of cytotoxic (37,38) or helper (38) T-cell lines with HTLV-I (37,38) or HTLV-II (38) results in a loss or modification of their cellular immunologic functions. These findings may explain the immune deficiency and the opportunistic infections observed in HTLV-associated malignancies.

Genetic Structure of HTLV

HTLV-I and HTLV-II carry the structures and genes characteristic to replication-competent retroviruses, i.e., two long terminal repeats (LTRs), and gag, pol and env genes (20,48,49). By sequence analysis of HTLV-II, we have found an open reading frame of 633 nucleotides, stretching from 30 nucleotides upstream to the 3' end of gag to 372 nucleotides downstream to the 5' end of pol (50). A similar open reading frame overlapping the gag and pol genes is found in BLV (51). This open reading frame likely encodes a viral protease. No data are available regarding such a protease-encoding sequence in HTLV-I. This protease is analogous to that found in many other retroviruses, but in most of the other viruses, this enzyme is expressed from the same reading frame as the gag gene and is generated as a cleavage product of the gag precursor protein (52).

In addition, like BLV, both HTLV-I and HTLV-II contain a fourth region with several possible reading frames located between the env gene and the 3' LTR. This region is called the X region (49) (also referred to as lor and x-lor) (53). Restriction enzyme and sequence analyses have demonstrated a remarkable homology between HTLV-I and HTLV-II genomes, with greatest homology in the X region (54,55).

LTR of HTLV

The LTR of HTLV-I and HTLV-II share only 30% homology; however, the homologous regions exhibit interesting features (56-58). Conserved sequences are found surrounding the cap site (21 out of 22 nucleotides), the TATA box (15 contiguous nucleotides), and the polyadenylation signal (AATAA). It is interesting to note that while in most cases the polyadenylation signal is located 15-20 nucleotides upstream to the polyadenylation site, in HTLV-I and HTLV-II LTRs this signal is located about 290 nucleotides upstream to the polyadenylation site. This has led us and others (57) to propose an unusual secondary structure for the RNA transcript, which would bring the polyadenylation signal closer to the polyadenylation site. In addition, a conserved stretch of 21 nucleotides with minor modifications is repeated three times in the U3 region upstream to the promoter region. This sequence probably plays a cis-acting regulatory role in the viral transcription control.

Using recombinant constructs for DNA transfection, the LTR of HTLV-I has been demonstrated as capable of activating gene expression in a wide variety of human and non-human cells (59). However, when tested by transfection of an infectious proviral clone, we found the LTR of HTLV-II to be active in human T and B lymphocytes, but not in fibroblasts (39). This restricted function of the HTLV-II LTR parallels the host range of this virus, which can readily infect T and B lymphocytes, but rarely fibroblasts (39). Later transfection experiments with recombinant constructs have shown that the LTRs of both viruses are transcriptionally more efficient in HTLV-I- or HTLV-II-infected than in uninfected cells (39,59), suggesting that their transcriptional function is regulated by a _trans_-acting viral gene product. As will be discussed in the following sections, this _trans_-acting factor is the viral χ gene-encoded protein.

The HTLV χ Gene

As noted above, HTLV-I and HTLV-II contain an X region which is analogous to a similar region in BLV. This region consists of four possible open reading frames in the case of HTLV-I (X-I, X-II, X-III, and X-IV) (49), and three open reading frames in HTLV-II (X-a, X-b, X-c) (53,60). There is a remarkable sequence homology (about 75%) between regions X-IV and X-c (49,53,60). These two open reading frames have been found to be active genes within the X region, called χ or _lor_ genes. We and others have found these genes to encode for a 40 kd protein in the case of HTLV-I ($p40x^{I}$), and a 37 kd protein in the case of HTLV-II ($p37x^{II}$) (61,62). We have determined the size of the χ gene mRNA of both viruses, and found it to be 2.1 kb (60,63). Furthermore, by S_1 nuclease and sequence analyses of this mRNA, we have shown that its transcription initiates from the cap site of the virus and that the _gag_ and _pol_ sequences are removed from the primary transcript by splicing the 5' leader segment to the methionine initiation codon of the _env_. Another splicing step links this methionine codon and the first nucleotide of the next codon of <u>env</u> to the second nucleotide of the second codon of the χ gene (64). Analogous processing has also been shown for the χ gene RNA of HTLV-I (65), and based on reported nucleotide sequences (66), the BLV χ mRNA is also believed to be similarly processed.

The χ gene-encoded protein has a relatively short half-life of about 120 minutes (67). It is found in all HTLV-I- and HTLV-II-infected cell lines, localized predominantly in the nucleus (67,68), and is immunoprecipitated by sera of patients with malignancies associated with these viruses (62). In addition to the χ gene, there is another sequence of 500-700 nucleotides in the X region of both HTLV-I and HTLV-II as well as of BLV, stretching between the <u>env</u> and the χ genes (49,60,66) with a function as yet unknown. Unlike the high sequence homology shared by the two HTLV viruses in the χ gene, they share only 30% homology in this part of the X region. It is likely that this region is important for <u>cis</u> regulation of the χ gene expression, or perhaps for virus assembly.

The Role of the χ Gene in HTLV Transcription

By introducing deletion mutations into the χ gene of infectious cloned DNA of HTLV-II and transfecting this DNA into EBV-transformed B-cell lines, we have demonstrated that the χ gene plays a critical role in viral replication by enhancing viral RNA transcription (69). Transfection of cells with the complete wild type HTLV-II genome resulted in a high level of viral RNA transcription. On the other hand, when HTLV-II genomes with various deletions in the χ gene were transfected, only low levels of viral RNA could be detected. However, RNA transcription from such mutant viral DNA was fully restored if the transfected cells were subsequently superinfected with a wild-type HTLV-II. Likewise, experiments with recombinant DNA constructs containing the LTR of HTLV-I or HTLV-II in front of the CAT gene have shown that these LTRs activate CAT gene transcription much more efficiently if these constructs have been transfected in HTLV-infected cells (39,59,70). These findings suggest the existence of a _trans_-acting factor in the infected cells which activates the LTR-mediated transcription. In order to identify this viral factor, we have co-transfected an LTR-CAT recombinant construct into cells together with the complete cloned HTLV-II genome in which the strong SV40 promoter has replaced the viral LTR, or with similar genomes that have carried various deletions (71). These experiments have shown that only deletions removing the χ gene or the _env_ methionine initiation codon (which, as noted above, is necessary for χ gene expression) abolished the viral _trans_-activation of the LTR-CAT expression. It is thus evident that the χ gene encodes for the _trans_-acting protein required for the LTR-mediated transcription. It is of interest to note that when the LTR-CAT was transfected with a complete genome containing the 5' LTR instead of the SV40 promoter, only low basal levels of CAT expression were detected. Since the viral LTR itself requires χ protein for its function, it was unable to mediate a high expression of viral genes, including the χ gene. Therefore, in the absence of virus-mediated synthesis of χ protein, only low CAT expression can take place. Taken together, these observations suggest that shortly after viral DNA is integrated into the cellular genome, there is only a low level of viral gene expression because of the absence of χ protein in the cells (analogous to the low CAT expression shown in our experiments under similar conditions (70), and to the low level of viral RNA transcription observed in cells transfected with HTLV-II genomes carrying deletion mutations in the χ gene (69)). Nevertheless, this low expression leads to the accumulation of some χ protein which, in turn, stimulates the function of the LTR so that the synthesis of χ protein is enhanced along with the expression of other viral genes. This proposed process is compatible with the delayed kinetics of HTLV replication observed in freshly infected cells (54).

Despite the structural and biological similarities between HTLV-I and HTLV-II, these two viruses differ in the nature and frequency of the T-cell malignancies with which they are associated. Furthermore, as noted above, their LTRs exhibit different target cell specificities. It was therefore of interest to investigate the interrelationships between their LTRs and the χ genes which govern their function. By co-transfection of various recombinant constructs aimed at elucidating these interrelationships, we have shown that the HTLV-II χ gene can activate the LTRs of both HTLV-I and HTLV-II. On the other hand, the HTLV-I χ gene can efficiently activate only the LTR of HTLV-I, but not that of HTLV-II (71, Shah et al., manuscript in preparation).

A Possible Role of the χ Gene in HTLV Leukemogenesis

Most animal retroviruses induce tumors either by an oncogene (v-onc) carried in their genome or by insertion of their proviral DNA near a cellular oncogene (c-onc) which is consequently activated by the viral transcription control elements located in the viral LTRs (72). Neither HTLV-I nor HTLV-II carry any of the known oncogenes. There is also no evidence that their leukemogenicity depends on integration of their genome into specific sites in the cellular DNA; therefore, the mechanism of their leukemogenicity evidently differs from that of most other retroviruses. It is of interest to note in this context that genes encoding for *trans*-acting proteins controlling viral transcription are present in various DNA tumor viruses such as adenoviruses, papova viruses and herpes viruses (73-77). The most extensively studied of these genes is the E1A immediate early gene of adenoviruses. The product of this gene is required for the expression of other adenovirus genes (73-75). It also activates several endogenous cellular genes (74) and newly transfected genes (78,79). Mutants of adenovirus lacking active E1A genes can still express E1A-dependent viral genes, although at a profoundly lower rate (74), but they completely lose their transforming capacity (80,81). Aberrant transcriptional activation has been proposed as a possible mechanism for cell transformation by viral and cellular transforming genes such as the adenovirus E1A gene (82), papova virus T antigen (83), and certain oncogenes (84). We have recently demonstrated functional relationships between the adenovirus E1A gene and the HTLV-II x gene; the HTLV-II x gene is capable of activating E1A-dependent adenovirus genes such as E-II and E-III, whereas the E1A gene is capable of weakly activating HTLV-II LTR-mediated transcription (85). In addition, it has been shown that the E1A gene product of the highly oncogenic adenovirus type 12 (Ad-12), but not of the non-oncogenic Ad-2 or Ad-5, suppresses the expression of the heavy chain of class I major histocompatibility (MHC) antigens (86). This suppression allows cells transformed by Ad 12 to escape the cytotoxic T-cell immune surveillence and to produce tumors in immunocompetent syngeneic animals (87). An interesting analogy may be pointed out between these findings and the above mentioned MHC modifications observed in HTLV-transformed T cells (12,21), although no direct data are available regarding the response of the immune system to these modifications. Nevertheless, the ability of the HTLV-II x gene to activate non-HTLV genes and its functional similarity to the adenovirus E1A gene strongly suggests the possibility that it may play an important role, analogous to that of the E1A gene, in cellular transformation, by regulating up or down the transcription of certain cellular genes. This possible attractive mechanism for HTLV-associated leukemogenesis still needs to be experimentally elaborated.

ACKNOWLEDGMENTS

We thank W. Aft for preparation of the manuscript.

REFERENCES

1. Weiss R, Teich N, Varmus H, Coffin J (eds) (1985). "RNA Tumor Viruses: Molecular Biology of Tumor Viruses." Cold Spring Harbor NY: Cold Spring Laboratory.
2. Sarngadharan MG, Sarin PS, Reitz MS, Gallo RC (1972). Reverse transcriptase activity in human acute leukemia cells: Purification of the enzyme, response to AMV 70S RNA and characterization of the DNA product. Nature, New Biol 240:67-72.
3. Baxt WG, Spiegelman S (1972). Nuclear DNA sequences present in human leukemic cells and absent in normal leukocytes. Proc Natl Acad Sci USA 69:3737-3741.
4. O'Brien SJ, Bonner TI, Cohen M, O'Connel C, Nash WG (1983). Mapping of an endogenous retroviral sequence to human chromosome 18. Nature (London) 300:74-77.
5. Rabson AB, Steel PE, Garon CF, Martin MA (1983). mRNA transcripts related to full-length of endogenous retroviral DNA in human cells. Nature (London) 306:604-607.
6. Repaske R, O'Neil RR, Steel PE, Martin MA (1983). Characterization and partial nucleotide sequence of endogenous type-C retrovirus segments in human chromosomal DNA. Proc Natl Acad Sci USA 80:678-682.
7. Ferrer JF, Abt DA, Bhatt DM, Marshak RR (1974). Studies on the relationship between infection with bovine C-type virus, leukemia and persistent lymphocytosis in cattle. Cancer Res 34:893-898.
8. Kettman R, Deschamps J, Cleuter Y, Couez O, Burny A, Marbaix G (1982). Leukemogenesis by bovine leukemia virus: Proviral DNA integration and lack of RNA expression of viral long terminal repeat and 3' proximate cellular sequences. Proc Natl Acad Sci USA 79:2465-2469.
9. Paul PA, Pomeroy KA, Joh DW, Muscoplat CC, Handwerger BS, Soper FS, Sorensen DK (1977). Evidence for replication of bovine leukemia virus in the B-lymphocytes. Am J Vet Res 38:873-876.
10. Burny A, Bruck C, Chantrenne H, Cleuter Y, Dekezel D, Ghysdael J, Kettemann R, Leclercq M, Leunen J,

Mammerickx M, Portelle D (1980). Bovine leukemia virus: Molecular biology and epidemiology. In Klein G (ed): "Viral Oncology," New York: Raven Press, pp 231-280.

11. Wong-Staal F, Gallo RC (1985). Human T-lymphotropic retroviruses. Nature 317:395-403.
12. Popovic M, Wong-Staal F, Sarin PS, Gallo RC (1984). Biology of human T-cell leukemia/lymphoma virus: Transformation of human T-cell in vivo and in vitro. In Klein G (ed): "Advances in Viral Oncology," New York: Raven Press, Vol 4, pp 45-70.
13. Gallo RC, Reitz MS Jr (1982). Human retroviruses and adult T-cell leukemia/lymphoma. J Natl Cancer Inst 69:1209-1212.
14. Yamada Y (1983). Phenotypic and functional analysis of leukemic cells from 16 patients with adult T-cell leukemia/lymphoma. Blood 61:192-199.
15. Gallo RC, Sarin PS, Gelmann EP, Robert-Guroff M, Richardson E, Kalyanaraman VS, Mann D, Sidhu GD, Stahl RE, Zolla-Pazner S, Leibowitch J, Popovic M (1983). Isolation of human T-cell leukemia virus in acquired immune deficiency syndrome (AIDS). Science 220:865-867.
16. Essex M, McLane ME, Lee TH, Falk L, Howe CWS, Mullins JI, Cabradilla C, Francis DP (1983). Antibodies to cell membrane antigens associated with human T-cell leukemia virus in patients with AIDS. Science 220:859-862.
17. Gelmann EP, Popovic M, Blayney D, Masur H, Sidhu G, Stahl ER, Gallo RC (1983). Proviral DNA of a retrovirus human T-cell leukemia virus in two patients with AIDS. Science 220:862-865.
18. Saxon A, Stevens RH, Golde DW (1978). T-lymphocyte variant of hairy cell leukemia. Ann Intern Med 88:323-326.
19. Kalyanaraman VS, Sarngadharan MG, Robert-Guroff M, Miyoshi I, Blayney D, Golde D, Gallo RC (1982). A new subtype of human T-cell leukemia virus (HTLV-II) associated with a T-cell variant of hairy cell leukemia. Science 218:571-573.

20. Chen ISY, McLaughlin J, Gasson JC, Clark SC, Golde D (1983). Molecular characterization of the genome of a novel human T-cell leukemia virus. Nature (London) 305:502-505.
21. Rosenblatt JD, Golde DW, Wachsman W, Jacobs A, Schmidt G, Quan S, Gasson JC, Chen ISY (1986). A new HTLV-II isolate associated with atypical hairy cell leukemia: Evidence for an etiological role. Manuscript submitted.
22. Wachsman W, Golde DW, Chen ISY (1984). Hairy cell leukemia and human T-cell leukemia virus. Semin Oncol 11:446-450.
23. Yoshida M, Seiki M, Yamaguchi K, Takatsuki K (1984). Monoclonal integration of human T-cell leukemia provirus in all primary tumors of adult T-cell leukemia suggests causative role of human T-cell leukemia virus in the disease. Proc Natl Acad Sci USA 81:2534-2537.
24. Sohn CC, Blayney DW, Misset JL, Mathe G, Flandrin G, Moran EM, Jensen FC, Winberg CD, Rappaport H (1986). Leukopenic chronic T-cell leukemia mimicking hairy cell leukemia. Blood, in press.
25. Hahn BH, Popovic M, Kalyanaraman VS, Shaw GM, Lomonico A, Weiss SH, Wong-Staal F, Gallo RC (1984). Detection and characterization of an HTLV-II provirus in a patient with AIDS. In Gottlieb MS, Groopman JE (eds): "Acquired Immune Deficiency Syndrome," New York: Alan R. Liss, pp 73-81.
26. Kalyanaraman VS, Narayanan P, Feorino P, Ramsey RB, Palmer EL, Chorba T, McDougal S, Getchell JP, Holloway B, Harrison AK, Cabradilla CD, Telfer M, Evatt B (1985). Isolation and characterization of a human T-cell leukemia virus type II from a hemophilia A patient with pancytopenia. EMBO J 4:1455-1460.
27. Popovic M, Sarngadharan MG, Read E, Gallo RC (1984). Detection, isolation and continuous production of cytopathic retrovirus (HTLV-III) from patients with AIDS and pre-AIDS. Science 224:497-500.
28. Gallo RC, Salahuddin SZ, Popovic M, Shearer GM, Kaplan M, Haynes BF, Parker TJ, Redfield R, Oleske J, Safai B, White G, Foster P, Markham PD (1984). Frequent

detection and isolation of cytopathic retroviruses (HTLV-III) from patients with AIDS and at risk for AIDS. Science 224:500-503.

29. SchÜpbach J, Popovic M, Gilden RV, Gond MA, Sarngadharan MG, Gallo RC (1984). Serological analysis of a subgroup of human T-lymphotropic retroviruses (HTLV-III) associated with AIDS. Science 224:503-506.
30. Sarngadharan MG, Popovic M, Bruch L, Schupach J, Gallo RC (1984). Antibodies reactive with human T-lymphotropic retroviruses (HTLV-III) in the serum of patients with AIDS. Science 224:506-508.
31. Chermann JC, Barré-Sinoussi F, Montagnier L (1984). Characterization and possible role in AIDS of a new human T-lymphotropic retrovirus. In Gottlieb MS, Groopman JE (eds): "Acquired Immune Deficiency Syndrome," New York: Alan R. Liss, pp 31-46.
32. Chen ISY, Quan SG, Golde DW (1983). Human T-cell leukemia virus type II transforms normal human lymphoctyes. Proc Natl Acad Sci USA 80:7006-7009.
33. Yamamoto N, Matsumoto T, Koyanagi M, Hinuma Y (1982). Unique cell lines harboring both Epstein-Barr virus and adult T-cell leukemia virus established from leukemia patients. Nature (London) 299:367-369.
34. Koyanagi Y, Yamamoto N, Kobayashi N, Hirai K, Konishi H, Takeuchi K, Tanaka Y, Hatanaka M, Hinuma Y (1984). Characterization of human B-cell lines harboring both adult T-cell leukemia (ATL) virus and Epstein-Barr virus derived from ATL patients. J Gen Virol 65:1781-1789.
35. Markham PD, Salahuddin SZ, Macchi B, Robert-Guroff M, Gallo RC (1984). Transformation of different phenotypic types of human bone marrow T-lymphocytes by HTLV-I. Int J Cancer 33:13-17.
36. Ruscetti FW, Robert-Guroff M, Ceccherini-Nelli L, Minowada J, Popovic M, Gallo RC (1983). Persistent in vitro infection by human T-cell leukemia-lymphoma virus (HTLV) of normal human T-lymphocytes from blood relatives of patients with HTLV-associated mature T-cell neoplasms. Int J Cancer 31:171-183.

37. Mitsuya H, Guo HG, Megson M, Trainor C, Reitz MS Jr, Broder S (1984). Transformation and cytopathogenic effect in an immune human T-cell clone infected by HTLV-I. Science 223:1293-1296.
38. Popovic M, Flomenberg N, Volkman DJ, Mann D, Fauci AS, Dupont B, Gallo RC (1984). Alteration of T-cell functions by infection with HTLV-I or HTLV-II. Science 226:459-462.
39. Chen ISY, McLaughlin J, Golde DW (1984). Long terminal repeats of human T-cell leukemia virus II genome determines target cell specificity. Nature (London) 309:276-279.
40. Popovic M, Kalyanaraman VS, Mann DS, Richardson E, Sarin P, Gallo RC (1984). Infection and transformation of T-cells by human T-cell leukemia/lymphoma virus and subgroups I and II. In Gallo RC, Essex M, Gross L (eds): "Human T-cell Leukemia/Lymphoma Virus," Cold Spring Harbor Laboratory, pp 217-228.
41. Hoxie JA, Matthews DM, Cines DB (1984). Infection of human endothelial cells by human T-cell leukemia virus type I. Proc Natl Acad Sci USA 81:7591-7595.
42. Ho DD, Rota TR, Hirsch MS (1984). Infection of human endothelial cells by human T-lymphotropic virus type I. Proc Natl Acad Sci USA 81:7588-7590.
43. Clapham P, Nagy K, Cheingsong-Popov R, Exley M, Weiss RA (1983). Productive infection and cell-free transmission of human T-cell leukemia virus in a nonlymphoid cell line. Science 222:1125-1127.
44. Miyoshi I, Taguchi H, Fujishita M, Yoshimoto S, Kuboishi I, Ohtsuki Y, Shiraishi Y, Akagi T (1982). Transformation of monkey lymphocytes with adult T-cell leukemia virus. Lancet i:1016.
45. Miyoshi I, Yoshimoto S, Taguchi H, Kubonishi I, Fujishita M, Ohtsuki Y, Shiraishi Y (1983). Transformation of rabbit lymphocytes with T-cell leukemia virus. Gann 74:1-4.
46. Hoshino H, Shimoyama M, Miwa M, Sugimura T (1983). Detection of lymphocytes producing a human retrovirus associated with adult T-cell leukemia by syncytia induction assay. Proc Natl Acad Sci USA 80:7337-7341.

47. Koeffler HP, Chen ISY, Golde DW (1984). Characterization of a novel HTLV-infected cell line. Blood 64:482-490.
48. Chen ISY, Golde DW, Slamon DJ, Wachsman W (1985). Comparative studies of HTLV-I and HTLV-II. In Gale RP, Golde DW (eds): "Recent Advances in Biology and Treatment," New York: Alan R. Liss, pp 137-149.
49. Seiki M, Hattori S, Hirayama Y, Yoshida M (1983). Human adult T-cell leukemia virus: Complete nucleotide sequence of provirus genome integrated in leukemia cell DNA. Proc Natl Acad Sci USA 80:3618-3622.
50. Shimotohno K, Takahashi Y, Shimizu N, Gojobori T, Golde DW, Chen ISY, Miwa M, Sugimura T (1985). Complete nucleotide sequence of an infectious clone of human T-cell leukemia virus type II: An open reading frame for the protease gene. Proc Natl Acad Sci USA 82:3101-3105.
51. Sagata N, Yasunaga T, Ohishi K, Tsuzuku-Kawamura J, Onuma M, Ikawa Y (1984). Comparison of the entire genomes of bovine leukemia virus and human T-cell leukemia virus and characterization of unidentfied open reading frames. EMBO J 3:3231-3237.
52. Schwartz DE, Tizard R, Gilbert W (1983). Nucleotide sequence of Rous sarcoma virus. Cell 32:853-869.
53. Haseltine WA, Sodroski J, Patarca R, Briggs D, Perkins D, Wong-Staal F (1984). Structure of 3' terminal region of type II human T lymphotropic virus: Evidence of new coding region. Science 225:419-421.
54. Rosenblatt JD, Cann AJ, Golde DW, Chen ISY (1986). Structure and function of human T-cell leukemia virus II genome. In Hinuma Y (ed.), "Cancer Review," Vol. 1, Copenhagen Denmark, Munksgaard, in press.
55. Gelmann EP, Franchini G, Manzari V, Wong-Staal F, Gallo RC (1984). Molecular cloning of a unique human T-cell leukemia virus (HTLV-II_{Mo}). Proc Natl Acad Sci USA 81:993-997.
56. Shimotohno K, Golde DW, Miwa M, Sugimura T, Chen ISY (1984). Nucleotide sequence analysis of the long terminal repeat of human T-cell leukemia virus type II. Proc Natl Acad Sci USA 81:1079-1083.

57. Seiki M, Hattori S, Yoshida M (1982). Human adult T-cell leukemia virus: Molecular cloning of the provirus DNA and the unique terminal structure. Proc Natl Acad Sci USA 79:6899-6902.
58. Sodroski J, Trus M, Perkins D, Patarca R, Wong-Staal F, Gelmann E, Gallo RC, Haseltine WA (1984). Repetitive structure in the long terminal repeat element of type II human T-cell leukemia virus. Proc Natl Acad Sci USA 81:4617-4621.
59. Sodroski JG, Rosen CA, Haseltine WA (1984). Trans-acting transcriptional activation of long terminal repeat of human T-lymphotropic viruses in infected cells. Science 225:381-385.
60. Shimotohno K, Wachsman W, Takahashi Y, Golde DW, Miwa M, Sugimura T, Chen ISY (1984). Nucleotide sequence of the 3' region of an infectious human T-cell leukemia virus type II genome. Proc Natl Acad Sci USA 81:6657-6661.
61. Slamon DJ, Shimotohno K, Cline MJ, Golde DW, Chen ISY (1984). Identification of the putative transforming protein of the human T-cell leukemia viruses HTLV-I and HTLV-II. Science 226:61-65.
62. Lee TH, Coligan JE, Sodroski JG, Haseltine WA, Salahuddin SZ, Wong-Staal F, Gallo RC, Essex M (1984). Antigens encoded by the 3' terminal region of human T-cell leukemia virus: Evidence for a functional gene. Science 226:57-61.
63. Wachsman W, Shimotohno K, Clark SC, Golde DW, Chen ISY (1984). Expression of the 3' terminal region of human T-cell leukemia viruses. Science 226:177-179.
64. Wachsman W, Golde DW, Temple PA, Orr EC, Clark SC, Chen ISY. HTLV χ gene product: Requirement for the env methionine initiation codon. Science 228:1534-1537.
65. Seiki M, Hikikoshi A, Tanigushi T, Yoshida M (1985). Expression of the pX gene of HTLV-I: General splicing mechanism in HTLV family. Science 228:1532-1535.
66. Rice NR, Stephens RM, Couez D, Deschmaps J, Kettmann R, Burny A, Gilden RV (1984). The nucleotide sequence of the env and post env region of bovine leukemia virus. Virology 138:82-93.

67. Slamon DJ, Press MF, Souza LM, Cline MJ, Golde DW, Gasson JC, Chen ISY (1985). Studies of the putative transforming protein of the the type I human T-cell leukemia virus. Science 228:1427-1430.
68. Goh WC, Sodroski J, Rosen C, Essex M, Haseltine WA (1984). Subcellular localization of the product of the long open reading frame of human T-cell leukemia virus type I. Science 227:1227-1228.
69. Chen ISY, Slamon DJ, Rosenblatt JD, Shah NP, Quan SG, Wachsman W (1985). The χ gene is essential for HTLV replication. Science 229:54-58.
70. Fujisawa JI, Seiki M, Kiyokawa T, Yoshida M (1985). Functional activation of the long terminal repeat of human T-cell leukemia virus type I by trans-acting factor. Proc Natl Acad Sci USA 82:2277-2281.
71. Cann AJ, Rosenblatt JD, Wachsman W, Shah N, Chen ISY (1985). Identification of the gene responsible for human T-cell leukemia virus transcriptional control. Nature (London) 318:571-574.
72. Bishop JM (1983). Cellular oncogenes and retroviruses. Ann Rev Biochem 52:301-354.
73. Jones N, Shenk T (1979). An adenovirus type 5 early gene function regulates expression of other early viral genes. Proc Natl Acad Sci USA 76:3665-3669.
74. Nevins JR (1981). Mechanism of activation of early viral transcription by the adenvoirus ElA gene product. Cell 26:213-220.
75. Berk AJ, Lee F, Harrison T, Williams J, Sharp PA (1979). Pre-early adenovirus 5 gene product regulates synthesis of early viral messenger RNAs. Cell 17:935-944.
76. El Karech A, Murphy AJM, Fichter T, Efstraitiadis A, Silverstein S (1985). Transactivation control signals in the promoter of herpes virus thymidine kinase gene. Proc Natl Acad Sci US 82:1002-1006.
77. Katinka M, Yanive M, Vasseur M, Blangy D (1980). Expression of polyoma early functions in mouse embryonal carcinoma cells depends on sequence rearrangements in the beginning of the late region. Cell 20:393-399.

78. Gaynor RB, Hillman D, Berk AJ (1984). Adenovirus early 1A protein activates transcription of a non viral gene introduced into mammalian cells by infection and transfection. Proc Natl Acad Sci USA 81:1193-1197.
79. Green MR, Treisman R, Maniatis T (1983). Transcriptional activation of cloned human β-globin genes by viral immediate-early gene products. Cell 35:137-148.
80. Montell C, Courtois G, Eng C, Berk A (1984). Complete transformation by adenovirus 2 requires both E1A proteins. Cell 36:951-961.
81. Montell C, Fisher EF, Caruthers MH, Berk A (1982). Resolving the functions of overlapping viral genes by site-specific mutagenesis at a mRNA splice site. Nature (London) 295:380-384.
82. Ruley HE (1983). Adenovirus early region 1A enables viral and cellular transforming genes to transform primary cells in culture. Nature (London) 304:602-606.
83. Land H, Parada LF, Weinberg RA (1983). Tumorigenic conversion of primary embryo fibroblasts require at least two cooperative oncogenes. Nature (London) 304:596.
84. Kingston RE, Baldwin AS, Sharp PA (1985). Transcription control by oncogenes. Cell 41:3-5.
85. Chen ISY, Cann AJ, Shah NP, Gaynor RB (1985). Functional relation between HTLV-II χ and adenovirus E1A proteins in transcriptional activation. Science 230:570-573.
86. Schrier PI, Bernards R, Vaessen RTMJ, Houweling A, van der Eb AJ (1983). Expression of class I major histocompatibility antigens switched off by highly oncogenic adenovirus 12 in transformed rat cells. Nature (London) 305:771-775.
87. Bernards R, Schrier PI, Houweling A, Bos JL, van der Eb AJ (1983). Tumorigenicity of cells transformed by adenovirus type 12 by evasion of T-cell immunity. Nature (London) 305:776-779.

Viruses and Human Cancer, pages 131–140

HTLV AND ATL[1]

Masanao Miwa, Kunitada Shimotohno, Toshio Kitamura,
Hiroshi Shima, Nobuaki Shimizu, Yuko Ootsuyama,
Atsumi Tsujimoto, Shaw Watanabe, Masanori Shimoyama,
and Takashi Sugimura

National Cancer Center Research Institute
1-1, Tsukiji 5-chome, Chuo-ku, Tokyo 104 Japan

Adult T-cell leukemia (ATL) is a well known disease characterized by onset in adulthood, subacute or chronic T-cell leukemia with rapidly progressive terminals, leukemic cells with frequently indented or lobulated nuclei, frequent skin involvement, lymphnode involvement and hepatosplenomegaly (1). In particular the distribution of the birth places of the patients with ATL is unique in that they are mainly localized in the south-west part of Japan and in the Carribean countries.

Human T-cell leukemia virus type I (HTLV-I) was first isolated by Dr. Gallo's group (2) and the total nucleotide sequence was determined by Dr. Yoshida's group (3). HTLV-I is considered to be closely associated with most ATL cases.

We have determined the total nucleotide sequence of infectious HTLV-II (4) isolated from cultured cells of a T cell variant of hairy cell leukemia (5).

Both HTLV-I and HTLV-II are related in that they are isolated from human T-cell leukemias. Also, normal human lymphocytes can be immortalized by cocultivation with cell lines producing HTLV-I or HTLV-II (6, 7).

Myristylation of gag proteins of HTLV-I and HTLV-II

Schultz et al. found that p19 of *gag* protein of HTLV-I was metabolically labeled with [^{3}H]myristic acid, suggesting that myristic acid was covalently bound to the protein (8).

[1] This work was supported by a Grant-in-Aid from the Ministry of Health and Welfare for the Comprehensive 10-Year Strategy for Cancer Control, Japan.

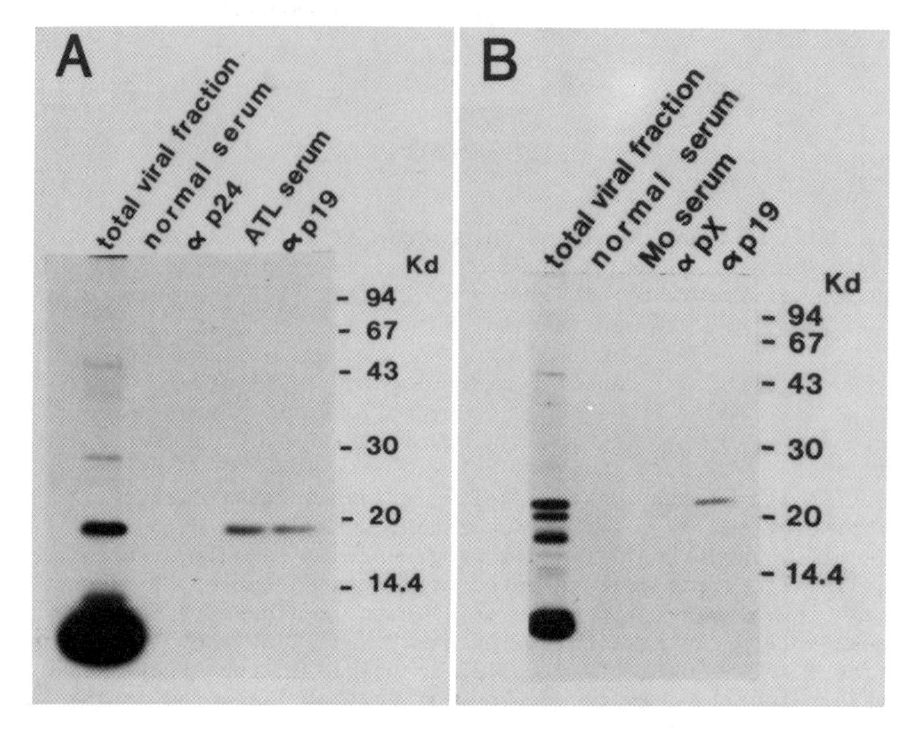

Fig. 1. Myristylation of gag protein (p19) of HTLV-I and HTLV-II.

Cells were labeled for 18 hr with [^{3}H]myristic acid in PRMI 1640 medium containing 10% fetal calf serum. Autoradiograms of immunoprecipitation products after SDS-polyacrylamide gel electrophoresis (9). A) HTLV-I lysate from MT-2 cell culture medium. Immunoprecipitates were prepared with either normal serum, or monoclonal antibody against, p24 serum from a patient with ATL or monoclonal antibody against p19. B) HTLV-II lysate from Tonl cell culture medium. Mo is a patient from whom HTLV-II producing Mo cells were originally established.

We studied myristylation of gag or gag related protein of HTLV-I and HTLV-II in more detail. HTLV-I producing MT-2 cells were incubated with [^{3}H]myristic acid and the virus fraction was prepared. With the monoclonal antibody against

$p19^{gag}$, the band corresponding to p19 was detected in HTLV-I lysate (Fig. 1A) (7).

HTLV-II producing Ton1 cells were also incubated with $[^3H]$myristic acid and the virus fraction was prepared. A radioactive protein with a molecular weight of 23 Kd was precipitated from HTLV-II lysate with the monoclonal antibody against $p19^{gag}$ of HTLV-I (Fig. 1B) (9). The apparent discrepancy between the observed molecular weight (23 Kd) and the expected molecular weight (19 Kd) of the gag protein of HTLV-II, may be explained by the high content of proline residues predicted by the amino acid sequence from the nucleotide sequence or by some unknown posttranslational modification other than myristylation. It is reported that v-src gene product, $p60^{src}$, is myristylated at its N-terminus (10, 11). Also v-abl and v-fes onc gene products are known to be fused proteins of the myristylated gag proteins (12). The significance of the myristylations to onc gene products in the transformation process is not well understood.

Protease gene coded by HTLV

A precursor gag protein of retrovirus is known to be cleaved by virally coded protease. In Rous sarcoma virus (RSV), the protease domain is at the 3' end of the gag frame (p15) (13). In murine leukemia virus (MuLV), this protease domain is at the 5' end of the pol frame (14). However there is no detectable homology between the amino acid sequence of these proteases and amino acid sequences coded from the 3' region of gag or the 5' region of pol of HTLV-II. There is an open reading frame covering 3' end of gag gene and 5' end of pol gene, which can code 178 amino acid residues in a different frame to gag and pol (4). The amino acid sequence from this open reading frame shows significant homology with the proteases of RSV and MuLV. In bovine leukosis virus (BLV), there is an open reading frame located in a similar position between gag and pol regions (15). The sequence of N-terminal portion of a purified protein from the virion, as determined by Dr. Oroszlan and his colleague, was consistent with the predicted amino acid sequence. Thus, the above open reading frame was shown to express a protein which probably is a protease of BLV (16) coded from the open reading frame. A similar amino acid sequence was observed in HTLV-I (3) and in HTLV-II (4). However, the sequence in HTLV-I was split by terminators and also showed a frame shift. The sequence in the protease portion in the published data (3) might be different from the sequence of the infectious HTLV-I which is not yet isolated.

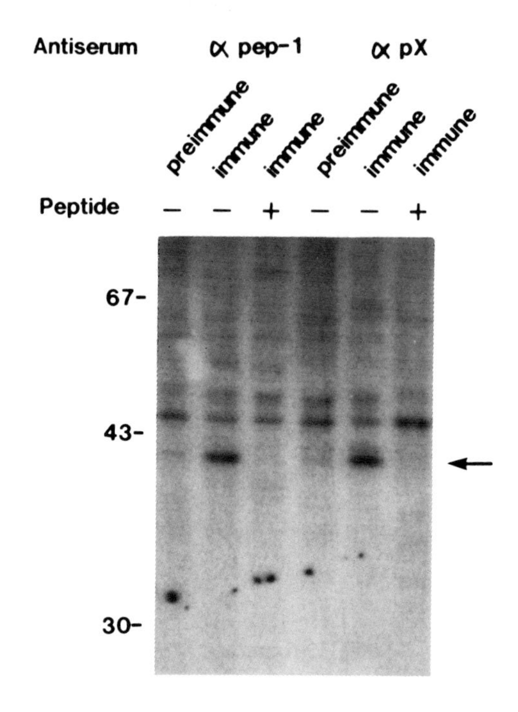

Fig. 2. Immunoprecipitation of monkey cell lysates with antisera against peptides corresponding to portions of the amino acid sequence of pX-IV gene product of HTLV-I. [^{35}S]Cysteine labeled cell lysate of BM5 cell line which harbors simian T-cell leukemia virus was immunoprecipitated with either preimmune rabbit serum, and immune rabbit serum against the synthetic tetradecapeptide CPEHQITWDPIDGR (α pep-1) or with preimmune or immune guinea pig serum against a fusion peptide of bovine growth hormone and pX-IV protein of HTLV-I (α pX) (22).

pX and X regions

Among the homologies between the amino acid sequences which were deduced from the nucleotide sequences of HTLV-I and HTLV-II proviruses, a higher homology was found between the open reading frame of pX-IV of HTLV-I and that of Xc of HTLV-II. The homology is more than 80%. We thought that these conserved sequences must be important for viral functions, or for cell immortalization or transformation. We identified proteins p41 and p38 coded by HTLV-I and HTLV-II, respectively, using the antibody against the synthetic dodecapeptide which corresponds to that near C-terminal portion of pX-IV (17, 18). These findings were also reported by other groups (19-21). Recently we identified a protein of 41 Kd as a pX product in simian lymphoid cell lines infected by simian T-cell leukemia virus which is closely related to HTLV-I (Fig. 2) (22).

Identification of Xb protein of HTLV-II

It is now well established that the pX (or X) regions are important for trans-acting transcriptional activation. Recently Dr. Yoshida's group reported the production of proteins from pX-III frame of HTLV-I (23). Until now, nothing has been known about the presence of other proteins coded by X region. We studied the presence of X protein coded from a frame other than Xc in HTLV-II infected cells. An antibody was raised against the synthetic octadecapeptide MPKTRRQRTRRARRNRPP corresponding to the N-terminal portion of Xb frame, taking into consideration the splicing of mRNA. Based on the assumed amino acid sequences, the size of the predicted protein would be about 20 Kd. The antibody precipitated a protein with a molecular weight of 24 Kd in [^{35}S]cysteine labeled Tonl cell extract. This band disappeared when the corresponding peptide was included as a competitor in immunoprecipitation reaction.

The protein could also be detected with Western blotting analysis. The protein was not recognized by normal guinea pig sera. The band disappeared when the corresponding peptide was added to the reaction, although the band still remained when another peptide was added to the reaction as a competitor. The protein was found in HTLV-II infected Mo cells, but not found in HTLV-II uninfected HL-60 cells. These observations suggest that the protein recognized by the anti-peptide antibody was coded by Xb frame of HTLV-II.

Monoclonal antibody against pX-IV protein of HTLV-I

To better understand the function of pX-IV protein, it will be helpful to make monoclonal antibody against pX-IV protein. For immunization, we used a fused protein, kindly given by Dr. Souza, consisting of a bovine growth hormone and C-terminal portion of pX-IV protein. We immunized BALB/C mice with this fused protein. Spleen cells from the immunized mice were fused to mouse myeloma cells. The hybridoma cells were screened for immunocytochemical staining

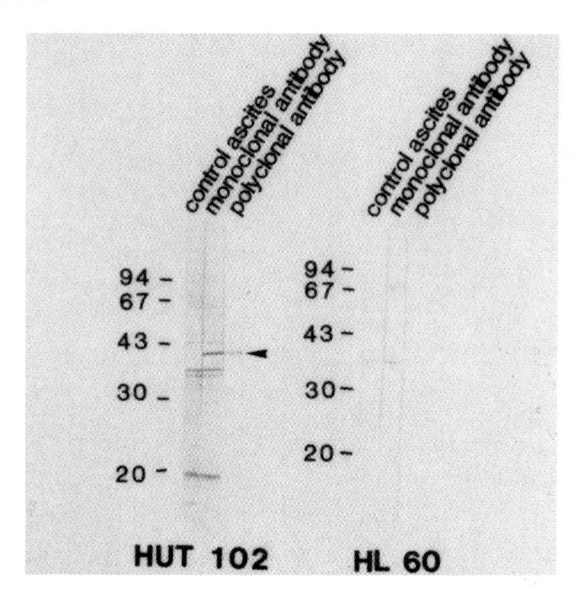

Fig. 3. Western blotting of HUT102 and HL60 cell extracts with, NCC-pX-1G, the monoclonal antibody against pX-IV protein of HTLV-I. The cell extracts were reacted with either control ascitic fluid, NCC-pX-1G in ascitic fluid or polyclonal guinea pig antiserum against pX-IV protein (24).

activity with HTLV-I infected HUT102 cells. A hybridoma cell line, NCC-pX-1G, was obtained (24). Western blotting analysis clearly revealed the protein band with a molecular weight around 41 Kd corresponding to pX-IV protein of HTLV-I (Fig. 3). For studying subcellular localization of pX-IV, HUT102 cell pellet was fixed with PLP, embedded in paraffin and sectioned. The nuclei were more strongly stained than cytoplasm. The staining pattern of the nuclei was diffuse and finely granular. The MT-2 and MT-1 cells were also stained. EBV transformed B-cell lines, HL-60 cells, and peripheral lymphocytes from healthy adults were not stained. Using this monoclonal antibody, it might be possible to detect pX-IV protein in certain organs of the patients with ATL or to check the physiological function of pX-IV protein by introducing the antibody into the HTLV-I infected cells.

ATL without Involvement of HTLV-I

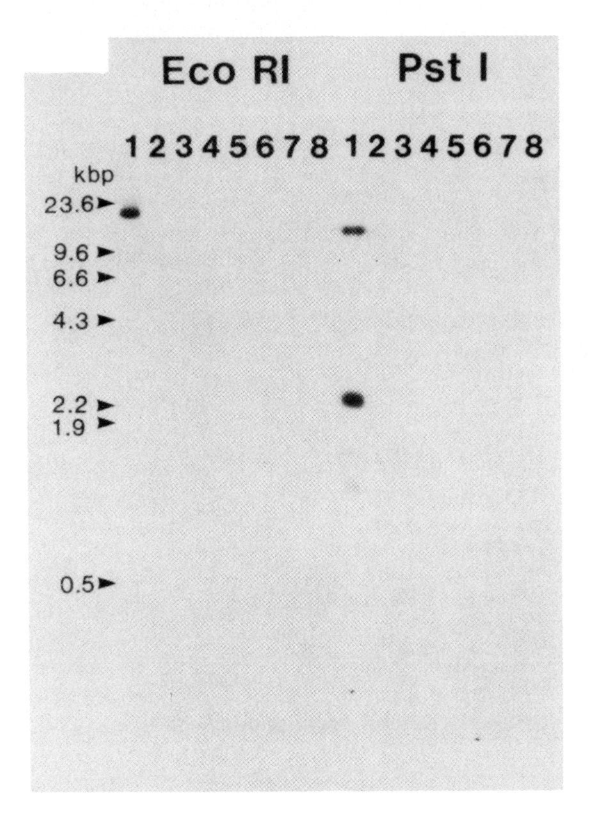

Fig. 4. Absence of the HTLV-I proviral genome in leukemia cells of patients with adult T-cell leukemia. Southern blot analysis with the ^{32}P-labeled probe of a mixture of nick-translated DNA fragments of gag-pol env-pX and LTR (25). Lanes 1: ATL-1K cell line. Lanes 2: HL-60 promyelocytic leukemia cell line. Lanes 3: peripheral blood mononuclear cells (PBMCs) from patient 1. Lanes 4: PBMCs from patient 2. Lanes 5: PBMCs from patient 3. Lanes 6: pericardial effusion cells from patient 3. Lanes 7: ascites cells from patient 4. Lanes 8: PBMCs from patient 5.

We have experienced 5 patients with ATL born in non-endemic areas of Japan. These patients did not have antibody against HTLV-I. We checked if the HTLV-I provirus was integrated in leukemic cells, with probes covering the whole area of HTLV-I provirus. We could not detect the provirus in DNA of the leukemic cells at any significant level in southern blot hybridization (Fig. 4) (25).

Although the possibility of deletion of once integrated HTLV-I from the leukemic cells is not excluded, these results suggest that some factor(s) other than HTLV-I is involved in pathogenesis of ATL in these patients. Epidemiological data suggest that HTLV-I infection occurs early in life but ATL become manifest in the fifties or sixties. The long latency between the viral infection and the appearance of ATL might be explained by multistep carcinogenesis which takes long term involvement. Therefore establishment of the cell lines from ATL patient not associated with HTLV-I will be useful for further clarification of the common features of transformation to ATL cells with or without HTLV-I.

ACKNOWLEDGMENTS

We are grateful to Dr. L. M. Souza for a fused protein of a part of pX protein. H. Shima is the recipient of a Research Resident Fellowship from the Foundation for Promotion of Cancer Research. A. Tsujimoto is the recipient of a fellowship from Sankyo Foundation of Life Science.

REFERENCES

1. Uchiyama T, Yodoi J, Sagawa K, Takatsuki K and Uchino H (1977). Blood 50: 481.
2. Poiesz BJ, Ruscetti FW, Gazdar AF, Bunn PA, Minna JD and Gallo RC (1980). Proc. Natl. Acad. Sci. USA 77: 7415.
3. Seiki M, Hattori S, Hirayama Y and Yoshida M (1983). Proc. Natl. Acad. Sci. USA 80: 3618.
4. Shimotohno K, Takahashi Y, Shimizu N, Gojobori T, Golde DW, Chen ISY, Miwa M and Sugimura T (1985). Proc. Natl. Acad Sci. USA 82: 3101.
5. Kalyanaraman VS, Sarngadharan MG, Robert-Guroff M, Miyoshi I, Blayney D, Golde D and Gallo RC (1982). Science 218: 571.

6. Miyoshi I, Kubonishi I, Yoshimoto S, Akagi T, Ohtsuki Y, Shiraishi Y, Nagata K and Hinuma Y (1981). Nature 294: 770.
7. Chen ISY, Quan SG and Golde DW (1983). Proc. Natl. Acad. Sci. USA 80: 7006.
8. Schultz AM, Henderson LE and Oroszlan S (1983). In "Leukemia Reviews International," (MA, Rich ed.) Vol.1. Marcell Dekker Press, New York, 304.
9. Ootsuyama Y, Shimotohno K, Miwa M, Oroszlan S and Sugimura T (1985). Jpn. J. Cancer Res. (Gann) 76: 1132.
10. Schultz AM, Henderson LE, Oroszlan S, Carber EA and Hanafusa H (1985). Science 227: 427.
11. Kamps MP, Buss JE and Sefton BM (1985). Proc. Natl. Acad. Sci. USA 82: 4625.
12. Schultz AM and Oroszlan S (1984). Virology 133: 431.
13. Schwartz DE, Tizard R and Gilbert W (1983). Cell 32: 853.
14. Yoshinaka Y, Katoh I, Copeland TD and Oroszlan S (1985). Proc. Natl. Acad. Sci. USA 82: 1618.
15. Sagata N, Yasunaga T, Tsuzuku-Kawamura J, Ohishi K, Ogawa Y and Ikawa Y (1985). Proc. Natl. Acad. Sci. USA 82: 677.
16. Oroszlan S, Copeland TD, Rice NR, Smythers GW, Tsai W-P, Yoshinaka Y and Shimotohno K (1985). In "Retroviruses in Human Lymphoma/Leukemia" (M. Miwa et al. eds.) Japan Sci. Soc. Press, Tokyo/VNU Science Press, Utrecht, 147.
17. Miwa M, Shimotohno K, Hoshino H, Fujino M and Sugimura T (1984). Gann (Jpn. J. Cancer Res.) 75: 752.
18. Shimotohno K, Miwa M, Slamon DJ, Chen ISY, Hoshino H, Takano M, Fujino M and Sugimura T (1985). Proc. Natl. Acad. Sci. USA 82: 302.
19. Lee TH, Coligan JE, Sodroski JG, Haseltine WA, Salahuddin SZ, Wong-Staal F, Gallo RC and Essex M (1984). Science 226: 57.
20. Slamon DJ, Shimotohno K, Cline MJ, Golde DW and Chen ISY (1984). Science 226: 61.
21. Kiyokawa T, Seiki M, Imagawa K, Shimizu F and Yoshida M (1984). Gann (Jpn. J. Cancer Res.) 75: 747.
22. Tsujimoto A, Tsujimoto H, Yanaihara N, Abe K, Hayami M, Miwa M and Shimotohno K (1986). FEBS Letters 196: 301.
23. Kiyokawa T, Seiki M, Iwashita S, Imagawa K, Shimizu F and Yoshida M (1985). Proc. Natl. Acad. Sci. USA 82: 8359.

24. Watanabe S, Sato Y, Shima H, Shimotohno K and Miwa M (1986). Jpn. J. Cancer Res. (Gann) 77: 338.
25. Shimoyama M, Kagami Y, Shimotohno K, Miwa M, Minato K, Tobinai K, Suemasu K and Sugimura T (1986). Proc. Natl. Acad. Sci. USA 83: 4524.

Viruses and Human Cancer, pages 141–150

AN ANIMAL MODEL OF HTLV-I INFECTION: INTRAVENOUS AND ORAL TRANSMISSION OF HTLV-I IN RABBITS[1]

I. Miyoshi, S. Yoshimoto, M. Fujishita, K. Yamato, S. Kotani, M. Yamashita, H. Taguchi, and Y. Ohtsuki

Departments of Medicine and Pathology, Kochi Medical School
Kochi 781-51, Japan

ABSTRACT Attempts were made to transmit human T-cell leukemia virus type I (HTLV-I) to rabbits with the purpose of investigating the alleged routes of virus transmission in humans. The virus could be transmitted to rabbits (12/12) by transfusion of 20 ml of whole blood or washed blood cell suspension (fresh or stored for 1-2 weeks at 4°C) but not to rabbits (0/2) by infusion of cell-free plasma from virus-infected rabbits. Seroconversion occurred 2-4 weeks after blood transfusion and virus-producing lymphoid cell lines were established from four of the seroconverted rabbits. Seroconversion likewise occurred in rabbits (3/3) transfused with blood that had been X-irradiated (6,000 rad) immediately before transfusion but not in rabbits (0/3) transfused with blood that had been irradiated and stored for 1-2 weeks at 4°C. Four rabbits were given twice weekly oral administration of $2\text{-}4 \times 10^7$ cells from an HTLV-I-producing rabbit lymphoid cell line. One of them seroconverted for HTLV-I after two months and the virus was isolated from this rabbit. These findings are discussed in terms of their clinical application for the prevention of HTLV-I infection in humans.

[1]This work was supported by a Grant-in-Aid (60218024) for Special Research Project, Cancer-Bioscience from the Ministry of Education, Science and Culture and a Grant-in-Aid (59shi-1) for Cancer Research from the Ministry of Health and Welfare of Japan.

INTRODUCTION

Human T-cell leukemia virus type I (HTLV-I) is a lymphotropic retrovirus which is associated with the etiology of adult T-cell leukemia (ATL). The virus is transmitted by blood transfusion (1) and by close family contact (2,3). Healthy persons seropositive for HTLV-I are all virus carriers (2,4) and the viral antigens have been detected in mothers' milk and semen (5,6). The seropositive rate in the general population is remarkably high in ATL-endemic areas such as southwest Japan, the Caribbean, and parts of Africa (7-9). It is, therefore, of urgent importance to screen seropositive individuals in blood banks and maternity clinics. We have found that HTLV-I is transmissible to rabbits serially by blood transfusion (10) and by oral administration of HTLV-I-infected cells. A method has been devised for preventing transfusion-related virus transmission in this animal model.

METHODS

Blood Transfusion

Japanese White rabbits weighing about 3 kg were used. Usually, 40 ml of heparinized blood were obtained from a seroconverted rabbit and later transfused into two normal rabbits, each receiving 20 ml as whole blood or washed blood cell suspension in physiological saline. Two rabbits were infused with 20 ml of cell-free plasma filtered through a Millipore (0.45 μm) membrane. In some pairs of animals, one was given blood that had been X-irradiated at 6,000 rad, while the other received non-irradiated blood from the same bleeding. The transfusion experiment was performed serially between rabbits of opposite sexes.

Detection of Antibodies to HTLV-I

Rabbits were bled immediately prior to and sequentially after blood transfusion or oral administration of HTLV-I-infected cells. Sera were titrated for the antibodies to HTLV-I antigens by indirect immunofluorescence as described previously (11).

Oral Administration of HTLV-I-infected Cells

Ra-1 cells were given orally twice a week to four female rabbits. Ra-1 is a male rabbit lymphoid cell line established by co-cultivation with MT-2 cells and is persistently infected with HTLV-I (12). The cells were washed and resuspended in physiological saline at 4-8 x 10^7/ml and 0.5 ml of the cell suspension was instilled into the oral cavity through a teflon-covered 16-gauge needle.

Cell Culture

After seroconversion, 10-20 ml of blood were obtained from rabbits and lymphocytes were separated on a Ficoll-Hypaque gradient. The cells were cultivated in medium RPMI 1640 supplemented with 10% human cord serum, 10% fetal calf serum, 10% crude T-cell growth factor (TCGF), and antibiotics. The cultures were incubated at 37°C in a humidified 7.5% CO_2 atmosphere and fed twice a week. After the cells had become TCGF-independent, they were adapted to growth in medium without human cord serum.

Surface and Antigen Markers

Surface and antigen markers were studied as described previously (11).

RESULTS

Transmission of HTLV-I by Blood Transfusion

Three female rabbits (A, B, and C) were first inoculated intravenously with Ra-1 cells. All three became seropositive within six weeks and HTLV-I was detected or isolated from these rabbits (11). Twenty ml of washed blood cell suspension from rabbit B were transfused into two female rabbits (D and E). Both recipients seroconverted for HTLV-I after three weeks and a virus-producing lymphoid cell line (Ra-2) was established from rabbit E.

Seven months after transfusion, a washed blood cell suspension prepared from rabbit E was transfused into two male rabbits (F and G), 20 ml each, after storage for two weeks at 4°C. Both of them became seropositive after four weeks and an HTLV-I-carrying cell line (Ra-3) was derived from rabbit F. In contrast, two male rabbits (H and I)

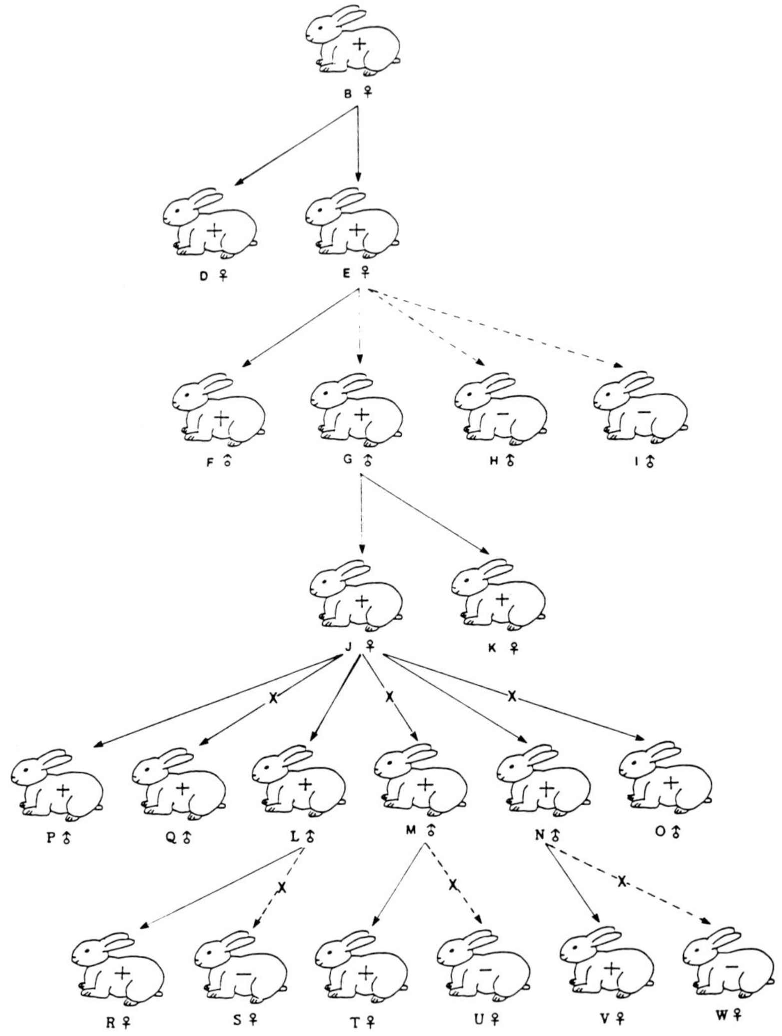

FIGURE 1. Solid lines, transfusion of whole blood or washed blood cell suspension (fresh or stored for 1-2 weeks at 4°C); dotted lines, infusion of cell-free plasma; crossed solid lines, transfusion of X-irradiated fresh blood; crossed dotted lines, transfusion of X-irradiated stored blood. Seroconversion is indicated by +. HTLV-I-producing lymphoid cell lines (Ra-2 to Ra-5) were estab-lished from rabbits E, F, J, and L, respectively. Reprinted with permission from Alan R. Liss, Inc. (Kotani et al. 1986).

infused with 20 ml of cell-free plasma from rabbit E remained seronegative.

Seven months after transfusion, 20 ml of blood from rabbit G were transfused into two female rabbits (J and K). These rabbits became anti-HTLV-I positive after four weeks and a virus-producing lymphoid cell line (Ra-4) was grown from rabbit J.

Prevention of Transfusion-related Transmission of HTLV-I by Use of X-irradiated Stored Blood

Two and half months after transfusion, 40 ml of blood were obtained from rabbit J; one half was transfused into rabbit L without irradiation and the other half was irradiated (6,000 rad) and immediately transfused into rabbit M. Both rabbits seroconverted for HTLV-I after two weeks and a virus-producing lymphoid cell line (Ra-5) was established from rabbit L. The same experiment was repeated with two more blood samples taken from rabbit J at an interval of one month; the non-irradiated half was transfused into rabbits N and P, and the irradiated half into rabbits O and Q. All four became anti-HTLV-I positive after 2-3 weeks.

Two to five months after transfusion, 40 ml of blood were taken from rabbits L, M, and N. One half of each blood sample was irradiated and was placed at 4°C together with the other non-irradiated half. After storage for 7, 10, and 14 days at 4°C, the non-irradiated half was transfused into rabbits R, T, and V, and the irradiated half into rabbits S, U, and W, respectively. Seroconversion occurred in rabbits R, T, and V after 2-4 weeks but not in rabbits S, U, and W. The serial transfusion experiment is schematically shown in Fig. 1.

Establishment and Characteristics of HTLV-I-producing Lymphoid Cell Lines.

Lymphocytes from four seroconverted rabbits were cultivated in the presence of TCGF and four lymphoid cell lines (Ra-2 to Ra-5) were established from rabbits E, F, J, and L, respectively, representing passages 1 to 4. Chromosome analysis of these cell lines showed normal rabbit karyotypes with sex chromosomes of the recipients. Three (Ra-2, Ra-3, and Ra-5) of the four cell lines became TCGF-independent after continuous cultivation for 2-12 months. The surface and antigen markers of the four cell lines were all identical to those of Ra-1. The cells were completely negative for sheep erythrocyte receptors and surface immunoglobulins and did not react with monoclonal antibodies to human T-cells (Leu-1, Leu-3a, and Leu-5) and to Ia antigens (OKIa1). The majority of cells were positive for HTLV-I antigens as indicated by positive fluorescence with sera from ATL patients and monoclonal antibodies to HTLV-I p19 and p24 core proteins (13,14). HTLV-I particles were demonstrated in all four cell lines by electron microscopy. These results are summarized in Table 1.

TABLE 1
CHARACTERISTICS OF RABBIT LYMPHOID CELL LINES[a]

	Ra-1	Ra-2	Ra-3	Ra-4	Ra-5
Sheep erythrocyte receptors	0	0	0	0	0
Surface immuno-globulins	0	0	0	0	0
Leu-1	0	0	0	0	0
Leu-3a	0	0	0	0	0
Leu-5	0	0	0	0	0
OKIa1	0	0	0	0	0
HTLV-I antigens[b]	90-100	90-100	70-80	70-80	70-80
p19	90-100	90-100	70-80	70-80	70-80
p24	90-100	90-100	70-80	70-80	70-80
HTLV particles	+	+	+	+	+
Karyotype	44,XY	44,XX	44,XY	44,XX	44,XY

[a]Figures indicate percentage of positive cells.
[b]Antigens detected with ATL patients' sera. Reprinted with permission from Alan R. Liss, Inc. (Kotani et al. 1986).

Oral Transmission of HTLV-I

One of four female rabbits given oral administration of Ra-1 cells became seropositive for HTLV-I after two months. Lymphocytes from the seroconverted rabbit were co-cultivated with irradiated (6,000 rad) lymphocytes from a seronegative male rabbit. This gave rise to an HTLV-I-producing lymphoid cell line derived from the female rabbit, indicating virus transmission via the oral route.

DISCUSSION

In the present experiment, HTLV-I could be serially transmitted for five passages from seropositive to seronegative rabbits by transfusion of 20 ml of whole blood or washed blood cell suspension but not by infusion of cell-free plasma. The rabbits regularly seroconverted for HTLV-I 2-4 weeks after blood transfusion and a virus-producing lymphoid cell line of recipient origin was established after each passage. The blood preparations stored for 1-2 weeks at 4°C were equally effective in transmittintg the virus. These results are consistent with a high seroconversion rate in recipients of blood and a lack of seroconversion in recipients of plasma from seropositive persons (1).

We then investigated the effect of X-irradiation (6,000 rad) on the transmission of HTLV-I by blood transfusion. Although virus transmission did occur when the blood was transfused immediately after irradiation, it was preventable when the irradiated blood was transfused after storage for 1-2 weeks at 4°C. These data suggest that irradiation per se does not inhibit virus induction from HTLV-I genome-carrying lymphocytes and that irradiated lymphocytes probably die within one week.

Transmission of HTLV-I by blood transfusion constitutes a major route of virus transmission in Japan. In Kyushu, the prevalence of seropositive blood donors is as high as 13% (15) and the number of seropositive adults is estimated to be about 500,000 (16). It is an urgent task, therefore, to screen seropositive blood donors to prevent the iatrogenic dissemination of HTLV-I. If exclusion of seropositive donors results in a shortage of blood supply, we recommend that these donors' blood be used after appropriate irradiation and storage.

Milk-borne transmission of HTLV-I from mother to infant is likely to occur in view of the presence of viral antigens in mothers' milk (5,6) and transmission of HTLV-I to a marmoset by feeding milk from seropositive women (17). This concept is further supported by the animal experiment of Yamanouchi et al (18) and ours in which a common marmoset and a rabbit were, respectively, infected with HTLV-I by oral administration of virus-infected cells. We have recently found that heating for 30 min at 56°C inactivates HTLV-I and virus-infected cells (19). Thus, it is of prophylactic importance to kill HTLV-I genome-carrying lymphocytes by heating breast milk for 30 min at 56°C before feeding babies. Alternatively, carrier mothers should refrain from breast feeding.

ACKNOWLEDGMENTS

We thank Dr. Robert C. Gallo and Dr. Barton F. Haynes for providing monoclonal antibodies to HTLV-I p19 and p24 and Miss Haruko Kawamura for preparing this manuscript.

REFERENCES

1. Okochi K, Sato J, Hinuma Y (1984). A retrospective study on transmission of adult T-cell leukemia virus by blood transfusion: seroconversion in recipients. Vox Sang 46: 245.
2. Miyoshi I, Taguchi H, Fujishita M, Niiya K, Kitagawa T, Ohtsuki Y, Akagi T (1982). Asymptomatic type C virus carriers in the family of an adult T-cell leukemia patient. Gann 73: 339.
3. Tajima K, Tominaga S, Suchi T, Kawagoe T, Komoda H, Hinuma Y, Oda T, Fujita K (1982). Epidemiological analysis of the distribution of antibody to adult T-cell leukemia virus-associated antigen: possible horizontal transmission of adult T-cell leukemia virus. Gann 73: 893.
4. Gotoh Y, Sugamura K, Hinuma Y (1982). Healthy carriers of a human retrovirus, adult T-cell leukemia virus (ATLV): demonstration by clonal culture of ATLV-carrying T-cells from peripheral blood. Proc Natl Acad Sci (USA) 79: 4780.
5. Kinoshita K, Hino S, Amagasaki T, Ikeda S, Yamada Y, Suzuyama J, Momita S, Toriya K, Kamihira S, Ichimaru M (1984). Demonstration of adult T-cell leukemia virus antigen in milk from 3 sero-positive mothers. Gann 75: 103.
6. Nakano S, Ando Y, Ichijo M, Moriyama I, Saito S, Sugamura K, Hinuma Y (1984). Search for possible routes of vertical and horizontal transmission of adult T-cell leukemia virus. Gann 75: 1044.

7. Hinuma Y, Nagata K, Hanaoka M, Nakai M, Matsumoto T, Kinoshita K, Shirakawa S, Miyoshi I (1981). Adult T-cell leukemia: antigen in an ATL cell line and detection of antibodies to the antigen in human sera. Proc Natl Acad Sci (USA) 78: 6476.
8. Clark H, Saxinger C, Gibbs WN, Lofters W, Lagranade L, Deceulaer K, Ensroth A, Robert-Guroff M, Gallo RC, Blattner WA (1985). Seroepidemiologic studies of human T-cell leukemia/lymphoma virus type I in Jamaica. Int J Cancer 36: 37.
9. Biggar RJ, Johnson BK, Oster C, Sarin PS, Ocheng D, Tuckei P, Nsanze H, Alexander S, Bodner AJ, Siongok TA, Gallo RC, Blattner WA (1985). Regional variation in prevalence of antibody against human T-lymphotropic virus types I and III in Kenya, East Africa. Int J Cancer 35: 763.
10. Kotani S, Yoshimoto S, Yamato K, Fujishita M, Yamashita M, Ohtsuki Y, Taguchi H, Miyoshi I (1986). Serial transmission of human T-cell leukemia virus type I by blood transfusion in rabbits and its prevention by use of X-irradiated stored blood. Int J Cancer 37: (in press).
11. Miyoshi I, Yoshimoto S, Kubonishi I, Fujishita M, Ohtsuki Y, Yamashita M, Yamato K, Hirose S, Taguchi H, Niiya K, Kobayashi M (1985). Infectious transmission of human T-cell leukemia virus to rabbits. Int J Cancer 35: 81.
12. Miyoshi I, Yoshimoto S, Taguchi H, Kubonishi I, Fujishita M, Ohtsuki Y, Shiraishi Y, Akagi T (1983). Transformation of rabbit lymphocytes with adult T-cell leukemia virus. Gann 74: 1.
13. Robert-Guroff M, Ruscetti FW, Posner LE, Poiesz BJ, Gallo RC (1981). Detection of the human T-cell lymphoma virus p19 in cells of some patients with cutaneous T-cell lymphoma and leukemia using a monoclonal antibody. J Exp Med 154: 1957.
14. Palker TJ, Scearce RM, Miller SE, Popovic M, Bolognesi DP, Gallo RC, Haynes BF (1984). Monoclonal antibodies against human T cell leukemia-lymphoma virus (HTLV) p24 internal core protein. J Exp Med 159: 1117.
15. Maeda Y, Furukawa M, Takehara Y, Yoshimura K, Miyamoto K, Matsuura T, Morishima Y, Tajima K, Okochi K, Hinuma Y (1984). Prevalence of possible adult T-cell leukemia virus-carriers among volunteer blood donors in Japan: a nation-wide study. Int J Cancer 33: 717.

16. Tajima K, Kuroishi T (1985). Estimation of rate of incidence of ATL among ATLV (HTLV-I) carriers in Kyushu, Japan. Jpn J Clin Oncol 15: 423.
17. Kinoshita K, Yamanouchi K, Ikeda S, Momita S, Amagasaki T, Soda H, Ichimaru M, Moriuchi R, Katamine S, Miyamoto T, Hino S (1985). Oral infection of a common marmoset with human T-cell leukemia virus type-I (HTLV-I) by inoculating fresh human milk of HTLV-I carrier mothers. Jpn J Cancer Res (Gann) 76: 1143.
18. Yamanouchi K, Kinoshita K, Moriuchi R, Katamine S, Amagasaki T, Ikeda S, Ichimaru M, Miyamoto T, Hino S (1985). Oral transmission of human T-cell leukemia virus type-I into a common marmoset (*Callithrix jacchus*) as an experimental model for milk-borne transmission. Jpn J Cancer Res (Gann) 76: 481.
19. Yamato K, Taguchi H, Yoshimoto S, Fujishita M, Yamashita M, Ohtsuki Y, Hoshino H, Miyoshi I (1986). Inactivation of lymphocyte-transforming activity of human T-cell leukemia virus type I by heat. Jpn J Cancer Res (Gann) 77: (in press).

Viruses and Human Cancer, pages 151–159

CHARACTERIZATION OF CONSERVED AND DIVERGENT REGIONS IN THE ENVELOPE GENES OF HTLV-III/LAV

B. Starcich[1], B. Hahn[2], G. Shaw[2], R. Gallo[1], and F. Wong-Staal[1]

[1]Laboratory of Tumor Cell Biology, DCT, National Cancer Institute, National Institutes of Health, Bethesda, MD 20892 USA;[2]Division of Haematology and Oncology, University of Alabama Medical Center, Birmingham, AL 35294, USA

ABSTRACT Genomic diversity is a prominent feature of different HTLV-III/LAV isolates and is likely to be fundamentally important in the virus's biology and pathogenicity. We determined the nucleotide sequence of an HTLV-III isolate from a Haitian man and analyzed the extent and nature of its genomic variation with respect to the published sequences of prototype HTLV-III/LAV viruses. This analysis demonstrated that nucleotide differences occur predominantly within the envelope gene. Furthermore, these differences clustered in regions corresponding to the extracellular portion, particularly in areas where antigenic sites are predicted. In contrast, certain other areas of the envelope gene, including parts of the extracellular domain and most of the transmembrane region were highly conserved. These findings suggest that the envelope glycoproteins of different HTLV-III/LAV isolates may vary substantially in their antigenic properties but also that certain conserved regions exist which may be useful in vaccine development.

INTRODUCTION

Human T-lymphotropic retrovirus type III (HTLV-III) and related viruses (LAV, ARV) are the presumed aetiological agents of acquired immunodeficiency syndrome

(AIDS) (1-4). The complete nucleotide sequences of HTLV-III/LAV and ARV have been reported (5-8). The sequenced proviral genomes contain at least six open reading frames which correspond to the replicative genes gag, pol, env, the transactivator gene tat-III, and two unique genes sor and 3'orf. Southern blot hybridization analysis of DNA from different HTLV-III isolates suggests that HTLVIII strains display considerable genomic diversity(9). In order to extend these studies we have sequenced part of the genome, including the complete envelope gene, of a divergent HTLV-III strain, HAT-3. This molecular clone was obtained from a Haitian AIDS patient (RF) in 1983. Previous comparisons of unintegrated DNA from this clone and a clone obtained at the same time from New York (BH10), have indicated that although restriction site differences occur throughout the genome, these differences, as analyzed by heteroduplex mapping, are most prominent in the envelope gene(10).

RESULTS

The complete nucleotide sequence of the envelope gene of HAT-3 is shown in Figure 1. The gene is 2619 nucleotides in length and has the coding potential for 873 amino acids. Its general stucture is similar to that of other HTLV-III isolates, encoding a signal peptide, extracellular domain and transmembrane region. The predicted amino acid sequence of the envelope of HAT-3 is shown in figure 2. The beginning of env is signified by a methionine codon at position 9 of the open reading frame. A leader sequence (signal peptide), comprised predominantly of hydrophobic amino acids, is located between positions 17 and 37. Immediately juxtaposed is the extracellular portion of the envelope which spans positions 38 to 527. This sequence is slightly hydrophilic and contains at least 27 potential sites for post-translational glycosylation (shown as open circles in figure 2). The cleavage site, at which the envelope precursor is presumed to be processed into the exterior and transmembrane regions, is located at position 527 (see second arrow, figure 2). It is immediately preceded by an arginine-rich sequence and marks the characteristic transition from the hydrophilic residues of the extracellular domain to the hydrophobic residues of the amino portion of the transmembrane region. The transmembrane region of HAT-3, like that of other HTLV-III isolates analyzed to date, is particularly long. It comprises 345 amino acids (from positions 528 to 873) and contains a highly hyrophobic stretch of residues and a hydrophilic region, in the amino and carboxy loci, respectively.

To examine the extent to which HAT-3 envelope resembles the envelope of other HTLV-III isolates, the predicted amino acid sequences of this clone was compared with clones BH10, LAV and ARV. As shown in figure 2, the overall structure of the envelope of HAT-3 is similar to that of BH10, LAV and ARV. The hydrophilicity profiles of all four isolates are also similar as shown in figure 3. However, differences in the predicted primary amino acid sequence of these genes, are evident. In the signal peptide and the extracellular regions of the envelope, differences between the isolates take the form of point mutations, insertions and deletions. This includes a 7 amino acid insertion in the extracellular domain of HAT-3 (positions 193 to 200) which is not evident in the other clones. It is perhaps worth noting that although the variation between clones BH10, LAV and ARV result in changes in the potential sites for glycosylation (11 of 27 sites in HAT-3 are different from BH10), the position and abundance of cysteine residues are preserved. This suggests that although selective pressures may operate to promote diversity in the envelope gene, the type of modifications that occur are probably restricted to those which do not greatly compromise protein structure. The transmembrane region of HAT-3 envelope shows less variability than the extracellular domain. The predominant differences between isolates in this region, occur as single point mutations in the carboxy portion. Variation in glycosylation is also less evident in the transmembrane, than extracellular domains of the envelope of HTLV-III.

To determine whether similarities/differences in nucleotide and amino acid sequence between the four HTLV-III isolates are likely to represent true similarities/

differences in protein structure and antigenicity, we adopted the Chou and Fasman (11), computer aided approach, to make rational predictions about the secondary structure of each envelope protein. Using this approach, the presence of alpha helicies, beta turns and beta sheets (shown as α , β and s, respectively, figure 2), were determined. We identified regions, common to all four isolates, which were similar both in terms of amino acid sequence and structure. These regions are indicated by shaded boxes in figure 2. Sixteen conserved regions in the extracellular, and ten conserved regions in the transmembrane domain were located. These included sequences which were highly hydrophilic, hydrophobic and of intermediary hydrophobicity (dark shading, pale shading and medium shading, figure 2), and contained alpha helicies, beta sheets and beta turn structures. These "conserved" sequences punctuate the variability seen within the envelope and raise the possibility that although the envelope proteins of HTLV-III isolates show divergence, antigenic epitopes common to all, exist. Of particular interest, in this respect, are the "conserved" regions which are hydrophilic and contain beta turns (conserved regions 3,5,7,15 and 16 in the extracellular domain), since the coincidence of hydrophilic regions and beta turns, has, in other model systems, been shown to associate with antigenic epitopes (12,13).

DISCUSSION

In summary, we have shown, by nucleotide sequence analysis, that HAT-3, an isolate obtained from a Haitian AIDS patient, differs significantly in the envelope gene from previously characterized isolates. This finding is consistent with previous observations that HAT-3 envelope differs from that of BH-10 in terms of both restriction enzyme analysis and heteroduplex analysis. Furthermore, by comparison to the published nucleotide sequences of three related viruses from AIDS or ARC patients, we were able to localize highly conserved and highly divergent regions within the viral envelope. It is probable that some of the conserved regions represent functionally important domains, and thus may be exploited to generate diagnostic and vaccine reagents which have broad cross-reactivity. Regions of variability may have arisen either from randon mutations in non-critical segments of the protein that tolerate such divergence, or they may have arisen from antigenic drift as a consequence of immunoselection. In this respect, other retroviruses such as visna (14) and equine infections anemia (15) virus are known to use changes on the envelope glycoprotein in order to escape immune surveillance by the host. These changes were observed after virus passage _in vivo_. To confirm whether modifications such as those described in this study are crucial for HTLV-III to escape immunosurveillance it will be necessary to evaluate the neutralizing properties of patient sera against different viral isolates. An important conclusion from the data presented in this study is the presence of highly conserved regions within the envelope of HTLV-III. It will now be possible to generate synthetic peptides corresponding to these regions, in attempts to develop a vaccine regime effective against different HTLV-III prototypes.

```
GAGAAAGAGCAGAAGACAGTGGCAATGAGAGTGATGGAGATGAGGAAGAATTGTCAGCAC   60
                                                  Pst I
TTGTGGAAATGGGGCACCATGCTCCTTGGGATGTTGATGATCTGTAGTGCTGCAGAGGAC   120
                           Kpn I
TTGTGGGTCACAGTCTATTATGGGGTACCTGTGTGGAAAGAAGCAACCACCACTCTATTT   180

TGTGCATCAGAAGCTAAAGCATATAAAACAGAGGTACATAATGTCTGGGCCAAACATGCT   240

TGTGTACCTACAGACCCCAACCCACAAGAAGTACTATTGGAAAATGTGACAGAAAATTTT   300

AACATGTGGAAAAATAACATGGTAGAACAGATGCATGAGGATATAATCAGTTTATGGGAT   360

CAAAGCCTAAAGCCATGTGTAAAATTAACCCCACTCTGTGTTACTTTAAATTGCACTGAT   420

GCTAACTTGAATGGTACTAATGTCACTAGTAGTAGCGGGGAACAATGATGGAGAACGGA   480

GAAATAAAAAACTGCTCTTTCCAGGTCACCACAAGTAGAAGAGATAAGACGCAGAAAAAA   540
                                Kpn I
TATGCACTTTTTTATAAACTTGATGTGGTACCAATAGAGAAGGGTAATATTAGCCCTAAG   600

AATAATACTAGCAATAATACTAGCTATGGTAACTATACATTGATACATTGTAATTCCTCA   660

GTCATTACACAGGCCTGTCCAAAGGTATCCTTTGAGCCAATTCCCATACATTATTGCACC   720

CCGGCTGGTTTTGCGATTCTAAAGTGTAATGATAAGAAGTTCAATGGAACAGGACCATGT   780

AAAAATGTCAGCACAGTACAATGTACACATGGAATTAGGCCAGTAGTGTCAACTCAACTG   840
                                          Bgl II
CTGTTAAATGGCAGTCTAGCAGAAGAAGAGGTAGTAATTAGATCTGAAAATTTCACGGAC   900

AATGTTAAAACCATAATAGTACAGCTGAATGCATCTGTACAAATTAATTGTACAAGACCC   960

AACAACAATACAAGAAAAAGTATAACTAAGGGACCAGGGAGAGTAATTTATGCAACAGGA   1020

CAAATAATAGGAGATATAAGAAAAGCACATTGTAACCTTAGTAGAGCACAATGGAATAAC   1080

ACTTTAAAACAGGTAGTTACAAAATTAAGAGAACAATTTGACAATAAAACAATAGTCTTT   1140

ACTTCATCCTCAGGAGGGGACCCAGAAATTGTACTTCACAGTTTTAATTGTGGAGGGGAA   1200

TTTTTCTACTGTAATACAACACAACTGTTTAATAGTACTTGGAATAGTACTGAAGGGTCA   1260

AATAACACTGGAGGAAATGACACAATCACACTCCCATGCAGAATAAAACAAATTGTAAAC   1320

ATGTGGCAGGAAGTAGGAAAAGCAATGTATGCCCCTCCCATCAGTGGACAAATTAAATGT   1380

ATATCAAATATTACAGGGCTACTATTAACAAGAGATGGGGGTGAAGATACAACTAATACT   1440
    Bgl II
ACAGAGATCTTCAGACTTGGAGGAGGAAATATGAGGGACAATTGGAGAAGTGAATTATAT   1500

AAATATAAAGTGGTAAGAATTGAACCATTAGGAGTAGCACCCACTAGGGCAAAGAGAAGA   1560

GTGGTGCAAAGAGAAAAAAGAGCAGTGGGAACAATAGGAGCTATGTTCCTTGGGTTCTTG   1620

GGAGCAGCAGGAAGCACTATGGGCGCAGGCTCAATAACGCTGACGGTACAGGCCAGACAC   1680

TTATTGTCTGGTATAGTGCAACAGCAAAACAATTTGCTGAGGGCTATTGAGGCGCAACAG   1740

CATCTGTTGCAACTCACGGTCTGGGGCATCAAACAGCTCCAGGCAAGAGTCCTAGCTGTG   1800

GAAAGATACCTAAGGGATCAACAGCTCCTAGGAATTTGGGGATGCTCTGGAAAACTCATT   1860

TGCACCACTACTGTGCCTTGGAATGCTAGTTGGAGTAATAAATCTCTGAATATGATTTGG   1920

AATAACATGACCTGGATGCAGTGGGAAAGAGAAATTGACAATTACACAGGCATAATATAC   1980

AACTTACTTGAAGAATCGCAGAACCAGCAAGAAAAGAATGAACAAGAATTATTGGAATTG   2040

GATAAATGGGCAAATTTGTGGAATTGGTTTGACATAACACAATGGCTGTGGTATATAAGA   2100

ATATTCATAATGATAGTAGGAGGCTTGGTAGGTCTAAAAATAGTTTTTGCTGTGCTTTCT   2160

ATAGTGAATAGAGTTAGGCAGGGATACTCACCATTATCATTTCAGACCCACCTCCCAGCC   2220
Ava I
CCGAGGGGACCCGACAGGCCCGAAGGAATCGAAGGAGAAGGTGGAGAGAGAGACAGAGAC   2280

AGATCCGGCGGTGCAGTGAATGGATTCTTGACACTTATCTGGGACGATCTGTGGACCCTG   2340

TGCAGCTTCAGCTACCACCGCTTGAGAGACTTACTCTTGATAGTAGTGAGGATTGTGGAA   2400

CTTCTGGGACGCAGGGGGTGGGAAGCCCTCAAGTATTGGTGGAATCTCCTGCAGTATTGG   2460

AGTCAGGAGCTAAAGAATAGTGCTGTTAGCTTGCTTAATACCACAGCAATAGCAGTAGCT   2520

GAAGGGACAGATAGGATTATAGAAGTAGCACAAAGAATTCTTAGAGCTTTTCTTCACATA   2580

CCTAGAAGAATAAGACAGGGCTTAGAAAGGCTTTGCTG                         2619
```

FIGURE 1. Nucleotide sequence of the envelope gene of the HAT-3 isolate

The nucleotide sequence of the envelope gene of HTLV-III isolate HAT-3, was deduced using the Maxam Gilbert (16) approach. The position of restriction enzyme sites, within the env open reading shown, are shown.

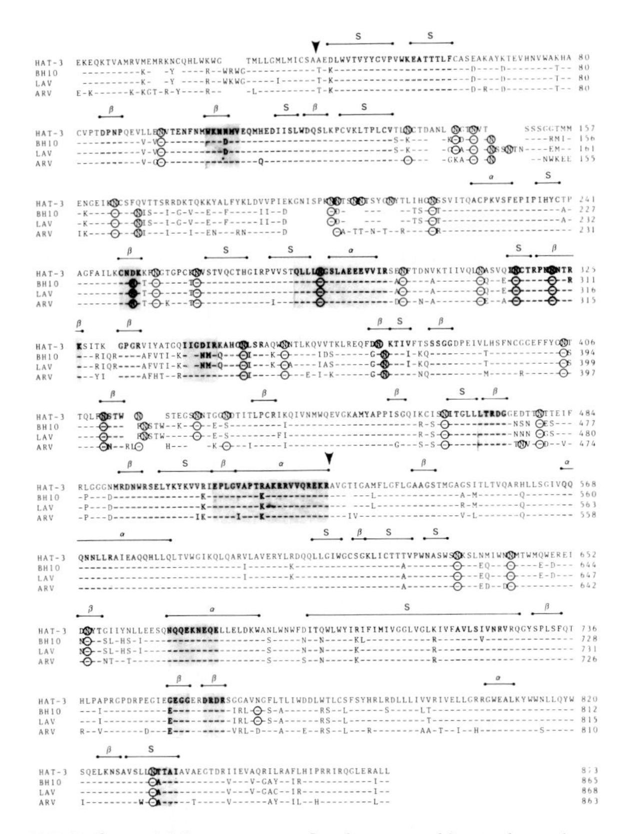

FIGURE 2. Alignment of the predicted amino acid sequence of the envelope proteins of HAT-3, BH10, LAV and ARV, HTLV-III isolates.

Using the Chou and Fasman (11) computer program, the predicted amino acid sequence, hydrophilicity and secondary structure of HAT-3 envelope was compared with that predicted for BH10, LAV and ARV isolates. Closed arrows mark the borders between the extracellular and transmembrane domains of the envelope. Amino acid identity between isolates are shown as -------. Beta sheets, beta turns and alpha helical structures are designated s, β and α , respectively. N - linked potential glycosylation sites are shown as open circles. 'Conserved' regions, common to all isolates are shown as shaded boxes. The intensity of shading indicates the hydrophilic properties of these conserved sequence where; hydrophilic = dark shading, hydrophobic = pale shading, intermediate hydrophilicity = intermediate shading. The amino acid position is shown in the extreme right hand column of the figure.

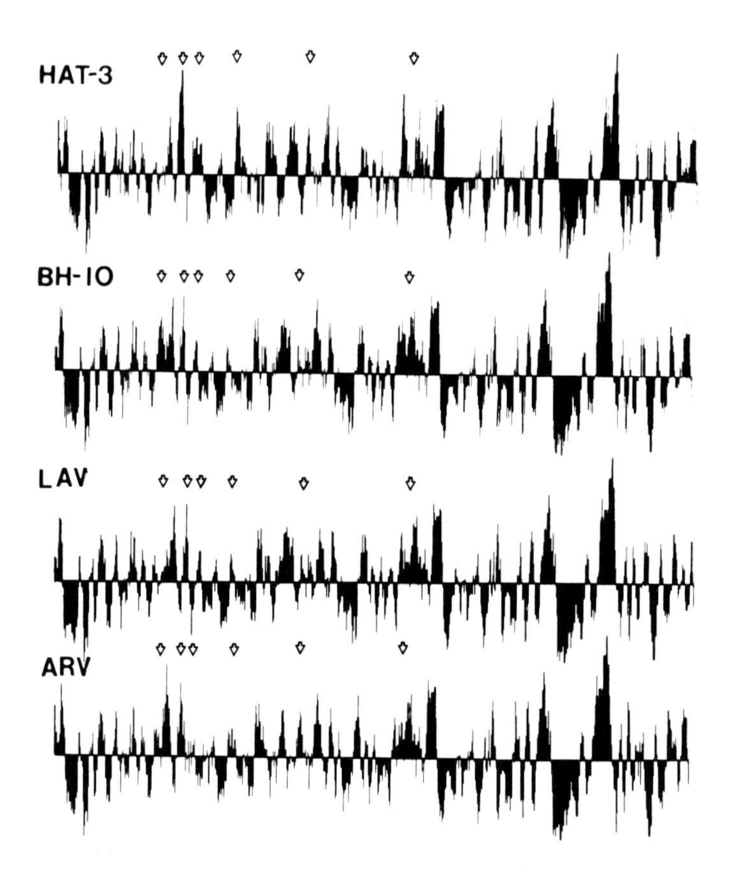

FIGURE 3. Hydrophilicity profiles of the predicted envelope proteins of HTLV-III isolates.

Using the approach of Hopp and Wood (12), the hydrophilicity profiles of the predicted envelope proteins of HAT-3, BH10, LAV and ARV, were assessed and aligned. 'Peaks' above the midline indicate hydrophilic regions. 'Peaks' below the midline indicate hydrophobic regions. Open arrows point to regions of dissimilarity between isolates.

1. Gallo RC, Salahuddin SZ, Popovic M, Shearer GM, Kaplan M, Haynes BF, Palker TJ, Redfield R, Oleske J, Safai B, White G, Foster P, Markham PD. (1984) Science 224:500.
2. Popovic M, Sarngadharan MG, Read E, Gallo RC. (1984) Science 224:937.
3. Schupbach J, Popovic M, Gilden RV, Gonda MA, Sarngadharan MG, Gallo RC. (1984) Science 224:503.
4. Sarngadharan MG, Popovic M, Bruch L, Schupbach J, Gallo RC (1984) Science 224:506.
5. Ratner L, Haseltine W, Patarca R, Livak KJ, Starcich B, Josephs SF, Doran ER, Rafalski JA, Whitehorn EA, Baumeister K et al. (1985) Nature 313:277.
6. Starcich B, Ratner L, Josephs SF, Okamoto T, Gallo RC, Wong-Staal F. (1985) Science 227:538.
7. Wain-Hobson S, Sonigo P, Danos O, Cole S, Alison M. (1985) Cell 40:9.
8. Sanchez-Pescador R, Power MD, Barr PJ, Steimer KS, Stempien MM, Brown-Shimer SL, Gee WW, Renard A, Randolph A, Levy JA et al (1986) Science 227:484.
9. Shaw GM, Hahn BH, Arya SK, Groopman JE, Gallo RC, Wong-Staal F. (1986) Science 226:1165.
10. Hahn BH, Gonda MA, Shaw GM, Popovic M, Hoxie JA, Gallo RC, Wong-Staal F. (1985) Proc Natl Acad Sci USA 82:4813.
11. Chou PY, Fasman GD (1976) Biochem 13:222
12. Hopp TP, Woods KR (1981) Proc Natl Acad Sci USA 78:3824.
13. Eisenberg RJ, Long D, Ponce de Leon M, Matthews JT, Spear PG, Gibson MG, Lasky LA, Berman P, Golub E, Cohen GH (1985) J Virol 53:634
14. Clements JE, Pederson FS, Narayan O, Haseltine WA (1980) Proc Natl Acad Sci USA 77:4454
15. Montelaro RC, Parekh B, Orrego A, Issel CJ (1986) J Biol Chem 259:10539.
16. Maxam AM, Gilbert W (1977) Proc Natl Acad Sci USA 74:560.

Viruses and Human Cancer, pages 161–176

HTLV-III/LAV INFECTION OF MONONUCLEAR PHAGOCYTE CELLS *IN VIVO* AND *IN VITRO*

M. Popovic, E. Read-Connole and S. Gartner

Laboratory of Tumor Cell Biology, National Institutes of Health, Bethesda, Maryland 20892

ABSTRACT

Cells exhibiting properties of mononuclear phagocytes were assessed for infectivity with different HTLV-III/LAV isolates. Mononuclear phagocytes recovered *in vitro* from brain and lung tissues of AIDS patients harbored the virus. Peripheral blood-derived macrophages from healthy donors were infected *in vitro* with four different HTLV-III/LAV isolates and found to be highly susceptible to and permissive for the virus. Unlike in HTLV-III/LAV infected T-cells, virus production in infected macrophages persisted longer and was independent of cell proliferation. Morphological examinations of the infected macrophage cell cultures revealed the presence of giant multinucleated cells and abundant numbers of virus particles frequently located in vacuole-like structures. The different virus isolates were compared quantitatively for their capacity to infect and replicate in macrophages and T-cells. Isolates recovered from brain and lung tissues and propagated in PB-derived macrophages exhibited 10-fold higher ability to infect these cells than T4+ lymphocytes. In contrast, the prototype HTLV-IIIB propagated in T-cells displayed a 10,000-fold lower capacity to infect macrophages than lymphocytes, and virus production in these HTLV-IIIB-infected macrophage cultures was 1/10th that in macrophage cultures infected with other isolates. These results indicate that different HTLV-III/LAV isolates may represent variants with increased tropism for macrophages or T-cells. In addition, the high susceptibility and permissivity of mononuclear phagocytes suggest that these cells may represent a primary target for virus infection and a reservoir for virus dissemination.

INTRODUCTION

It is generally accepted that the T4+ lymphocytes are the targets for the human T cell lymphotropic retrovirus or lymphadenopathy associated virus (HTLV-III/LAV) (the etiological agent of the acquired immune deficiency syndrome (AIDS)) (1-4). The selective depletion of this population of cells, particularly the helper/inducer subset of T4+ lymphocytes, is considered to be the crucial event in the development of profound immunodeficiency in AIDS patients (5). Because of the central role of this T cell subset in regulation of immune processes, its depletion may indeed explain many aspects of the immune disorders occurring at terminal stages of the disease. However, the mechanism(s) that account for development of immune deficiency during the early stages of the viral infection are still obscure. Limited in vivo as well as in vitro studies suggest that, in addition to T4+ lymphocytes, other susceptible and permissive cell types for the virus may be involved in the development of immunodeficiency at early stages of infection (6,7). A notion that cells of the mononuclear phagocyte series may represent sensitive and permissive targets for the virus is based on the following observations: 1) Detection of HTLV-III/LAV DNA sequences by Southern blot analysis and the viral mRNA by in situ hybridization in brain tissues from neurosymtomatic AIDS patients showed that cells harboring the virus are not T lymphocytes (8). Further characterization of primary cultures from brain tissues of a neuroosymptomatic AIDS patient revealed that the virus-harboring cells belong to the mononuclear phagocyte series (9). 2) Morphogical and immunohistological studies in patients with persistent generalized lymphadenopathy representing early stages of the disease (PGL) revealed that the germinal centers within the lymph nodes were characterized by an expansion of follicular dendritic reticulum (FDR) cells and B-cell blasts (7,10). The presence of retrovirus particles and HTLV-III core proteins have been demonstrated within the network of these FDR cells (7,11). 3) In addition, the defect in antigen presentation that has been described in AIDS patients could also be due to the virus infection of cells from the mononuclear phagocyte series (12,13). 4) Finally, there are several lines of evidence including morphology of the virus particles, genomic organization and some nucleotide sequence homology that suggests a relationship between the lentiviruses, visna, and HTLV-III/LAV (14,15). The major targets for the visna virus are macrophage cells (15). To establish that these types of cells are targets for HTLV-III/LAV, we have carried out numerous infections of macrophages recovered from the peripheral blood of healthy donors with four different virus isolates. In this paper, we report that macrophages are highly susceptible to and permissive for the virus and the isolates recovered from cells of the mononuclear phagocyte series of patients with AIDS exhibited a higher tropism for macrophages than T-cells.

MATERIALS AND METHODS

Cells and Virus Isolates

Cultures of pure populations of macrophage cells were obtained by methods previously described (16). Briefly, 3×10^7 mononuclear cells separated by ficall-hypaque gradients were seeded into T25 plastic flasks (Corning) in RPMI1640 supplemented with 20% heat-inactivated (h.i.) fetal bovine serum (FBS), 10% h.i. pooled human serum (Advanced Biotechnology) and antibotics. Nonadherent cells were removed by extensive washing with phosphate buffered saline (PBS) 3 to 5 days following initiation of the cultures. Macrophage cell cultures were routinely maintained in RPMI1640 supplemented with 20% h.i. FBS and antibiotics. The HTLV-IIIRC-br isolate was recovered from adherent cells of primary brain cultures grown in Dulbecco's hi-gluocse MEM supplemented with 15% FBS and antibiotics. HTLV-IIIRC-PB was recovered from the same patients peripheral blood (PB) T-cells and propagated in PB-derived T cells as previously described (17). Both isolates were obtained from a 50-year old bisexual male with neurological symptoms who subsequently developed AIDS (9). The prototype of HTLV-IIIB was propagated in H9 cells (2). The HTLV-IIIBa-L isolate was recovered from a primary lung culture grown in RPMI-1640 complete medium. The post-mortem specimen was obtained from the lung of a 7-month-old boy who died of AIDS (18). Both isolates, HTLV-IIIRC-br and HTLV-IIIBa-L, were propagated in PB-derived macrophage cells.

Cultures were subjected to trypsinization by exposing the adherent cells to CA^{++}-Mg^{++} free PBS containing 0.05% trypsin (Gibco) and 0.05% EDTA for 1-5 minutes at 37°C.

Virus Assays

Undiluted culture fluids harvested from virus-infected cell cultures were used in the transmission studies. The virus incubation (culture fluid) of HTLV-IIIB contained 10^6cpm reverse transcriptase (RT) activity per milliliter (ml), while all others contained 10^5 cpm RT activity /ml. The infection of macrophage or T cells followed the well-established procedure previously described (2,18). RT activity was followed in culture fluids harvested from cell cultures at 3-5 day intervals and expressed in counts per minute (cpm) per one ml of harvested culture fluid. RT assays were performed as previously described (19). Both infected and noninfected Ca^{++}-Mg^{++}-free macrophage cells were detached from the surface of plastic flasks by incubation of the flasks containing Hanks' balanced salt solution on ice for 15 to 30 minutes and then the cells were gently scraped into the Hanks' solution and processed similarly as T-cells for indirect immunofluorescence (IF) assay as described by Robert-Guroff et al. (20). Highly specific mouse monoclonal antibody directed against the viral core protein HTLV-IIIp17 was used for detection of the virus positive cells by IF assay (21).

Southern blot analysis.

High molecular DNAs were prepared using standard methods (22). The samples were electrophoresed in 0.7% agarose gels, transfered to nitro cellulose and hybridized as described previously (23). The probe used was either the SST1-SST1 region from the BH-10 clone (24) or a mixture of the 5'and 3' ends of the HxB-2 clone (25), each representing an 8.9 kb-long fragment of the HTLV-IIIB genome.

Functional, cytochemical and phagocytic assays.

Phagocytic activity of macrophage cells was demonstrated by incubating of these cell cultures at 37°C for 14 hrs in the presence of 1.05μ latex beads (Poly-sciences, Inc.). Following the incubation period, the cultures were washed with RPMI1640 culture medium to remove the excess beads and examined microscopically. Cytochemical staining of macrophage cultures for alpha naphthyl acetate estrase [nonspecific estrase] (NSE) was performed on in situ-fixed cells using Sigma kit #90. For morphological examinations cells were stained with Wright's-Giemsa, and for electron microscopic examinations a culture was fixed in situ and processed by well-established procedure previously described (19).

RESULTS

HTLV-III/LAV Infection of Macrophage Cells.

PB-derived macrophage cells have been used for HTLV-III/LAV infection with four different isolates. These plastic adherent cells were trypsin resistant and positive for NSE staining up to 100%. (See Table 1). Trypsin resistant plastic adherence and NSE positivity proved to be reliable markers for characterization of macrophage cell cultures, because of the consistent positive correlation between these properties of macrophage cells and their functional (phagocytic activity) and cell surface characteristics determined by antimacrophage cell antibodies. In addition, the former properties of these cells could be more easily and accurately applied than the later ones (functional and cell surface markers) for identification of these cells in primary cultures derived from brain and lung tissues where cells of the mononuclear phagocyte series represent a minority. As shown in Table 1, only primary cultures of brain and lung tissues that were NSE positive and not susceptible to trypsinization expressed the virus. Attempts to infect these subcultured cells (NSE-negative cultures) with two different isolates, HTLV-IIIRC-br and HTLV-IIIBa-L, gave consistently negative results.

On the contrary, infection of PB-derived macrophage cells were highly susceptible to and permissive for the virus. The percentage of HTLV-III-infected macrophage cells as determined by IF assay using highly specific mouse monoclonal antibodies recognizing the viral core protein, HTLV-IIIp17, were in the range of from 5-21%. The virus production detected in culture fluids by reverse transcriptase (RT) assay harvested from these cell cultures was in the range of from 0.2 - 8 x 10^5 cpm/ml. In some experiments, using the brain or lung-derived HTLV-IIIBa-L isolates, RT activity in macrophage cultures peaked to levels of 2 x 10^6 cpm/ml culture fluid. These results indicate that cells of the mononuclear phagocyte series are susceptible to and permissive for HTLV-III/LAV.

Detection of HTLV-III/LAV DNA Sequences in Macrophage Cells.

To substantiate further that HTLV-III/LAV replicates in macrophage cells, Southern blot analysis was performed using DNAs prepared from these cells as well as from T lymphocytes both infected with a brain isolate, HTLV-IIIRC-br and from macrophage cells infected with an isolate from lung tissue, HTLV-IIIBa-L. Both these isolates were recovered from cells of the mononuclear phagocyte series of AIDS patients and had been propagated *in vitro* only on PB-derived macrophages. The DNAs were digested separately with several restriction enzymes. Figure 1 depicts a representative Southern blot of a hybridization analysis obtained by Hind III digestion. First of all, as the biological assays indicated, subcultured brain and lung cells (see lanes e and g) from the primary cultures established from AIDS patient's tissues contained no detectable HTLV-III/LAV sequences. A comparison of lanes c and d reveals that the isolate from brain (HTLV-IIIRC-br) transmitted to PB-derived macrophage cells is identical to that transmitted into T cells. Lane f contained the isolate from lung (HTLV-IIIBa-L) which exhibited a restriction pattern different from the brain-derived isolate. Moreover, both two isolates, HTLV-IIIRC-br and HTLV-IIIBa-L, are clearly distinct from the prototype HTLV-IIIB commonly used in our laboratory (compare lanes a, c, d and f).

Table 1. Infection of Macrophage Cells with HTLV-III/LAV Isolates*

Virus isolate	Source of tissue	No. of posit./ No. of tested	Characteristics of Cultured Cells: response to trypsin	Characteristics of Cultured Cells: % NSE** positive	Virus expression: HTLV-IIIp17 % positive	Virus expression: RT activity cpm/ml
HTLV-IIIRC-br						
	brain (primary)	1/1	resistant	1	NT	$2x10^4$
	brain (subcultured)	0/1	sensitive	0	0	0
	skin (subcultured)	0/1	sensitive	0	0	0
	peripheral blood	11/11	resistant	>95	5-21	$0.02-2x10^6$
HTLV-IIIRC-PB						
	peripheral blood	3/3	resistant	>95	14	$0.24-1.3x10^6$
HTLV-IIIB						
	peripheral blood	3/3	resistant	>95	5-10	$0.3-1x10^5$
HTLV-IIIBa-L						
	lung (primary)	1/1	resistant	10	NT	$0.65x10^4$
	lung (subcultured)	0/1	sensitive	0	0	0
	peripheral blood	8/8	resistant	>95	10-20	$1.3-2x10^6$

*For details of the HTLV-III isolates and experimental procedures see Materials and Methods. **Nonspecific esterase

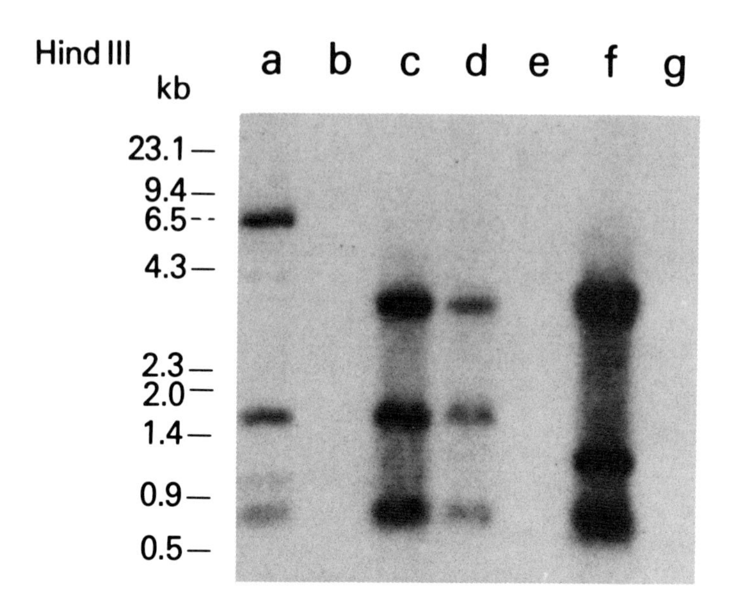

Fig. 1. Southern blot analysis of Hind III digested DNA from a) H9/HTLV-IIIB cells, b) H9 uninfected cells, c) normal PB-derived macrophages infected with HTLV-III/RCbr, d) normal PHA-stimulated TCGF-dependent T lymphocytes infected with HTLV-IIIRC-br, e) subcultured glial-like cells derived from the same biopsy from which HTLV-III RC-br isolate was recovered, f) normal PB-derived macrophages infected with HTLV-IIIBa-L and g) subcultured fibroblast-like cells derived from the same lung specimen from which HTLV-IIIBa-L was isolated. Note differences in restriction patterns of these isolates.

Comparison of the Susceptibility and Permissivity of Macrophage vs T cells to HTLV-III/LAV Isolates.

The susceptibility of macrophage vs T cells was compared quantitatively using the four different HTLV-III/LAV isolates. In parallel experiments, approximately equal numbers of both types of cells derived from PB of the same donor were exposed to various dilutions of the virus inoculum. Virus production was followed by RT assays in culture fluids harvested from these cultures at 3-day intervals for 1.5 months. The results of these experiments comparing the susceptibility of macrophages vs T-cells to infection with these four isolates are shown in Table 2. The susceptibility is expressed in relative values as the ratio of the dilution of virus that was able to infect the macrophage cells

TABLE 2. RELATIVE SUSCEPTIBILITY OF MACROPHAGES VS. T CELLS TO INFECTION WITH DIFFERENT HTLV-III/LAV ISOLATES

Virus Isolate	Macrophages/T cells*
HTLV-III_B	0.0001
HTLV-III_{RC-br}	10.0
HTLV-III_{RC-PB}	1.0
HTLV-III_{Ba-L}	10.0

*Data represents a ratio of end point dilutions of the virus inoculum that was able to productively infect the macrophages compared to that able to infect the T cells. For details of procedures used, see "Materials and Methods"

compared to that able to infect the T cells. Macrophages and T cells exhibited the same susceptibility to infection with the HTLV-IIIRC-PB isolate recovered from a patient's T-cells, while the isolates HTLV-IIIRC-br and HTLV-IIIBa-L, both recovered and propagated in macrophage cells were ten-fold more efficient in infecting macrophages than T-cells. In contrast, the prototype HTLV-IIIB, which has been propagated for a long time only in T-cells, exhibited 10,000-100,000 fold higher efficiency in infecting T cells then the other three isolates. These results suggest that the preferential tropism of the virus for macrophage or T cells may be an acquired behavior resulting from propagation of a particular isolate in one or the other type of cells both _in vivo_ or _in vitro_.

Similarly, the permissivity of macrophages vs T-cells to infection with HTLV-III/LAV was compared quantitatively using autologous PB-derived cells of healthy donors. Approximately 2 x 10^6 macrophages and T-cells were exposed to the same virus inoculum (10^4cpm/ml) of the HTLV-III RC-br isolate, respectively. The longevity and magnitude of virus production were followed in harvested culture fluids by RT assay at 3 to 5 day intervals for 2 to 4 months. The results of these experiments of long-term production of HTLV-III/LAV by macrophage vs T cells are shown in Figure 3. The values are expressed in cpm of RT activity per milliliter of culture fluid. The virus production persisted for at least 40 days longer and its peak was 10-fold higher in the HTLV-IIIRC-br infected macrophage cultures than in the infected T cell cultures. Although comparable numbers of cells were used for the virus infection, it should be noted that the macrophage cell cultures were essentially non-proliferating cell populations, while the T cells, maintained in TCGF, continued to proliferate, resulting in substantial increases in the total number of cells.

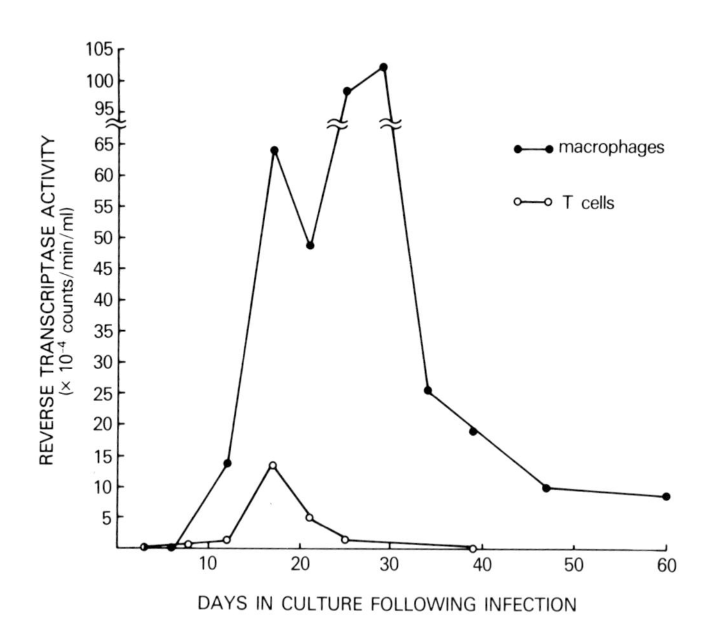

Fig. 2. Comparison of HTLV-IIIRC-br production by macrophages and T-cells during long-term *in vitro* cultivation. The values are expressed as cpm per milliliter of culture fluids. At the peak points of virus production, 10% of macrophages and 6% of T cells were positive for the core protein HTLV-III p17 as determined by IF assay.

In this experiment 10% of macrophage cells and 6% of T cells were positive for the core protein HTLV-III p17 which corresponded to approximately 2.5 x 10^4 macrophage cells and 6 x 10^4 T-cells per milliliter of culture fluid determined at the peak of RT activity, respectively. Therefore, the HTLV-IIIRC-br production in macrophage cultures was even greater than in the T cell cultures if expressed per equal number of the virus positive cells. The same pattern of long-term virus production of macrophage cells and transient production in T cell cultures was observed in all four isolates used in this study. There was, however, a differnce in the magnitude of virus production by macrophage cells infected with these isolates. Under the same culture conditions the virus production of HTLV-IIIB-infected macrophage cells was 10-fold lower compared to the other three isolates (See Table 1).

Morphology of HTLV-III/LAV infected Macrophage Cells.

Wright-Giemsa staining of HTLV-III/LAV-infected macrophage cell cultures frequently revealed the presence of giant multinucleated cells, some containing over one hundred nuclei (Fig. 3.A & B). In general, the virus-induced cytopathic effect on these cultured macrophage cells was less profound as that seen in cultures of virus-infected T-cells. Electron microscopic examinations of the virus-infected macrophage cell cultures showed the presence of abundant numbers of virus particles (Fig.3.C) which, as evidenced by ruthenium red staining, were frequently located within vacuole-like structures (26).

DISCUSSION

The results described in this paper clearly demonstrate that cells of the mononuclear phagocyte series are highly susceptible to and permissive for HTLV-III/LAV. The cells with mononuclear phagocyte characteristics recovered from brain and lung tissues of AIDS patients harbored the virus. In addition, PB-derived macrophages from 25 speciments of 20 different donors consistently exhibited *in vitro* susceptibility to infection with four different HTLV-III/LAV isolates. The virus production

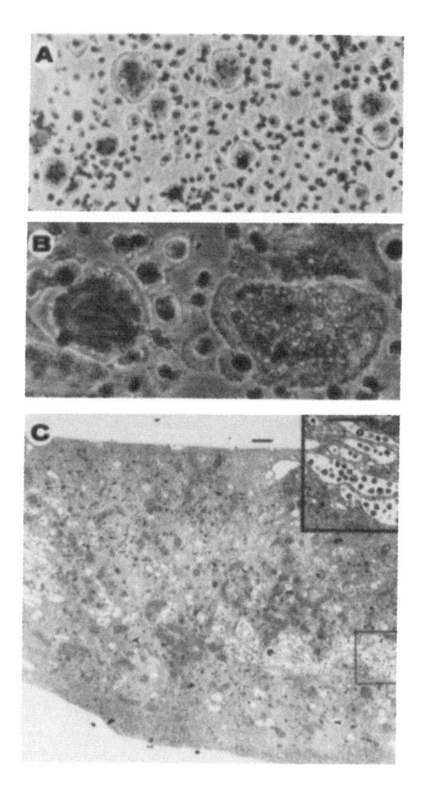

Fig. 3. (A and B). Wright's-Giemsa staining of HTLV-IIIRC-br infected PB-derived macrophages two weeks following infection (Ax80, Bx640). Note the presence of numerous multinucleated giant cells. C) Thin section of electron micrograph depicting a macrophage infected with HTLV-IIIRC-br taken from a culture fixed *in situ*. The top of the photograph represents part of the cell adhering to the plastic flask. The insert shows an abundance of virus particles located within vacuale-like structures (size bar= 1 micron).

by these cells persisted for a considerably longer period of time and the cytopathic effect was less pronounced than in HTLV-III/LAV infected T-cell cultures where the virus expression is transient and followed by rapid cell death. Physiological function(s) of cells of the mononuclear phagocyte series and the pattern of long-term virus expression in these cells indicate that they should be regarded as a reservoir and source of the virus dissemination to target organs (e.g., brain, lung, etc.) *in vivo*. Moreover, the viral budding and presence of mature virions in vacuole-like structures on the infected macrophages suggest that HTLV-III/LAV may escape from immunological reaction(s) directed against the virus *in vivo* and persist in an infectious form for a long time. Furthermore, it should be emphasized that the first type of cells encountering a foreign material (e.g. virus) in the body are phagocytes. Since these types of cells are highly susceptible to and permissive for the virus, it is conceivable that the mononuclear phagocytes represent a primary target for the virus. In recent preliminary experiments performed in our laboratory, freshly separated blood mononuclear cells were exposed to HTLV-IIIBa-L (lung isolate) and then used to establish cultures of macrophages and PHA-stimulated TCGF-dependent T cells. The fact that the macrophage cell cultures representing a minority (less than 5%) of the total PB-derived mononuclear cells preferentially expressed the virus as compared to T cell cultures further suggests that macrophage cells may be the first cells to become infected *in vivo*.

There are several lines of evidence including nucleotide sequence homology and morphology of the virus particles, that suggest a relationship between the lentivirus, visna, and HTLV-III/LAV (14,15,26). The findings presented in this paper provide additional evidence of a similarity, since macrophages have been shown to be the major targets for visna virus (15). However, visna virus replication in macrophages has been shown to be non-productive or minimally productive while HTLV-III/LAV replication in mononuclear phagocytes was fully productive in the experiments we assessed. In addition, the presence of the 3'orf gene and the ability to acquire a tropism for T4+ lymphocytes and infect them productively appear to be features unique to HTLV-III/LAV (26,27). It should be noted that the prototype HTLV-IIIB which infects macrophages only at a high multiplicity of infection has been selected for growth on normal and neoplastic T4+ lymphocytes. Quantitative titration of the prototype on PB-derived lymphocytes and macrophages showed a 10,000 fold greater susceptibility of T cells than macrophage cells. Moreover, a binding assay exhibited high affinity binding of the HTLV-IIIB virus particles to the T4 antigen positive cells (M. Popovic, *et al*., unpublished) suggesting that through infection and growth in T4+ lymphocytes, HTLV-III/LAV can increase its tropism to these cells. Since T4+ lymphocytes (helper/inducer) are the first type of cells that interact with antigen presenting cells (APC, the cells of the mononuclear phagocyte series) following antigen uptake, it is conceivable that HTLV-III/LAV infection *in vivo* initially follows the same general pathway as any other antigen which enters the human body. Because of the close association between APC (macrophages) and T4+ lymphocytes during antigen recognition, HTLV-III/LAV transmission from the virus positive macrophages to T4+ lymphocytes may occur very efficiently even if the particular variant has a low tropism for T4+ lymphocytes. However, because of the ability of the virus to acquire a high affinity for and also chronically infect T lymphocytes (28), it is likely that *in vivo* transmission of the virus from T-cells to T-cells and from T-cells to macrophage cells also occurs. It should be taken into consideration that the ability of the virus to develop into a particular variant with preferential tropism for mononuclear phagocytes or T cells may partly determine both the appearance of certain symptoms and the clinical course of the disease.

ACKNOWLEDGMENTS

We thank S. Pahwa, M. H. Kaplan, D. M. Markovitz for clinical material, B. Kramarsky for the electron micrographs, M. Niclas for technical assistance, and Dee Goodrich for editorial assistance.

REFERENCES

1. Barre-Sinoussi F, Chermann JC, Rey F, Nugeyre MT, Chamaret S, Gruest T, Dauguet C, Axler-Blin C, Vezinet-Brun F, Rouzioux C, Rozenbaum W, Montagnier L. (1983). Science, 220:868.
2. Popovic M, Sarngadharan MG, Read E, Gallo, RC. (1984). Science 224:497.
3. Gallo RC, Salahuddin SZ, Popovic M, Shearer GM, Kaplan M, Haynes BF, Palker TJ, Redfield R, Oleske T, Safai B, White G, Foster P, Markham PD. (1984). Science 224:500.
4. Sarngadharan MS, Popovic M, Bruch L, Schupbach J, Gallo RC. (1984). Science 224:506.
5. Klatzmann D, Barre-Sinoussi F, Nugeyre MT, Dauguet CH, Vilmer E, Brun-Vezinet CGF, Rouzioux C, Gluckman JC, Chermann TC, Montagnier L (1984). Science 225:59.
6. Montagnier L, Gruest T, Chamaret S, Dauguet C, Axler C, Guertard D, Nugyre MT, Barre-Sinoussi F, Chermann J-C, Bruent JB, Klatzmann D, Gluckman JC (1984). Science 225:63.
7. Tenner-Racz K, Racz P, Bofill M, Schulz-Meyer A, Dietrich M, Kern P, Weber T, Pinching AJ, Veronese-DiMarzo F, Popovic M, Klatzmann D, Gluckman JC, Janossy G. (1986). Am J Pathol 123:1
8. Shaw GM, Harper ME, Hahn BH, Epstein LG, Gajdusek DC, Price RW, Navia BA, Petito C, O'Hara CJ, Groopman JE, Cho E-S, Oleske JM, Wong-Staal F, Gallo RC. (1985). Science 227:177.
9. Gartner S, Markovits P, Markovitz DM, Betts RF, Popovic M (1986). JAMA, in press.
10. Racz P, Tenner-Racz K, Kahl C, Feller AC, Kern P, Dietrich M (1985). Progress Allergy 37:81.
11. Armstrong JA, Horne R (1984). Lancet 2:370.
12. Lane HC, Depper GM, Greene WC, Whalen G, Waldman PA. (1985). N Engl J Med 313:79.
13. Prince ME, Moody DI, Shubin BI, Fahey JH (1985). J Immunol 5:21
14. Gonda MA, Wong-Staal F, Gallo RC, Clements JE, Narayan O, Gilden RV (1985). Science 227:173.
15. Narayan O, Wolinsky JS, Clements JE, Strandberg JD, Griffin DE, Cork LC (1982). J Gen Virol 59:345.
16. Kaplan HS, Gartner S (1977). Int J Cancer 19:511.

17. Morgan DA, Ruscetti FW, Gallo RC (1976). Science 193:1007.
18. Gartner S, Markovits P, Markovitz DM, Kaplan MH, Gallo RC, Popovic M (1986). Science, in press.
19. Poiesz BJ, Ruscetti FW, Gazdar AF, Bunn PA, Minna JD, Gallo RC (1980) Proc Natl Acad Sci, USA 77:7415.
20. Robert-Guroff M, Ruscetti FW, Posner LE, Poiez BJ, Gallo RC. (1981). J Exp Med 154:1957.
21. Di Marzo Veronese F, Sarngadharan MG, Rahman R, Markham PD, Popovic M, Bodner AJ, Gallo RC (1985). Proc Natl Acad Sci, USA 82:5199.
22. Gross-Bellard M, Oudet P, Chambon P (1973). Eur J Biochem 36:32.
23. Southern E (1975). J Mol Biol 98:503.
24. Hahn BH, Shaw GM, Arya SK, Popovic M, Gallo RC, Wong-Staal F (1984). Nature 312:166.
25. Shaw GM, Hahn BH, Arya SK, Groopman JE, Gallo RC (1984). Science 226:1165.
26. Popovic M, Kramarsky B, Niclas M, Read-Connole E, Gartner S., in preparation.
27. Wain-Hobson S, Sonigo P, Danos O, Cole S, Alizon M (1985) Cell 40:9.
28. Hoxie T, Haggarty B, Rackowski J, Pillsbury N, Levy J (1985). Science 229:1400.

Viruses and Human Cancer, pages 177–190

EXPRESSION MECHANISMS OF THE X GENES OF HUMAN T-CELL LEUKEMIA AND BOVINE LEUKEMIA VIRUSES: ALTERNATIVE TRANSLATION OF A DOUBLY SPLICED X mRNA[1]

Noriyuki Sagata, Teruo Yasunaga,[2] Kazue Ohishi,[3] and Yoji Ikawa

Laboratory of Molecular Oncology
The Institute of Physical and Chemical Research
Wako, Saitama 351-01, Japan

ABSTRACT Bovine leukemia virus (BLV) and human T-cell leukemia virus (HTLV) have a potential transforming gene, termed X. The major long open reading frames of these X genes encode proteins of 38-42K, which appears to be nuclear transcriptional activators of the long terminal repeats. In addition to the major open reading frame, the X genes commonly harbor another short open reading frame that overlaps this major one. Both of these open reading frames are found on a single spliced X mRNA in a potentially functional form. Circumstantial evidence strongly suggests that they are both translated from the single X mRNA molecule by alternative translation mechanism. We note that the short open reading frame has the capability to encode a putative nuclear protein with structural features similar to those of an AIDS virus trans-acting protein. Thus, we propose that the X genes of HTLV and BLV are both an overlapping gene encoding two distinct polypeptides, both of which may be involved in viral replication, cellular transformation, or both, possibly interacting with each other.

[1] This work was supported partly by Research Grant of the Princess Takamatsu Cancer Research Fund and partly by a grant from the Ministry of Education, Science and Culture of Japan.
[2] Computation Center, The Institute of Physical and Chemical Research, Wako, Saitama 351-01.
[3] Present address: Division of Life Science, Tonen R. & D. Laboratories, Ohi, Irima County, Saitama 354, Japan.

INTRODUCTION

Human T-cell leukemia virus type I (HTLV-I) and bovine leukemia virus (BLV) cause adult T-cell leukemia (1) and enzootic bovine leukosis (2), respectively. These retroviruses and their close relatives simian T-cell leukemia virus (STLV) and human T-cell leukemia virus type II (HTLV-II) all have a potential transforming gene, termed X, between the env gene and the 3' long terminal repeat (LTR) (3-6). The major long open reading frames in the HTLV X genes encode proteins of 38-42 kDa (7-9). We have recently shown that an analogous open reading frame of the BLV X gene also encodes a 38 kDa protein, which is localized in the nucleus of an infected cell (10). Probably, these proteins are responsible for viral replication, cellular transformation, or both (11-13).

In addition to the major long open reading frame, the X genes of these retroviruses commonly contain a short open reading frame which overlaps this major one (3-6). We show here the expression mechanism of the X genes and propose that both the long and short open reading frames are read from a single spliced X mRNA by alternative translation. A part of this paper has already been published elsewhere (14).

MATERIALS AND METHODS

Northern Blotting Analysis of BLV RNA

Poly(A)$^+$RNA was obtained from BLV-producing cells (FLK-BLV (15)) (10) and subjected to Northern blotting analysis (16) using various portions of the cloned BLV DNA as probes (4).

Calculation of Nucleotide Difference at Synonymous Sites

For each pair of aligned nucleotide sequences of homologous genes, the nucleotide difference at synonymous sites was calculated as described (17).

RESULTS AND DISCUSSION

Structure of the X Gene: A Potential Overlapping Gene

Existence of two overlapping open reading frames, X-I and X-II. Fig. 1 schematically shows the structure of the X genes of BLV (4), HTLV-I (3), STLV (5), and HTLV-II (6): in addition to the functional major open reading frame (designated here X-I frame), all the X genes commonly contain another short open reading frame (designated X-II frame) which overlaps the 5' half of the X-I frame. In all X genes, neither X-I nor X-II frame has an initiator ATG codon. However, a splice acceptor sequence occurs at the 5' end of each X-I frame (Fig. 1), suggesting that at least this frame is expressed as a spliced mRNA.

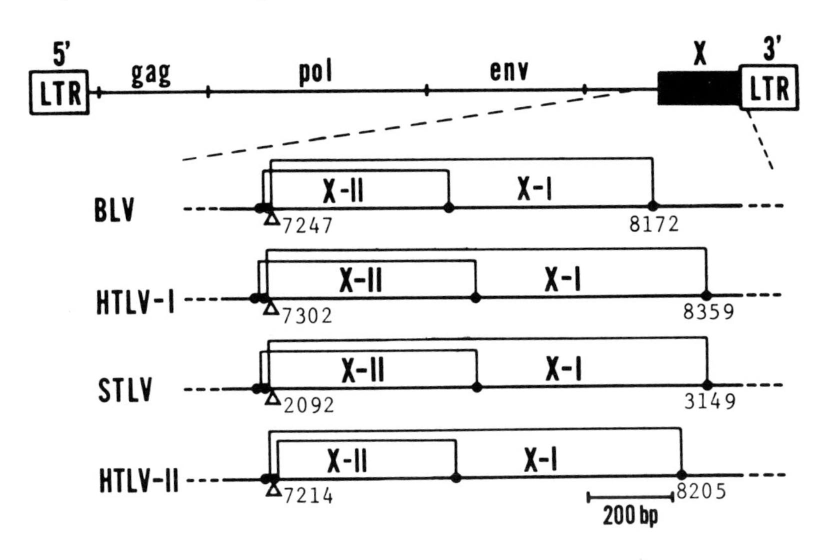

FIGURE 1. Structures of the X genes of the HTLV/BLV family. References are from 4(BLV), 3(HTLV-I), 5(STLV), and 6(HTLV-II). Closed circle, termination codon; triangle, splice acceptor site.

Theoretical analysis of the protein-coding capability of the X-II frame. To assess whether the overlapping X-II frame also encodes a functional protein, we compared the nucleotide sequences of the *gag*, *pol*, *env*, and X genes for three pairs of viruses: BLV-J (Japanese isolate (4,18)) vs. BLV-B (Belgian isolate (19,20)), HTLV-I vs. STLV, and HTLV-I vs. HTLV-II. From these comparisons, we calculated the nucleotide differences at synonymous sites of the respective genes (14,17); for the X gene, the calculation was done separately for overlapping (i.e., X-II region) and non-overlapping regions of the X-I frame. Fig. 2 shows that the nucleotide differences at synonymous sites of all the genes or regions, except for the X-II region, give nearly identical values in a given pair of viruses; this is consistent with the notion that evolutionary rate of the nucleotide substitution of synonymous sites is constant among different genes (21). In contrast, in any pair of viruses the nucleotide difference at synonymous sites of the X-II region is extremely reduced as compared with those of the other genes or regions. Since extreme reduction in synonymous substitutions is typical of overlapping genes of certain prokaryotic and eukaryotic viruses (21,22) and since the X-II region contains an open

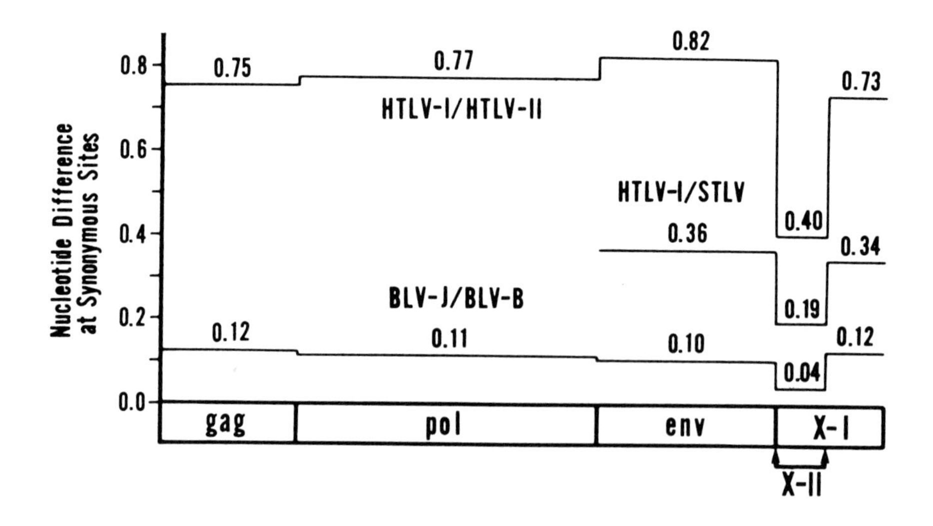

FIGURE 2. Theoretical analysis of the protein-coding capability of the X-II frame. Calculation according to Miyata and Yasunaga (17). See text for detailed explanation.

reading frame commonly in all members of the HTLV/BLV family (Fig. 1), it is most likely that the X-II region, or the X-II frame, encodes an unknown protein. Thus, we propose that the X gene is an overlapping gene.

Structure of the X mRNA: A Bicistronic mRNA

Generation of the X mRNA by double splicing. We previously predicted the existence of a spliced X mRNA of BLV (18). To verify this, we performed Northern blotting analyses of the poly(A)$^+$RNA isolated from BLV-producing FLK cells (14), using various portions of the BLV proviral DNA as probes (Fig. 3A). Fig. 3B shows that, of the three BLV-specific mRNAs detected, a subgenomic 1.8 kb mRNA preferentially hybridizes with an X gene-specific probe (lane 8), indicating that it is an X mRNA. This subgenomic X mRNA also hybridizes, although very weakly, with both 5'LTR and pol-env junction probes (lanes 1 and 5), which implies that small portions of these are also exons of the X mRNA. Thus, the BLV X mRNA has a doubly spliced structure, △5'LTR-pol/env junction-X. Based on our published BLV DNA sequence (4,24), two potential splice donor sites are located in the R region of the 5'LTR and immediately upstream of the 3' end of the pol gene, while two potential splice acceptor sites are located about 170 bp upstream of the env gene and at the 5' end of the X-I frame (Fig. 3C); most of these splice donor/acceptor sites are consistent with our earlier predictions (4,18). Recently, an almost identical structure of the X mRNA was accurately determined for HTLV-I (24) and HTLV-II (25). HTLVs and BLV therefore have a common and novel splicing mechanism of RNA for expression of their X genes.

The Spliced X mRNA Has a Bicistronic Structure, Containing Both X-I And X-II Frames in a Functional Form.

Fig. 4 shows a deduced structure of the spliced BLV X mRNA (14), along with those of the other viruses (24,25). The BLV X-I frame, which originally had no initiator ATG codon in the proviral genome (Fig. 1), now displays a potentially functional form on the spliced mRNA: an AUG triplet, initially located just upstream of the second splice donor site or the 3' end of the pol gene but in a different reading frame from those of both pol and env genes (Fig. 3C),

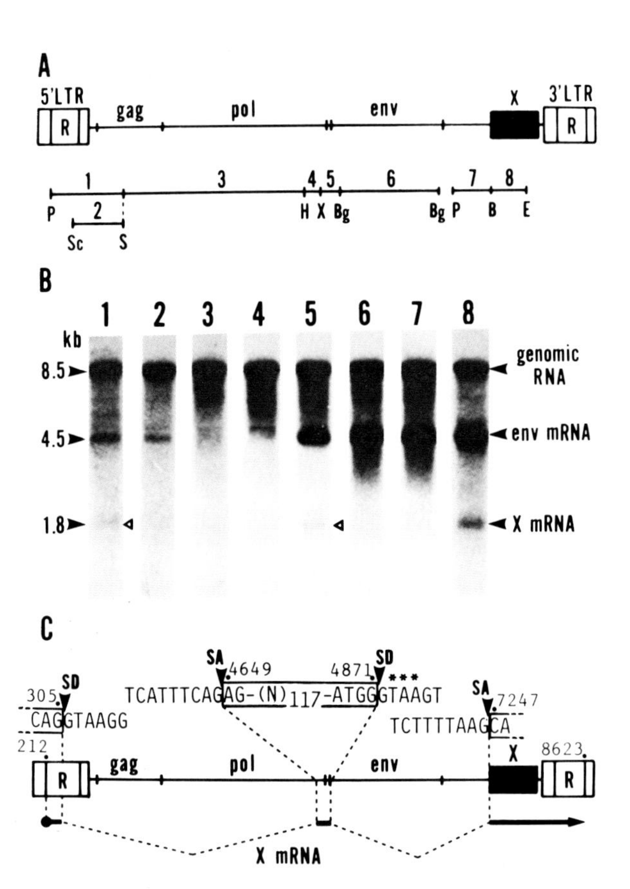

FIGURE 3. Identification and structure of the BLV X mRNA. (A) Location of probes used for Northern blotting analyses. P, Pvu II; Sc, Sac I; S, Sal I; H, Hind III; X, Xba I; Bg, Bgl II; B, Bam HI; E, Eco RI. (B) Northern blotting analyses of the FLK-BLV poly(A)$^+$RNA (lane numbers coincide with probe numbers in A). Triangles (lanes 1 and 5) denote weak hybridization with the X mRNA. Genomic RNA and env RNA are also indicated. (C) Spliced structure of the X mRNA. SD, splice donor; SA, splice acceptor. TAA with stars is a termination codon of the pol gene (4).

is joined to the 5' end of the original X-I frame (Fig. 4). The resulting X-I frame, supplied with only one initiator methionine codon, can thus encode a 309-amino acid residue protein, which has recently been identified as p38(X_{BL}) (10). As shown in Fig. 4, X mRNAs of HTLV-I, STLV, and HTLV-II have similarly organized X-I frames, each being supplied with an initiator AUG codon, which is also located just upstream of the second splice site; it is notable that these AUG codons, unlike that of BLV, are themselves env start codons (24,25).

A closer examination of the nucleotide sequence of the BLV X mRNA reveals another AUG triplet which is located 44 nucleotides upstream of the X-I AUG start codon, (Fig. 4). Interestingly, this is an env start codon (4,19), is a 5'-proximal AUG triplet, and is in frame with the X-II frame. Thus, the BLV X-II frame also has a potentially functional form on the spliced mRNA, with the first 51 env-coding nucleotides at its 5' end (Fig. 4); it should be noted that this 51-nucleotide sequence apparently encodes an NH_2-terminal half of an env signal peptide (4). Inspection of the nucleotide sequences of HTLV-I and STLV X mRNAs also reveals an AUG triplet which is located 53 nucleotides upstream of the respective X-I AUG start codons (Fig. 4). To our surprise, these AUG triplets also are 5'-proximal AUG codons and are in frame with the respective X-II frames, although unlike the putative BLV X-II start codon, they are within, but in a different reading frame from, the pol gene (3,5). Thus, in all these viruses, both X-I and X-II frames are on a single X mRNA in an apparently functional form.

The HTLV-II X mRNA also contains an AUG triplet, 56 nucleotides upstream of the X-I AUG start codon; again, this is a 5'-proximal AUG codon and is in frame with the original X-II frame (Fig. 4). In this case, however, the resulting X-II frame is disrupted by an in-frame AUG terminator codon located at the 5' end of the original X-II frame. The reason why the HTLV-II X-II frame is prematurely terminated in spite of its strong protein-coding capability (Fig. 2) is not known at present.

A Possible Translation Mechanism of the X mRNA: Alternative Translation.

Evidence for translatability of the X-II frame. A plasmid construct containing a cDNA clone of the HTLV-I X mRNA has recently been shown to direct synthesis of the major X-I product $p40^X$ in the transfected cells (24), clearly showing

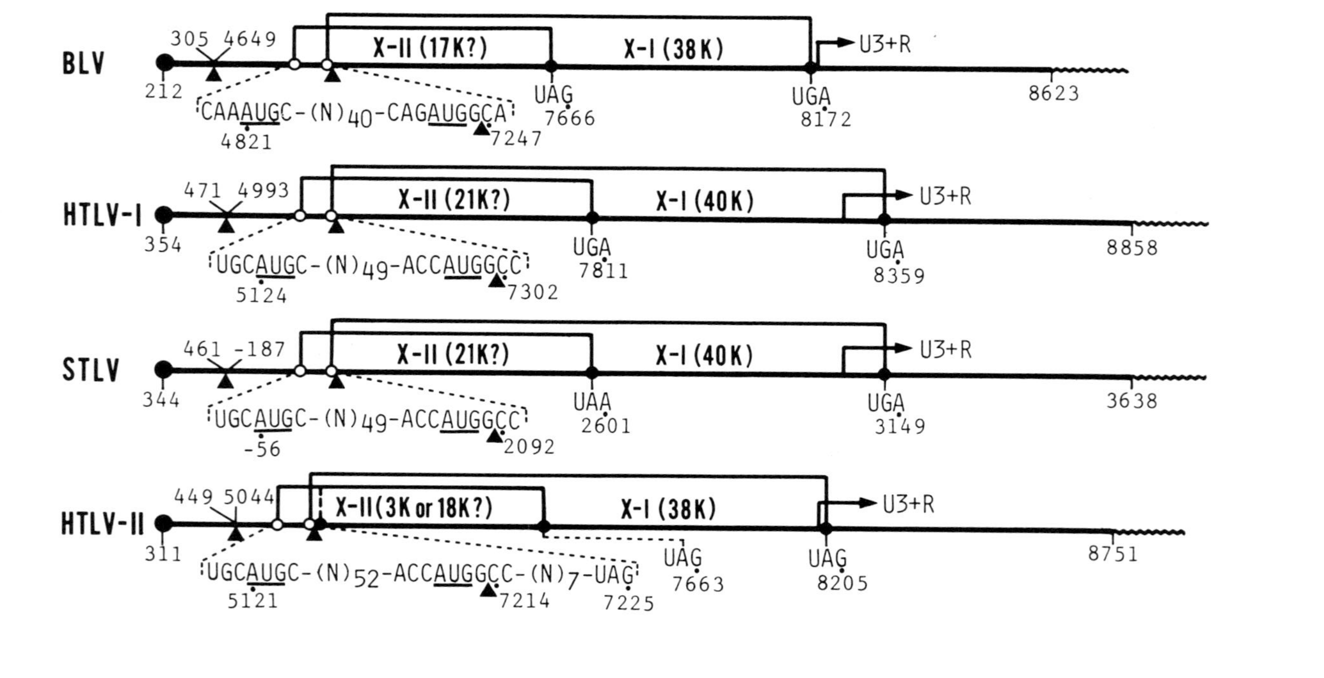

FIGURE 4. Fine structures of the X mRNAs of the HTLV/BLV family. Large closed circle, cap; closed triangle, splice site; small open circle, initiation codon; small closed circle, termination codon; wavy line, poly(A) tail.

that the X-I frame is translatable from the X mRNA *in vivo*. In contrast, no direct experimental data on the translatability of the X-II frame from the same X mRNA are available. Recently, however, sera from BLV-infected cattle have been shown to react with an *in vitro* translated X-II-related protein, suggesting the existence of an XII product *in vivo* (26). More recently, using a peptide antiserum against the HTLV-I X-II frame, a polypeptide with a size similar to that of the predicted HTLV-I X-II product (21 kDa, Fig. 4) has been detected in HTLV-I-producing cell lines (27). We believe that these putative X-II products are translated from the above mentioned BLV and HTLV-I X mRNAs, respectively, since no other mRNAs which could encode them have so far been observed. Thus, these observations and the common structure of the X mRNA in all members of HTLV/BLV family (Fig. 4) strongly suggest that in all these viruses both the X-I and X-II frames are read from the single X mRNA *in vivo* by alternative translation. We are currently substantiating this highly likely possibility by hybrid-arrested translation both *in vivo* and *in vitro*.

Possible mechanism of the alternative translation. If the unusual alternative translation of the X mRNA truly occurs, then how can ribosomes recognize both the X-I and X-II AUG triplets as start codons on the same single X mRNA? For the selection by eukaryotic ribosomes of an initiator AUG codon, both proximity of the AUG codon to the 5' end of the mRNA and the sequences that flank the AUG codon are important (modified scanning model (28)). Fig. 5 shows that the X-II AUG start codon of every X mRNA is a 5'-proximal AUG and is in a suboptimal sequence context, whereas the X-I AUG codon

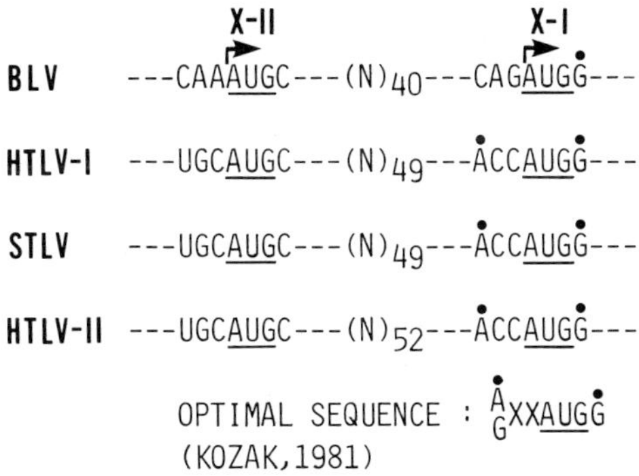

X-II X-I

BLV ---CAAAUGC---$(N)_{40}$---CAGAUGG---

HTLV-I ---UGCAUGC---$(N)_{49}$---ACCAUGG---

STLV ---UGCAUGC---$(N)_{49}$---ACCAUGG---

HTLV-II ---UGCAUGC---$(N)_{52}$---ACCAUGG---

OPTIMAL SEQUENCE : A/G XXAUGG
(KOZAK,1981)

FIGURE 5. Comparison of the sequences that flank the X-I and X-II AUG start codons with those in the optimal sequence context proposed by Kozak (28).

is a second AUG and is in an optimal or nearly optimal sequence context. Thus, according to the modified scanning model, some 40S ribosome subunits would stop and initiate at the 5'-proximal X-II AUG codon, but some would bypass that site because of its suboptimal sequence context and initiate at the downstream X-I AUG codon which is in an optimal sequence context, thereby resulting in the alternative translation of the single X mRNA as illustrated in Fig. 6. There are indeed several other examples of eukaryotic viral and cellular mRNAs that are translated from more than one AUG codon (29-30); among these, the most notable example is a reovirus sl gene mRNA, which also has both suboptimal and optimal AUG start codons in different reading frames and encodes two distinct proteins (32).

Nature of the Putative X-II Product: A Nuclear Regulatory Protein?

The X-II frames are capable of encoding proteins of 17-21 kDa (for BLV, HTLV-I, and STLV) or only 3 kDa (for HTLV-II, if premature termination truly occurs) (Fig. 4). These putative X-II products apparently do not show overall amino acid sequence homology between BLV and the other viruses (data not shown). However, they commonly share a very high proline residue content (20-23% of total residues). More interestingly, their NH_2-terminal 20-residue sequences, which are derived from second exons or pol-env junctions (Fig. 4 and 7), are commonly highly basic with many arginine residues

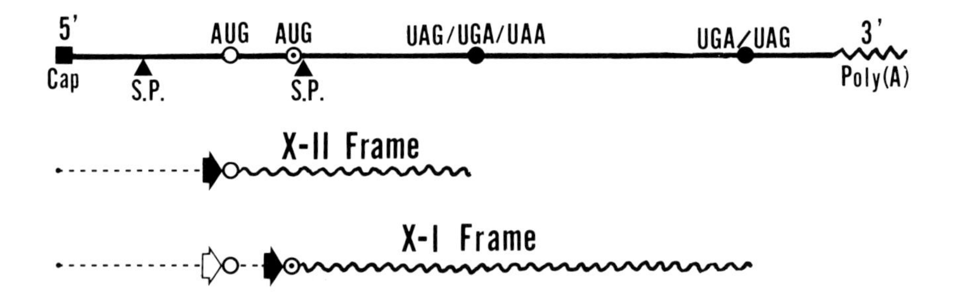

FIGURE 6. Alternative translation mechanism of the X mRNA. See text for details.

and show even sequence homologies between BLV and the other viruses (Fig. 7): These sequence homologies are significantly higher than those observed for the pol reading frames at the corresponding regions (Fig. 7). These findings suggest that the highly basic NH_2-terminal sequences are functional domains of the respective X-II products. In certain cellular and viral proteins, such a highly basic amino acid sequence has DNA-binding activity (33) and is a nuclear transport signal (34), raising the possibility that the putative X-II product is a nuclear protein.

Very recently, acquired immune deficiency syndrome (AIDS)-associated retrovirus (HTLV-III) has been shown to encode a nuclear trans-acting protein which also carries a number of proline residues and a striking cluster of arginine-lysine residues (35,36). This highly basic amino acid cluster, Arg-Lys-Lys-Arg-Arg-Gln-Arg-Arg-Arg-Pro-Pro, is also read from the pol-env junction of the proviral genome (35,36) and appears to have significant sequence homology with the basic NH_2-terminal sequences of the putative X-II products (cf. Fig. 7), suggesting analogous function between

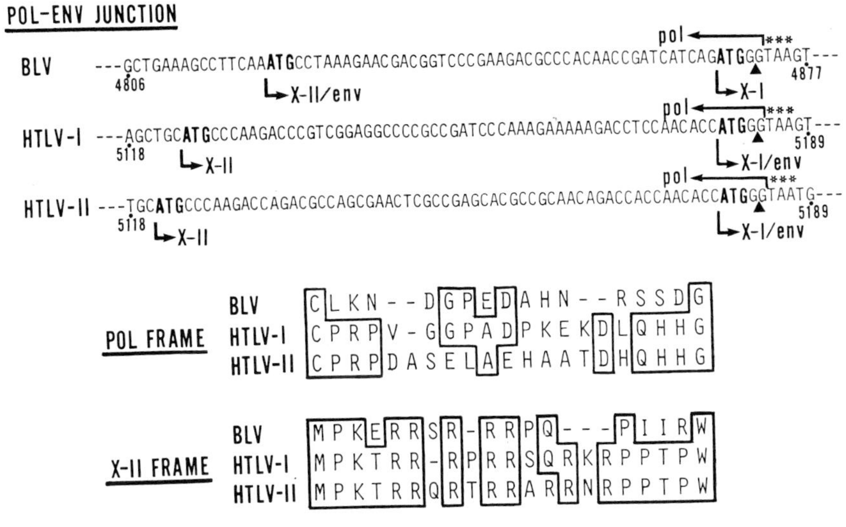

POL-ENV JUNCTION

BLV ---GCTGAAAGCCTTCAA**ATG**CCTAAAGAACGACGGTCCCGAAGACGCCCACAACCGATCATCAG**ATG**GGTAAGT--- (4806, 4877; X-II/env; X-I; pol ***)

HTLV-I ---AGCTGC**ATG**CCCAAGACCCGTCGGAGGCCCCGCCGATCCCAAAGAAAAAGACCTCCAACACC**ATG**GGTAAGT--- (5118, 5189; X-II; X-I/env; pol ***)

HTLV-II ---TGC**ATG**CCCAAGACCAGACGCCAGCGAACTCGCCGAGCACGCCGCAACAGACCACCAACACC**ATG**GGTAATG--- (5118, 5189; X-II; X-I/env; pol ***)

POL FRAME

BLV C L K N - - D G P E D A H N - - R S S D G
HTLV-I C P R P V - G G P A D P K E K D L Q H H G
HTLV-II C P R P D A S E L A E H A A T D H Q H H G

X-II FRAME

BLV M P K E R R S R - R R P Q - - - P I I R W
HTLV-I M P K T R R - R P R R S Q R K R P P T P W
HTLV-II M P K T R R Q R T R R A R R N R P P T P W

FIGURE 7. Comparison of the NH_2-terminal sequences of the putative X-II products of the HTLV/BLV family. In the upper pannel, pol-env junctions of BLV, HTLV-I and HTLV-II are shown (closed triangle; splice sife). In the lower pannels, sequence alignments are performed for the pol and X-II frames and homologous residues are boxed.

the HTLV-III trans-acting protein and the HTLV/BLV X-II products. Thus, we propose that the X-II product is a nuclear regulatory protein which is involved in viral replication, cellular transformation, or both, possibly interacting with the known X-I product.

ADDENDUM

Recently it was reported that the antisera to the synthesized oligopeptides of the c terminus of HTLV-I pX_{III} (corresponding to our X-II) product could precipitate 27K and 21K proteins in HTLV-I producing cells and in short-term cultures of adult T-cell leukemia tissue, and that this 27K protein was phosphorylated and located in the nuclei (Yoshida, M., personal communication).

A transacting factor encoded by HTLV-III/LAV genome has turned out to control translation of the gene product by targeting nucleotide sequences residing in the R region of LTR (G. Pavlakis, personal communication) or those in R-U_5 region (W. Haseltine, personal communication).

ACKNOWLEDGEMENT

We thank Mrs. Michiko Kimura for manuscript preparation.

REFERENCES

1. Gallo RC, Essex ME, Gross L (eds) (1984). Human T-cell Leukemia/Lymphoma Virus, Cold Spring Harbor Laboratory, Cold Spring Harbor, New York.
2. Burny A, Bruck C, Chantrenne H, Cleuter Y, Dekegel D, Ghysdael J, Kettmann R, Leclercq M, Leunen J, Mammerickx M, Portetelle D (1980). In Klein G (ed): "Viral Oncology," New York: Raven, p 231.
3. Seiki M, Hattori S, Hirayama Y, Yoshida M (1983). Proc Natl Acad Sci USA 80:3618.
4. Sagata N, Yasunaga T, Tsuzuku-Kawamura J, Ohishi K, Ogawa Y, Ikawa Y (1985). Proc Natl Acad Sci USA 82:677.
5. Watanabe T, Seiki M, Tsujimoto H, Miyoshi I, Hayami M, Yoshida M (1985). Virology 144:59.
6. Shimotohno K, Takahashi Y, Shimizu N, Gojobori T, Golde DW, Chen ISY, Miwa M, Sugimura T (1985). Proc Natl Acad Sci USA 82:3101.

7. Kiyokawa T, Seiki M, Imagawa K, Shimizu F, Yoshida M (1984). Gann 75:747.
8. Miwa M, Shimotohno K, Hoshino H, Fujino M, Sugimura T (1984). Gann 75:752.
9. Lee TH, Coligan JE, Sodroski JG, Haseltine WA, Salahuddin SZ, Wong-Staal F, Gallo RC, Essex M (1984). Science 226:57.
10. Sagata N, Tsuzuku-Kawamura J, Nagayoshi-Aida M, Shimizu F, Imagawa K, Ikawa Y (1985). Proc Natl Acad Sci USA 82:7879.
11. Sodroski JG, Rosen CA, Haseltine WA (1984). Science 225:381.
12. Rosen CA, Sodroski JG, Kettmann R, Burny A, Haseltine WA (1985). Science 227:320.
13. Fujisawa J, Seiki M, Kiyokawa T, Yoshida M (1985). Proc Natl Acad Sci USA 82:2277.
14. Sagata N, Yasunaga T, Ikawa Y (1985). FEBS Lett 192:37.
15. Van der Maaten MJ, Miller JM (1976). Biblio Haematol 43:360.
16. Thomas PS (1980). Proc Natl Acad Sci USA 77:5201.
17. Miyata T, Yasunaga T (1980). J Mol Evol 16:23.
18. Sagata N, Yasunaga T, Ohishi K, Tsuzuku-Kawamura J, Onuma M, Ikawa Y (1984). EMBO J 3:3231.
19. Rice NR, Stephens RM, Couez D, Deschamps J, Kettmann R, Burny A, Gilden RV (1984). Virology 138:82.
20. Rice NR, Stephens RM, Burny A, Gilden RV (1985). Virology 142:357.
21. Miyata T, Yasunaga T, Nishida T (1980). Proc Natl Acad Sci USA 77:7328.
22. Yasunaga T, Miyata T (1982). J Mol Evol 19:72.
23. Sagata N, Yasunaga T, Ogawa Y, Tsuzuku-Kawamura J, Ikawa Y (1984). Proc Natl Acad Sci USA 81:4741.
24. Seiki M, Hikikoshi A, Taniguchi T, Yoshida M (1985). Science 228:1532.
25. Wachsman W. Golde DW, Temple PA, Orr EC, Clark SC, Chen ISY (1985). Science 228:1534.
26. Yoshinaka Y, Oroszlan S (1985). Biochem Biophys Res Commun 131:347.
27. Kiyokawa T, Seiki M, Iwashita S, Imagawa K, Shimizu F, Yoshida M (1985). Proc Natl Acad Sci USA 82:8359.
28. Kozak M (1983). Microbiol Rev 47:1.
29. Kozak M (1981). Curr Topics Microbiol Immunol 93:81.
30. Mardon G, Varmus HE (1983). Cell 32:871.
31. Bos JL, Polder LJ, Bernards R, Schrier PI, van den Elsen PJ, van der Eb A, van Ormondt H (1981). Cell 27:121.

32. Ernst H, Shatkin AJ (1985). Proc Natl Acad Sci USA 82:48.
33. Saenger W (1983). "Principles of Nucleic Acid Structure." New York: Verlag, p 385.
34. Kalderon D, Roberts BL, Richardson WD, Smith AE (1984). Cell 39:499.
35. Arya SK, Guo C, Josephs SF, Wong-Staal F (1985). Science 229:69.
36. Sodroski J, Patara R, Rosen C, Wong-Staal F, Haseltine W (1985). Science 229:74.

Viruses and Human Cancer, pages 191–204

EXPRESSION OF A SECRETED FORM OF THE AIDS RETROVIRUS ENVELOPE IN MAMMALIAN CELLS

Laurence A. Lasky, Christopher Fennie, Donald Dowbenko, Wylla Nunes, and Phillip Berman

Department of Molecular Biology, Genentech, Inc.
460 Pt. San Bruno Blvd, South San Francisco, CA 94080

ABSTRACT The envelope antigen of the AIDS retrovirus is hypothesized to play a major role in the induction of virus neutralizing antibodies. In order to study its potential as a vaccine, this protein has been produced using a permanent mammalian cell line expression system. The protein has been truncated to enable secretion of the antigen from the cell line. The resultant retrovirus envelope protein is found in the medium of transfected cells as a highly glycosylated, soluble, stable protein. The antigen is terminally sialated and is apparently secreted from these cells with a half-life of approximately seven hours. Future studies will be directed towards determining the effectiveness of this secreted antigen as a subunit vaccine for AIDS.

INTRODUCTION

The acquired immune deficiency syndrome (AIDS) is a physically devastating disorder which has reached epidemic levels in several parts of the world (1). The disease, which generally manifests itself in the form of opportunistic infections, neoplasms, and a variety of other symptoms, has been shown to be associated with a retrovirus which has been isolated in several laboratories (2-4). The AIDS retrovirus appears to cause immune disfunction by infecting the helper subset of T cells, an infection which inevitably results in cell death and a depletion of this T cell subset from the immune system (5). At present, the major high risk groups for infection with the AIDS retrovirus include homosexual men, intravenous drug users, and hemophiliacs (1). However, recent evidence has shown that heterosexual transmission can occur as a result of sexual contact, suggesting that a much larger percentage of the population may eventually be at risk for infection with this virus (6-8). The potential magnitude of this epidemic cannot, therefore, be overestimated.

While alterations in the lifestyles of certain high risk groups may have an impact on the spread of AIDS, the development of a vaccine which protects against infection by the AIDS retrovirus would undoubtedly profoundly affect the current epidemic (1,9). Recent evidence which demonstrates neutralizing antibodies in many patients (10,11), together with other reports which show that the majority of people infected with the AIDS retrovirus produce antibodies against the envelope surface antigen, suggest that this protein may potentially induce the formation of neutralizing antibodies (12-14). Thus, with these data in mind, and by analogy to previous work on the induction of neutralizing antibodies by other viral envelope proteins (15-17), we have chosen to concentrate on the AIDS retrovirus envelope antigen as a potential AIDS vaccine.

In order to study the efficacy of this protein in inducing a neutralizing response, it is convenient to produce large quantities of this material. Although the envelope antigen of the AIDS retrovirus may be isolated from virus infected cells, the difficulties in producing significant amounts of antigen in this way, as well as the obvious dangers of growing and inactivating the virus, suggested that this antigen could be more practically produced using recombinant DNA technology. In this paper, we report the synthesis of the AIDS retrovirus antigen as a secreted, glycosylated molecule in a permanent mammalian cell line.

The AIDS Retrovirus Envelope Protein

Previous work has indicated that the AIDS retrovirus envelope is apparently synthesized in the form of a 160,000 dalton precursor protein which is subsequently cleaved into a 120,000 dalton protein (gp120) and a 41,000 dalton putative transmembrane binding protein (gp41) (18). It is thought that the larger, N-terminal protein is bound to the viral surface by interaction with the smaller envelope protein which, by virtue of a hydrophobic domain, is bound to the plasma membrane surrounding the virus. While it has not yet been conclusively demonstrated, analogies with other viral systems suggest that one or both of these envelope antigens may be good candidates for inclusion in an AIDS vaccine (9). This hypothesis stems from the belief that at least one of these viral surface antigens is responsible for binding of the retrovirus to a receptor on the surface of the T cell, and that antibodies directed against envelope antigens of the virus should disrupt this binding.

The DNA sequence of the AIDS retrovirus genome has revealed, in some part, the nature of the envelope antigen (19-22). This sequence has demonstrated an open reading frame of 856 amino acids which is located in the last third of the viral genome and which encodes a protein which contains a region identical to that determined for the processed N-terminus of the 120,000 dalton virus envelope antigen isolated from AIDS retrovirus infected cells (23). The encoded protein contains approximately 30 potential N-linked glycosylation sites (Asn-X-Ser/Thr) (depending upon the strain of the virus) which could contribute, in large part, to the apparent size of the mature envelope antigen. In addition, as can be seen in Figure 1, several additional features of the virus envelope antigen may be noted from a hydropathy analysis of the envelope protein sequence (24). The N-terminus of the protein appears to contain a hydrophobic stretch of amino acids which could function as a signal sequence, although this sequence is preceeded by several charged amino acid residues. Interestingly, previous work has shown that the signal cleavage site for the AIDS retrovirus envelope appears to occur before the N-terminal hydrophobic region, suggesting that this region may function as an internal signal sequence which may remain bound to the lipid bilayer (18). In addition, figure 1 shows that two other major hydrophobic domains are found in the envelope antigen. The first, from amino acids 512-539, contains only hydrophobic residues and is found immediately following the processing site which clips the 160,000 dalton precursor protein into the 120,000 dalton protein and the 41,000 dalton transmembrane protein (18). A second hydrophobic domain is found from amino acids 684-705; however, this domain invariably contains a charged arginine residue at position 695 in all strains examined.

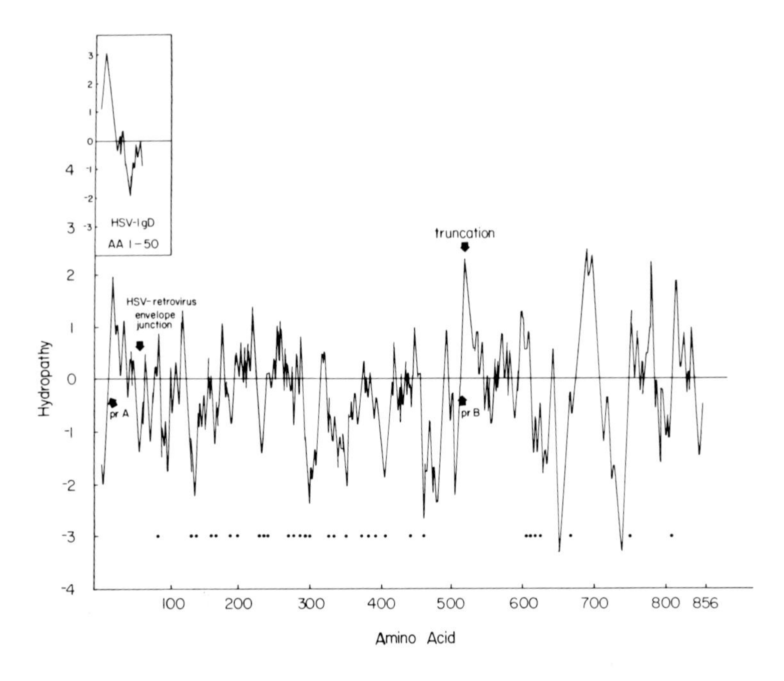

FIGURE 1. Hydropathy Analysis of the AIDS Retrovirus Envelope Protein. The envelope antigen of the AIDS retrovirus was analyzed for hydrophilic and hydrophobic regions using the hydropathy program of Hopp and Woods (24). Regions above the zero line correspond to potentially hydrophobic areas while regins below this line correspond to potentially hydrophilic regions, with the distance above this line corresponding to degree of hydrophobicity or hydrophilicity. The dots at -3 show potential N-linked glycosylation sites. "pr A" refers to the N-terminal signal sequence processing site (23). "pr B" refers to the processing site which clips the 160,000 dalton envelope precursor protein into the N-terminal 120,000 dalton protein and the C-terminal 41,000 dalton transmembrane protein (18). "truncation" illustrates the site at which the recombinant, secreted form of the envelope antigen was truncated here. "HSV-retrovirus envelope junction" illustrates the site at whch the Herpes Simplex Virus glycoprotein D(gD) signal sequence was joined to the signalsequence-lacking AIDS retrovirus envelope gene. The inset illustrates a hydropathy analysis of the first 50 amino acids of gD, including the highly hydrophobic N-terminal signal sequence.

Expression of A Secreted, Glycosylated Form of the AIDS Retrovirus Envelope in Mammalian Cells

Previously, we demonstrated the usefulness of mammalian cell expression systems for the production of a safe and efficacious subunit vaccine for herpes simplex virus types 1 and 2 infections (15,16). In this work, we demonstrated that the glycoprotein D (gD) of this virus could be secreted from mammalian cell lines by transfection of a gD gene which was truncated in such a way that it lacked the transmembrane binding domain which normally serves to bind the protein to the viral or cellular surface. This resulted in the synthesis of a glycosylated, immunogenic antigen which, because it was found to be secreted in the media of transfected cells, was relatively easy to purify and whose purification did not require extraction of the cells. We felt that a similiar approach could be utilized to produce a potential AIDS subunit vaccine. Several differences in the natures of the gD and AIDS retrovirus envelope antigens had to be addressed, however. The first issue was the nature of the signal sequence of the retroviral envelope. As noted above, this envelope contains a hydrophobic domain at the N-terminus which is preceeded by a highly charged region. This is in contrast to other viral envelope antigens, such as gD (see inset, figure 1), which contain an N-terminal signal sequence which usually possesses a fifteen to twenty residue hydrophobic domain containing a single charged amino acid. A second problem revolved around the nature of the putative transmembrane binding protein at the C-terminus of the retroviral envelope antigen. While a well-defined, C-terminal transmembrane domain is commonly found in other viral envelope proteins, the complex nature of the C-terminal gp41 protein of the AIDS retrovirus made the choice of a truncation site for secretion more difficult.

In order to efficiently express a secreted form of the AIDS retrovirus envelope antigen, a truncated form of the gene was constructed which utilized the gD signal sequence. The use of the gD signal sequence was necessitated by initial envelope expression experiments, utilizing the normal, retroviral signal sequence, which demonstrated that the protein was very inefficiently expressed. It was felt that this inefficient expression was due, in part, to the unusual nature of the retrovirus envelope signal sequence. The construction of this chimaeric, truncated form of the retrovirus envelope is illustrated in figures 1 and 2. The gD signal sequence was included in a 50 amino acid coding fragment which was joined in frame to amino acid 61 of the retrovirus envelope. Correct cellular processing of the gD signal sequence would result in a retroviral envelope antigen containing 25 amino acids of gD at its N-terminus. Expression of a full-length retrovirus envelope-gD signal sequence chimaeric gene in transfected chinese hamster ovary (CHO) cells resulted in the production of significant amounts of a highly glycosylated retrovirus envelope antigen which appeared to be bound to the membranes of one of the cellular compartments with no evidence of proteolytic processing or surface transport (data not shown). In order to secrete the antigen from the cell, the entire putative transmembrane binding domain (ie. gp41) was deleted from the envelope gene. As can be seen in figure 1 and 2, the truncation site was at amino acid 531 of the envelope, 20 amino acids C-terminal to the actual cellular proteolytic processing site (18), resulting in a truncated chimaeric envelope protein of 520 amino acids in length. This truncated envelope gene was under the transcriptional control of an SV40 early promoter, with the 3' terminal polyadenylation signals derived

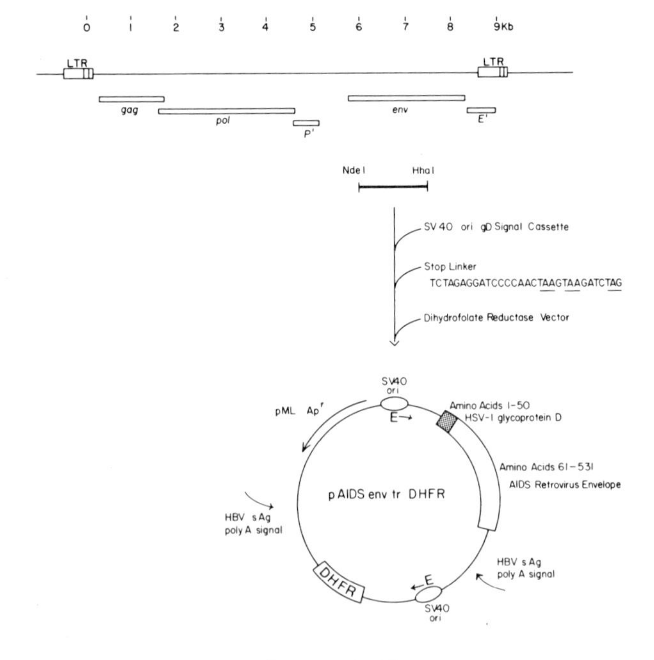

(FIGURE 2)

from the hepatitis B virus surface antigen gene (figure 2). In addition, the plasmid contained a cDNA clone of the murine dihydrofolate reductase gene (dhfr) which served as a selectable and amplifiable marker for transfection into CHO cells deficient for dihydrofolate reductase activity (dhfr) (25,26).

Transfection of this plasmid onto dhfr minus CHO cells and subsequent selection (25,26) resulted in the appearance of several stable colonies which contained the retrovirus expression plasmid. Several of the clones were labeled with 35S methionine, and the intracellular and extracellular protein was immunoprecipitated using antiserum from a healthy homosexual male who had high antibody titres against the native envelope antigen of the AIDS retrovirus. The results of these experiments are illustrated in figure 3. As can be seen in figure 3d,

---

FIGURE 2. Construction of a Plasmid for the Synthesis of a Secreted Form of the AIDS Retrovirus Envelope in Mammalian Cells. The top of the figure illustrates the viral genome with the location of each of the five major open reading frames of the virus (19-22). The env reading frame encodes the viral envelope antigen utilized here (23). An envelope fragment encoding amino acid residues residues 61-531 was used for expression of a secreted envelope protein by in-frame ligation to a herpes simplex virus glycoprotein D (gD) signal sequence fragment (15). These fragments were incorporated into a vector which contained the components necessary for proper expression of this integrated envelope gene as well as for selection in mammalian cells. "SV40 ori" contains an early promoter from the SV40 virus ("E") which is utilized to drive transcription of either the AIDS retrovirus envelope or the dihydrofolate reductase (dhfr) cDNA clone (25,26). Transcription termination and message polyadenylation are accomplished using signals derived from the 3 prime non-translated region of the hepatitis B virus surface antigen gene ("HBVs Ag poly A signal"). Growth in the bacteria E.coli is accomplished by the inclusion of the ampicillin resistance gene ("Ap R") of the plasmid pML as well as the origin of replication of this plasmid. Finally, the murine dhfr cDNA clone is utilized as a selectable marker transfection and selection in chinese hamster ovary (CHO) cells which lack this gene product (25,26).

immunoprecipitation of the intracellular contents of one transfected cell line shows an AIDS retrovirus antibody specific band at approximately 100,000 daltons. This band was absent when using control serum as well as in non transfected cells (figure 3b). When the extracellular (secreted) contents of this cell line were analyzed, a

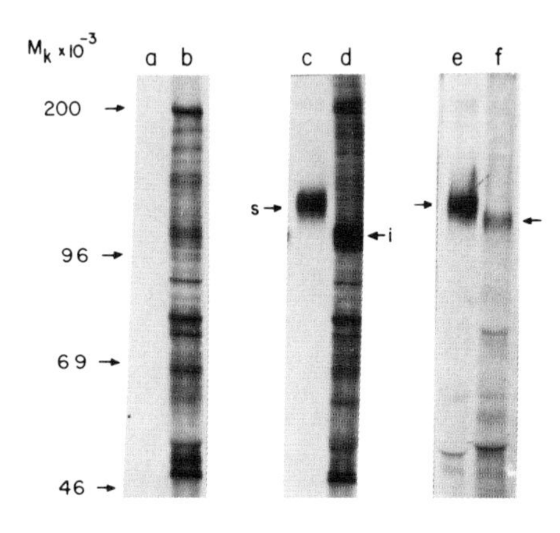

FIGURE 3. Immunoprecipitation of the Intracellular and Secreted Forms of the AIDS Retrovirus Envelope From Transfected Cells. Control (CHO, dhfr minus) and p AIDS env tr DHFR-transfected cells were labelled with 35S methionine, and the intracellular and secreted proteins were immunoprecipitated with serum from an AIDS retrovirus seropositive patient. The precipitated proteins were then resolved by SDS polyacrylamide gel electrophoresis by autoradiography: a. material secreted from control cells, b. intracellular material from control cells, c. material secreted from cells transfected with the AIDS retrovirus envelope expression plasmid, d. intracellular material from cells transfected with the AIDS retrovirus envelope expression plasmid. s: secreted, retrovirus envelope antigen, i: intracellular, retrovirus envelope antigen. The nature of the glycosylation of the secreted envelope antigen was investigated by neuraminadase digestion and immunoprecipitation. e. untreated material secreted from envelope-producing cell lines, f. neuraminadase-treated material secreted from envelope-producing cell lines.

larger, more diffuse immunoprecipitable band which migrates at approximately 130,000 daltons was found. This band was immunoprecipitable by a variety of AIDS retrovirus seropositive sera, and it was only precipitable from cell lines transfected with this plasmid (figure 3c). These results, thus, demonstrated that the cell lines transfected with this expression plasmid are secreting a highly glycosylated form of the AIDS retrovirus envelope protein which is capable of reacting with antibodies directed against the native, viral protein.

The nature of the glycosylation of this secreted protein was further analyzed in neuraminadase and pulse-chase experiments. Figure 3f shows that neuraminadase treatment of the secreted envelope protein reduced its size to approximately 110,000 daltons, suggesting that this protein is terminally glycosylated with sialic acid residues (27). As shown in figure 4, pulse-chase experiments support this view and demonstrate that the 100,000 dalton intracellular precursor protein is fully glycosylated as it is secreted from this cell line with a half life of approximately 7 hours. The apparent slight decrease in molecular weight of the intracellular precursor may be due to trimming of the core mannose residues before sialic acid addition and secretion (27). In agreement with these results, the intracellular form of the envelope protein is sensitive

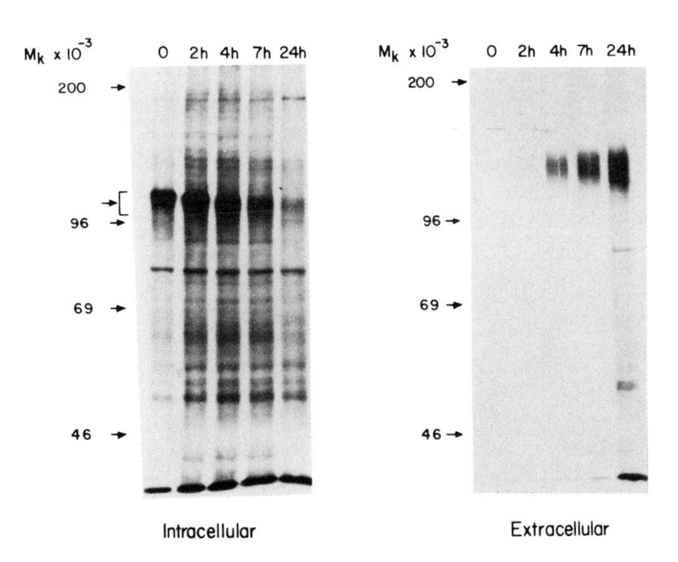

to endoglycosidase H digestion (27), suggesting that this protein corresponds to the high mannose precursor of the secreted antigen (data not shown).

DISCUSSION

In summary, these results demonstrate that the AIDS retrovirus envelope can be produced in a heterologous, mammalian cell expression system in a secreted, glycosylated form. In addition, and most importantly, the fact that the antigen can be immunoprecipitated by antibodies from patients exposed to the AIDS retrovirus suggests that at least some of the natural epitopes are maintained in the secreted antigen produced in these transfected cell lines. The use of transfected, permanent mammalian cell lines for the production of a potential AIDS retrovirus subunit vaccine has several advantages over other microbial systems, such as bacteria or yeast. While undoubtedly a more difficult approach, the mammalian cell system allows for the production of a recombinant envelope antigen which has many of the characterisics of the native antigen. For example, the envelope protein produced in mammalian cells is glycosylated. While the role(s) of the sugar residues found in most virus surface antigens is unclear, it seems likely that these residues must contribute to the overall tertiary structure of the antigen (28). Thus, the protein produced in this system will probably represent an antigen which is very similiar to that found on the surface of the virus. While very large quantities of envelope antigens may be produced in bacteria or yeast, the antigens produced in these systems are neither glycosylated nor is it clear that their three-dimensional structure is entirely correct (29). Thus, it would appear, from the results reported here, that the synthesis of the envelope antigen in an environment in

---

FIGURE 4. Kinetics of Secretion of the Truncated From of the AIDS Retrovirus Envelope Antigen. Envelope-producing CHO cell lines were pulse-labelled for 10 minutes with 70 microcuries per ml of 35S methionine. The cells were washed and chased in cold-methionine containing media for the indicated times. Intracellular and extracellular materials were analyzed at these times using immunoprecipitation as described in figure 3.

which it is normally made, ie. the mammalian cell, yet in the absence of infectious virus, is the best method to insure an envelope antigen which is both highly immunogenic and safe.

An additional benefit of the mammalian cell system is that it results in the secretion of proteins by utilizing the constituitive secretory pathway of the cell. This allows for the secretion of a truncated form of the AIDS retrovirus envelope antigen into the medium, resulting in a purification procedure which does not require the extraction of the envelope-producing cells. This enables the continued production of the secreted envelope for a longer period of time as well as an easier purification procedure, since far fewer cellular proteins have to be removed. This should ultimately result in the purification of large quantities of homogeneous secreted antigen for use in vaccine trials.

In conclusion, this paper presents one of the possible routes towards the production of a safe vaccine which may be efficacious in the prevention of transmission of the AIDS retrovirus. While the preliminary results reported here suggest that the antigenic structure of the recombinant protein may be representative of that found in the native antigen, vaccination studies in a variety of animals will have to be completed in order to demonstrate the potential efficacy of this protein as an AIDS retrovirus vaccine.

ACKNOWLEDGMENTS

We thank Rebecca Cazares for excellent manuscript preparation and Dr. Jerome Groopman for the gift of high titre anti-AIDS retrovirus antiserum.

REFERENCES

1. Fauci, A, Macker, A, Longo, D, Rook, A, Masur, H, Gelman, E (1984). Acquired Immunodeficiency Syndrome: Epidemiological, Clinical, Immunologic and Therapeutic Consideration. Ann Int Med 100:92.
2. Barre-Sinoussi, F, Chermann, J, Rey, F, Nugeyre, M, Chamarat, S, Gruest, J, Dauguet, L, Axler Blin, C, Vezinet-Brun, F, Rouzioux, C, Rozenbaum, W, and Montaignier, L (1983). Isolation of a T-lymphotrophic Retrovirus from a Patient at Risk for Acquired Immunodeficiency Syndrome. Science 220:868.
3. Gallo, R, Salahuddin, S, Popovic, M, Shearer, G, Kaplan, M, Haynes, B, Paiker, T, Redfield, R, Olesko, J, Safai, B, White, G, Foster, P, and Markham, P (1984). Frequent Detection and Isolation of Cytopathic Retroviruses (HTLV III) from Patients with AIDS and at Risk for AIDS. Science 224:500.
4. Levy, J, Hoffman, A, Kramer, S, Landis, J, Shimabukuro, J, and Levy, J (1984). Isolation of Lymphocytopathic Retroviruses from San Francisco Patients with AIDS. Science 225:840.
5. Klatzman, D, Barre-Sinoussi, F, Nugeyre, M, Dauguet, C, Vilmer, E, Guiscelli, C, Vezinet-Brun, F, Rouzioux, C, Gluckman, J, Cherman, J, and Montaignier, L (1984). Selective Tropism of Lymphadenopathy Associated Virus (LAV) for Helper-Inducer T Lymphocytes. Science 225:59.
6. Harris, C, Small, C, Klein, R, Friedland, G, Moll, B, Emerson, E, Spigland, I, and Steigligel, N (1983). Immunodeficiency in Female Sexual Partners of Men with the Acquired Immunodeficiency Syndrome. New Engl J Med 308:1181.
7. Pitchenik, A, Shafren, R, Glasser, R, and Spira, T (1984). The Acquired Immunodeficiency Syndrome in the Wife of a Hemophiliac. Ann Intern Med 100:62.
8. Groopman, J, Sarngadharan, M, Salahuddin, S, Buchsbaum, R, Huberman, M, Kinniburgh, J, Sliski, A, McLean, M, Essex, M, Gallo, R (1985). Apparent Transmission of HTLV III to a Heterosexual Woman with AIDS. Ann Int Med 102:63.
9. Francis, D, Pettricianni, J (1985). The prospects and pathways toward a vaccine for AIDS. New Engl J Med 313:1586.

10. Weiss, R, Clapham, R, Cheingsong-Popov, R, Dalgleish, A, Carne, C, Weller, I, Tedder, R (1985). Neutralization of human T-lymphotrophic virus Type III by Sera of AIDS and AIDS-Risk Patients. Nature 316:69.
11. Robert-Guroff, M, Brown, M, Gallo, R (1985). HTLV-III Neutralizing Antibodies in Patients with AIDS and AIDS-related Complex. Nature 316:72.
12. Barin, F, McClane, M, Allan, J, Lee, T, Groopman, J, Essex, M (1985). Virus Envelope Protein of HTLV-III Represents Major Target Antigen for Antibodies in AIDS Patients. Science 228:1094.
13. Chang, T, Kato, I, McKinney, S, Chanda, P, Barone, A, Wong-Staal, F, Gallo, R, Chang, N (1985). Detection of Antibodies to Human T-Cell Lymphotropic Virus (HTLV-III) with an Immunoassay Employing a Recombinant Escherichia coli-Derived Viral Antigenic Peptide. Bio/Technology 3:905.
14. Cabradilla, C, Groopman, J, Lannigan, J, Renz, M, Lasky, L, Capon, D (1986). Serodiagnosis of Antibodies to the Human AIDS Retrovirus with a Bacterially Synthesized Env Polypeptide. Bio/Technology in press.
15. Lasky, L, Dowbenko, D, Simonsen, C, and Berman, P (1984). Protection of Mice From Lethal Herpes Simplex Virus Infection by Vaccination with a Secreted Form of Cloned Glycoprotein D. Bio/Technology 2:527.
16. Berman, P, Gregory, T, Crase, D, and Lasky, L (1985). Protection From Genital Herpes Simplex Virus Type 2 Infection by Vaccination with Cloned Type 1 Glycoprotein D. Science 227:1490.
17. Patzer, E, Gregory, T, Nakamura, G, Simonsen, C, Eichberg, J, Levinson, A, Hershberg, R (1986) Recombinant Hepatitis B Surface Antigen Vaccine from a Continuous Cell Line. Bio/Technology, in press.
18. Veronese, F, Devico, A, Copeland, T, Oroszlan, S, Gallo, R, Sarngadharan, M (1985). Characterization of gp41 as the Transmembrane Protein Coded by the HTLV-III/LAV Envelope Gene. Science 229:1402.
19. Muesing, M, Smith, D, Cabradilla, C, Lasky, L, Capon, D (1984). Nucleic Acid Structure and Expression of the Human AIDS/lymphadenopathy virus. Nature 313:450.

20. Ratner, L, Haseltine, W, Pataria, R, Livak, K, Starcich, B, Josephs, S, Doran, E, Ratalski, J, Whitehorn, E, Baumeister, K, Ivanoff, L, Pettevag, S, Pearson, M, Lautenberger, J, Papas, T, Ghrageb, J, Chang, N, Gallo, R, Wong-Staal, F (1985). Complete Nucleotide Sequence of the AIDS virus, HTLV III. Nature 313:277.
21. Sanchez-Pescador, R, Power, M, Barr, P, Steimer, K, Stemplem, M, Brown Shimer, S, Gee, W, Renard, A, Randolph, A, Levy, J, Dino, D, Luciw, P (1985). Nucleotide Sequence and Expression of an AIDS Associated Retrovirus (ARV-2). Science 227:484.
22. Wain-Hobson, S, Sonigo, P, Danas, O, Cole, S, Alizon, M (1985). Nucleotide Sequence of the AIDS virus, LAV, Cell 40:9.
23. Allen, J, Coligan, J, Barin, F, McLane, M, Sodovoski, J, Rosen, C, Haseltine, W, Lee, T, Essex, M (1985). Major Glycoprotein Antigens that Induce Antibodies in AIDS Patients are Encoded by HTLV-III. Science 228:1091.
24. Hopp, T, Woods, K (1981). Prediction of Antigenic Determinants from Amino Acid Sequences. Proc Natl Acad Sci USA: 78,1099.
25. Urlaub, G, Chasin, L (1980). Isolation of Chinese hamster cell mutants deficient in dihydrofolate reductase activity. Proc Natl Acad Sci USA 77:4216.
26. Simonsen, C, Levinson, A (1983). Isolation and expression of an altered mouse dihydrofolate reductase cDNA. Proc Natl Acad Sci USA 80:2495.
27. Hubbard, S, Ivatt, R (1981). Synthesis and processing of Asparagine-linked oligosacharrides, Ann Rev Biochem 50:555.
28. Berman, P, Lasky, L (1985). Engineering Glycoproteins for use as pharmaceuticals. Trends in Biotech 3:51.
29. Yelverton, E, Norton, S, Obijeski, J, Goeddel, D (1983). Rabies Virus Glycoprotein Analogues: Biosythesis in Eschericia coli. Science 219:614.

Viruses and Human Cancer, pages 205–219

EXPRESSION OF FUNCTIONAL DOMAINS OF THE AIDS ASSOCIATED RETROVIRUS (ARV) IN RECOMBINANT MICROORGANISMS

Philip J. Barr, Debbie Parkes, Elizabeth A. Sabin, Michael D. Power, Helen L. Gibson, Chun Ting Lee-Ng, James C. Stephans, Carlos George-Nascimento, Ray Sanchez-Pescador, Robert A. Hallewell, Kathelyn S. Steimer and Paul A. Luciw

Chiron Corporation, 4560 Horton Street, Emeryville, California 94608

ABSTRACT Using genetically engineered microorganisms, we have expressed functional domains of each of the major genes of ARV-2, a retrovirus associated with Acquired Immune Deficiency Syndrome (AIDS). Recombinant proteins corresponding to precursors and individual subunits of the *gag-pol* precursor, together with *N* and *C*-terminal regions of the *env* precursor were analyzed. Immunoblotting procedures using proteins derived from these heterologous expression systems showed that antibodies in AIDS sera react with domains within each of the *gag*, *pol* and *env* genes. A53kD unglycosylated *env* protein representing the polypeptide moeity of viral envelope gp120 was isolated from yeast and used to immunize rabbits; antibodies from these animals reacted with viral *env* glycoproteins both in virus and in infected cells. Recombinant proteins from yeast and bacteria together with glycosylated proteins produced in mammalian cell systems will be useful in the diagnosis of AIDS and AIDS-Related Complex (ARC) and also for studies directed towards immunopreventive therapies for AIDS. In addition, active recombinant viral enzymes may be useful in the identification of inhibitors as potential antiviral chemotherapeutic agents.

INTRODUCTION

Several closely related human retroviruses that are considered to be the etiologic agents of the acquired immune deficiency syndrome (AIDS) have been molecularly cloned and sequenced (1, 2, 3). This information, and the availability of cloned viral DNA, possible detailed studies of the molecular biology of AIDS viral infection at the nucleic acid level (4). In addition, expression of viral proteins in recombinant microorganisms has offered a means to further assess certain aspects of the interaction of the AIDS virus with the infected individual. In this report we describe the production, by genetic engineering techniques, of proteins corresponding to each of the major genes of ARV-2 (a San Francisco isolate) (3, 5) in yeast and bacteria and their use in the study of immunological and clinical issues associated with this disease.

GENOMIC ORGANIZATION OF ARV-2

Similar structures have been reported for the genomes of several isolates of the retrovirus responsible for AIDS (1, 2, 3, 4). Despite complex polymorphism patterns (4, 6) which may represent antigenic variation (4), ARV-2 shows an identical overall genomic organization to HTLV-III and LAV. As with other well-studied retroviruses, the AIDS viruses contain large open reading frames corresponding to group-specific antigen (*gag*) or core protein genes, *pol* genes coding for the viral enzymatic activities, and the surface protein or envelope (*env*) gene. These genes are situated between the long terminal repeats (LTRs) (Figure 1), elements of the virus that regulate transcription. Each of these three viral genes encode precursor polyproteins that are proteolytically processed to give structural or enzymatic subunits of the mature virion. In addition to these major genes, each virus strain also contains two open reading frames (*orfs*) predicted to encode proteins of as yet unascribed function. A protein, designated p27, arising from translation of the 3'-orf of HTLV-III has been detected in infected cell lines using sera from AIDS patients (7). Finally, a recently discovered protein encoded by two exons, one immediately 5' of the env gene and the second within the *env* gene is responsible for an elevated levels of viral gene expression. Demonstration of the existence of this transactivating factor or *tat* gene product has allowed analyses of transcriptional regulation mechanisms which may be peculiar to certain retroviruses, including HTLV-I, -II (4), BLV (8) and RSV (9).

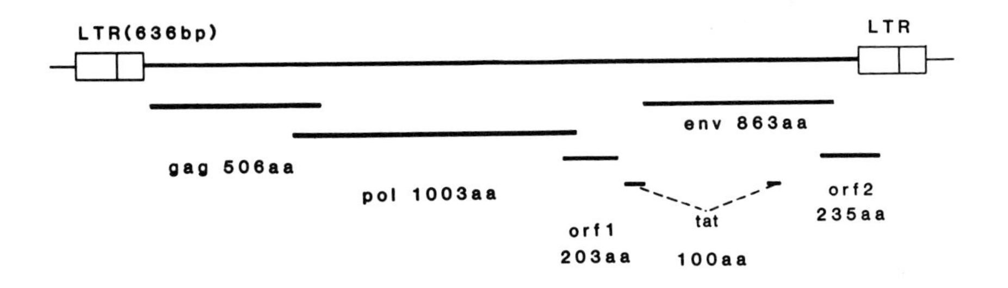

FIGURE 1. Genomic organization of the AIDS virus.

EXPRESSION AND USES OF RECOMBINANT ARV PROTEINS

Rationale.

There are several reasons for the production of ARV proteins by recombinant DNA methodologies. First, difficulties associated with large scale propagation of the virus make the production of viral polypeptides through non-infectious reagents more attractive. These techniques can also give virtually unlimited supplies of the protein desired. Furthermore, accurate knowledge of the DNA sequence encoding each protein precursor makes possible the construction of vectors for the expression of any desired region of the AIDS virus genome. This may include the expression of known or predicted functional domains of the virus or polyprotein precursors of these domains. Also, to further dissect the biological role of each protein, one may express smaller segments of each polypeptide.

We have developed systems for the high level constitutive or regulatable expression of heterologous proteins in bacteria and the yeast S. cerevisiae. We have also used mammalian cell expression systems for the production of heterologous proteins in which post-translational modification (e.g., glycosylation) may be advantageous. Using these systems, together with cloned proviral, fully synthetic or semi-synthetic ARV-2 coding sequences we have expressed and studied proteins from each region of the AIDS virus genome.

The gag Precursor Gene.

Both DNA sequence and analysis of purified viral proteins have defined the gag proteins of ARV-2 (3). The major core protein p25 is flanked by p12 and p16, all detectable by antibodies in human AIDS sera. p25, a major immunogenic component of the virus (10,11), was selected initially for expression as a useful diagnostic reagent. Using a semi-synthetic gene for p25 and a ribosome binding site randomization approach (12) we expressed large quantities of p25 in E. coli using the tac-1 promoter (13). Further derivatization of this gene using synthetic oligonucleotide adapters and standard yeast expression vectors (14), allowed either high level constitutive expression of p25 from the glyceraldehyde-3-phosphate dehydrogenase (GAPDH) promoter (14) or glucose regulatable expression from an alchohol dehydrogenase-2 (ADH-2)/GAPDH hybrid promoter (15) in S. cerevisiae. Using this latter promoter, we were also able to produce the gag precursor polypeptide p53 in yeast. This could be detected by Western blotting techniques using AIDS sera or monoclonal antibodies to p25 (figure 2). The presence of several immunoreactive gag-related bands is presumably due to specific proteolytic cleavage by yeast enzymes. Further studies of this form of the gag gene product both as a diagnostic tool and as the natural substrate for the AIDS viral protease (see below) are currently in progress. Both the bacterial and yeast derived p25 proteins have been purified to homogeneity and shown to react with many AIDS sera on immunoblots (figure 2). However, all sera with antiviral antibodies were not necessarily reactive with p25. Using the E. coli derived recombinant gag protein in a sensitive ELISA assay, it was found that antibodies to p25 were only rarely detected in patients with frank AIDS (16%) (12). Antibodies to this protein were found to be much more prevalent in patients with AIDS related complex (ARC) (48%) or clinically healthy individuals with documented exposure to the virus through sexual contact with AIDS or ARC patients (71%). This finding extends a semi-quantitative study in which recombinant p25 related antigens from an HTLV-III like virus were described to detect all serum samples which had previously been characterized as seropositive in an ELISA for viral antibodies using disrupted whole virus (16). Our results suggest a potential use in predicting the prognosis or stage of the disease using this particular functional domain of the virus.

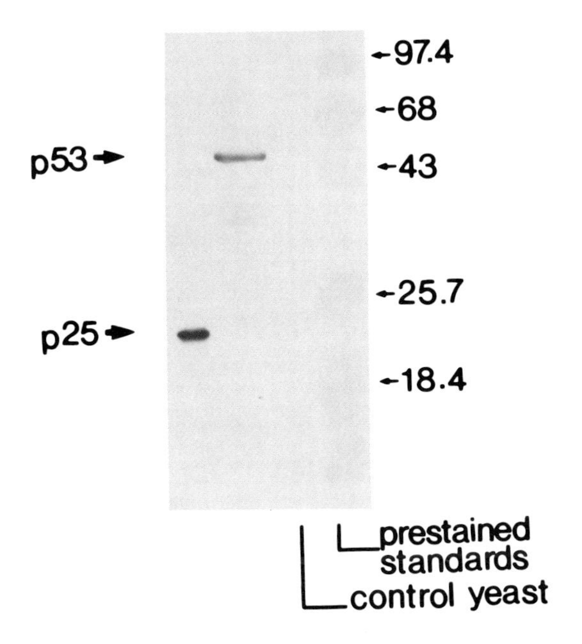

FIGURE 2. Western blot of yeast derived p25 and p53 using monoclonal antibody to p25.

The *pol* Gene.

The *pol* gene, encodes the enzymes of the AIDS retrovirus, in one separate open reading frame. These enzymes are presumed to be biosynthesized initially as a large *gag-pol* precursor polyprotein (17). Mechanisms for circumventing the shift in reading frame between the gag and pol genes of retroviruses have been proposed to involve either RNA splicing (18), or translational frameshifting. This latter mechanism has been shown to exist in the case of Rous sarcoma virus (RSV) (19), and most probably occurs with the AIDS virus (20). This mechanism of regulation of enzyme levels relative to structural protein levels poses several problems for the expression of each functional subunit in recombinant microorganisms. First, accurate localization of the N-terminus of the pol gene product can not be determined from the DNA sequence alone. Second, this method of biogenesis inevitably leads to a much lower level of proteins from the pol gene relative to proteins from the gag region. Accurate knowledge of both the N and C termini of each enzyme can be obtained only from protein sequencing.

Analysis of homologies of the AIDS retrovirus pol genes with those of other retroviruses allows the prediction of protein subunits of the gag-pol precursor. In addition to the gag proteins, one can also localize protease, reverse transcriptase and endonuclease (integrase) domains. N and C termini of each domain can be estimated from these homologies (21). Also, single cleavage sites can be predicted based on the substrate specificities of the protease activity itself, which is probably responsible for processing of its own precursor (22). Using these principles, we have attempted to express potentially active enzymes of the AIDS virus in both bacteria and yeast. Active recombinant enzymes may be useful in the screening of inhibitors as potential AIDS chemotherapeutics, in addition to their use as diagnostic reagents.

Protease. For expression of AIDS protease regions, we have utilized the yeast α-factor secretion system. We and others have previously demonstrated high level expression of fully active growth factors and lymphokines using this system (23, 24), and more recently this method of heterologous secretion has been extended to the protease carboxypeptidase B (25). In the absence of N and C terminal protein sequence for the ARV-2 protease we have used synthetic oligonucleotides to fuse regions of viral DNA predicted to code for this protein to the leader sequence of the yeast mating pheromone α-factor. Three such regions were designed to encompass protease molecules. These are designated prt-1, prt-2 and prt-3. We have previously used prt as a nomenclature to define the open (and closed) reading frame coding for the protease region of Simian AIDS-associated retrovirus-1 (SRV-1) a retrovirus causing immunodeficiency in rhesus monkeys (21). prt1 (amino acids 79 to 192 of the pol precursor) prt2 (aa 34 to 192) and prt3 (aa 34 to 163) (figure 3) were each secreted into the yeast media in varying yields. Activity of these polypeptides is currently being studied using the gag p53 precursor as a substrate.

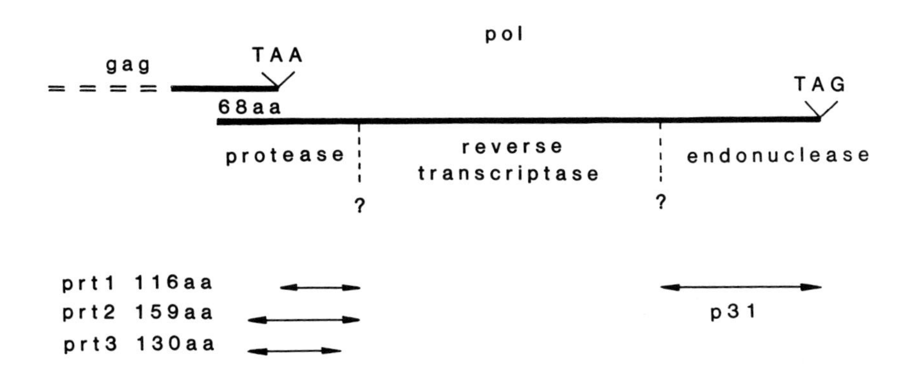

FIGURE 3. Schematic representation of putative protease regions expressed in yeast.

Reverse Transcriptase (RT). Several vectors for reverse transcriptase expression were constructed for both E. coli and S. cerevisiae. The most likely cleavage site defining the N-terminus of RT is between amino acids Phe 155 and Pro 156 of the pol precursor. Accordingly, using synthetic adaptors, we expressed an RT molecule from Pro 156 to an estimated C-terminus at Ala 693. Moderate levels of RT were produced in E. coli using the tac-1 promoter (13), and in yeast using the GAPDH promoter. This was shown to be active by a standard reverse transcriptase assay using polydT and polydG primed synthesis on polyA and polyC templates. We are currently attempting to optimize expression of RT in both bacteria and yeast to produce significant quantities of active enzyme for biochemical analysis.

Endonuclease (p31). Surprisingly, although presumably produced in equimolar quantities to protease and RT, the AIDS virus endonuclease region is highly immunogenic and detects antibodies in most infected individuals. Chang *et al.* (26), produced a *ca.* 15 kD protein corresponding to approximately the *C*-terminal half of this 31 kD protein. We have used a combination of site-directed mutagenesis and DNA synthesis to construct vectors for the expression of the majority of the protein (29 kD) in bacteria and yeast (27). Direct expression in each of these systems gave relatively low but immunoblot detectable yields. However, fusion of the endonuclease gene to that of human superoxide dismutase, an extremely stable protein in both bacteria and yeast (13), gave high levels of the hybrid protein (27). Using the SOD-p31 fusion protein produced in *E. coli* in an ELISA, we were able to accurately detect antibodies in 95% of patients seropositive in a whole virus ELISA (27). Unfortunately, some virus seropositive individuals clearly lacked antibody to p31, thus precluding the use of recombinant p31 derivatives as the sole antigen in a screening assay.

The *env* gene.

The envelope proteins of the AIDS virus are considered to be of extreme importance in both the diagnosis of disease and also in its possible prevention. The product of the *env* gene has been well characterized for HTLV-III at the protein level (28, 29). By analogy with other retroviruses, the highly glycosylated *N*-terminal domain (gp120) forms the exterior portion of the envelope protein complex. The *C*-terminal domain (gp41), containing two extremely hydrophobic transmembrane regions, serves as an anchor for gp120, (Figure 4).

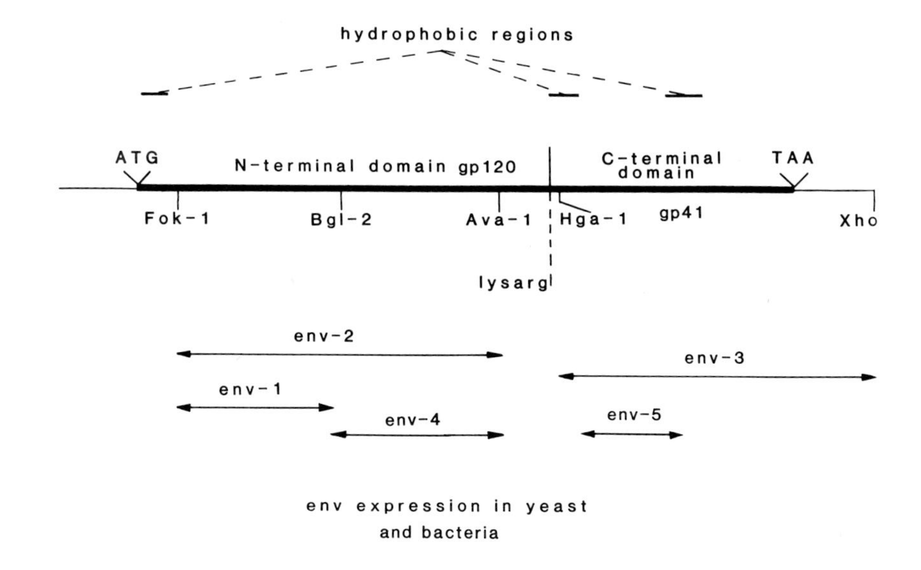

FIGURE 4. Schematic representation of AIDS virus envelope protein and regions expressed.

Using direct yeast expression systems with the glycolytic enzyme promoters for pyruvate kinase (pyk), GAPDH and the ADH-2/GAPDH hybrid promoter described above, we have expressed both domains of the AIDS virus (ARV-2) envelope protein in the yeast *S. cerevisiae*. Yeast systems have previously been shown to be useful for the production of viral coat proteins. For example, the *env* protein of the retrovirus associated with feline leukemia (FeLV) was expressed at high levels in yeast (30). More importantly, yeast derived hepatitis B surface antigen (HBsAg) is currently being used as an effective vaccine in humans (31).

Expression of a major portion (residues 26-490) of the envelope protein analogous to gp120 (env-2) was achieved in yeast using pyk or GAPDH promoters. This 53kD unglycosylated protein was readily detected on Western blots using sera from AIDS patients (Figure 5(a)). Use of these constitutive promoters for expression of the gp41 equivalent (env-3) was compromised by toxicity to the yeast cells. However, regulated expression from the ADH-2/GAPDH hybrid promoter (15) gave production of a major portion of gp41, detectable as several bands on immunoblots with AIDS sera (Figure 5(b)). That this heterogeneity was due to glycosylation of the yeast recombinant protein env-3 was demonstrated by endoglycosidase H treatment prior to immunoblotting (Figure 5(c)). We attribute this yeast glycosylation to the presence of most of the first hydrophobic membrane spanning sequence of gp41 in this particular construction. This hydrophobic region presumably targets env-3 to the yeast secretion/glycosylation pathway. However, as one might anticipate, env-3 was not detected in the yeast media.

Extensive serological analysis using these recombinant antigens (K.S. Steimer et al., in preparation) has shown that the of the vast majority of whole virus seropositive specimens have antibodies which react with both envelope domains. 98% of positive sera were detected with env-2 using Western blot assays. Similarly, 98% were detected with env-3.

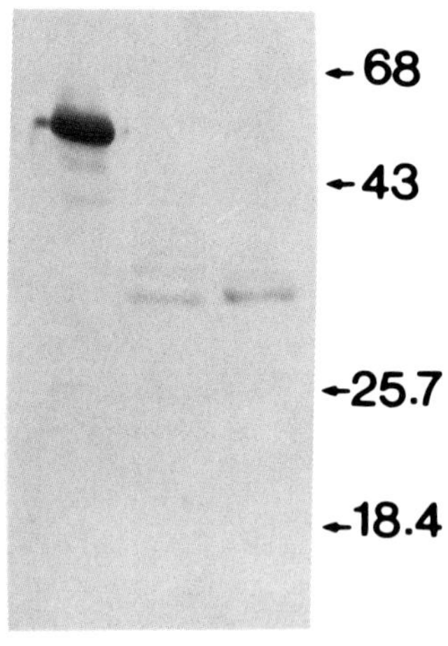

FIGURE 5. Western blot of AIDS env protein synthesized in yeast.

In order to improve expression yields from gp41 coding regions and also to further characterize smaller regions of gp120, we have made expression constructions corresponding to env-1, env-4 and env-5 (Figure 4). Env-1 was constitutively expressed at high levels in yeast, however env-4 and env-5 were best expressed as fusion proteins with β-galactosidase in *E. coli* (M. Truett *et al.*, unpublished) and with superoxide dismutase in *E. coli*.

Finally, recombinant env-2 was purified and used to raise antibodies in rabbits and rhesus monkeys. Hyperimmune sera from these two species reacted with the mature env glycoprotein, gp120. Several monoclonal antibodies reacting with viral gp120 resulted from fusion of spleen cells from mice immunized with env-2. Specificity of these antibodies was analyzed using env-1 and env-4 (J.C. Stephans *et al.*, this volume).

SUMMARY

We have shown the utility of genetic engineering techniques for the production of individual proteins and subunits of the AIDS virus ARV2. Recombinant products of each of the major viral genes (*gag*, *pol* and *env*) have been demonstrated to be effective for the detection of antibodies in AIDS sera and in some cases, for potential stage specific analysis of disease.

The availability of active recombinant enzymes from the AIDS virus may be an effective strategy for identifying or designing inhibitors as potential antiviral therapeutic agents. Detailed studies of the enzymatic mechanisms and structures of the viral protease and RT will be feasible with large quantities of these enzymes produced in genetically engineered microorganisms. The structural properties of precursor polypeptides such as gag p53, the substrate for the viral protease, can be analyzed with the view of both understanding the mechanism of the protease and designing substrate analogs for protease inhibition. This concept may also be extended to the less well characterized proteins of the virus encoded by the 5'- and 3'- open reading frames and the trans-activating factor. This however, will no doubt be secondary to genetic analyses of the function of these proteins through DNA transfection in various mammalian cells systems.

Finally, high level expression of domains of the env gene in yeast has provided material for the analysis of antiviral immune responses in heterologous hosts including mice, rabbits and rhesus monkeys. Strategies untilizing recombinant antigens will be essential for the development of an effective subunit vaccine for humans. Antigenic variation between known strains of the virus is most pronounced in the gp120 region of the virus. This is compounded by an even greater variation in distribution of potential N-linked glycosylation sites in this region. The relevance of this observation may be assessed by comparing the immunogenicities of non-glycosylated material from yeast and glycoslated protein from mammalian cells. To this end, we have also constructed expression vectors for the production of gp120 in several mammalian cell expression systems. Mammalian cell expression systems are expected to yield env glycoprotein that is modified (by proteolytic processing and glycosylation) in a fashion similar to that observed in cells infected with virus; the importance of modification and conformation can be evaluated by comparing recombinant envelope proteins derived from mammalian cells and yeast.

ACKNOWLEDGEMENTS

The scientific contributions of Ian Bathurst, Amy Dennis, John Puma and Jim Merryweather; and the help of Diana Wyles and Toni Jones in the preparation of the manuscript are gratefully acknowledged. We also thank our colleagues at Chiron, particularly Pablo Valenzuela, Dino Dina and Lacy Overby for many helpful discussions.

REFERENCES

1. Wain-Hobson S, Sonigo P, Danos O, Cole S, Alizon M (1985). Nucleotide sequence of the AIDS virus, LAV. Cell 40:9
2. Ratner L, Haseltine W, Patarca R, Livak K, Stavcich B, Josephs SF, Doran ER, Rafalski JA, Whitehom EA, Baumeister K, Ivanoff L, Petteway Jr. SR, Pearson ML, Lautenberger JA, Papas TS, Ghrayeb J, Chang NT, Gallo RC, Wong-Staal F (1985). Complete nucleotide sequence of the AIDS virus, HTLV-III. Nature 313:277
3. Sanchez-Pescador R, Power MD, Barr PJ, Steimer KS,

Stempien MM, Brown-Shime SL, Gee WW, Renard A, Randolph A, Levy JA, Dina D, Luciw P (1985). Nucleotide sequence nd expression of an AIDS-associated retrovirus (ARV-2). Science 227:484

4. Wong-Staal F, Gallo RC (1985). Human T-lymphotropic retroviruses. Nature 317:395
5. Luciw PA, Potter SJ, Steimer K, Dina D, Levy JA (1984). Molecular cloning of AIDS-associated retrovirus. Nature 312:760
6. Rabson AB, Martin MA (1985). Molecular organization of the AIDS retrovirus. Cell 40:477
7. Allan JS, Coligan JE, Lee T-H, McLane MF, Kanki PJ, Groopman JE, Essex M (1985). A new HTLV-III/LAV encoded antigen detected by antibodies from AIDS patients. Science 230:810
8. Rosen CA, Sodroski JG, Kettman R, Burny A, Haseltine W (1984). Trans activation of the Bovine Leukemia Virus long terminal repeat in BLV-infected cells. Science 227:320
9. Broome S, Gilbert W (1985). Rous Sarcoma Virus encodes a transcriptional activator. Cell 40:537
10. Kitchen LW, Barin F, Sullivan JL, McLane MF, Brettler DB, Levine PH, Essex M (1984). Aetiology of AIDS-antibodies to human T-cell leukaemia virus (type III) in haemophiliacs. Nature 312:367
11. Brun Vezinet F, Rouzioux C, Montagnier L, Chamaret S, Gruest J, Barré-Sinoussi F, Geroldi D, Chermann JC, McCormick J, Mitchell S, Piot P, Taelman H, Mirlanger KB, Mbendi N, Kayembe Kalambayi MP, Bridts C, Desmyter J, Feinsod FN, Quinn TC (1984). Prevalence of antibodies to lymphadenopathy-associated retrovirus in African patients with AIDS. Science 226:453
12. Steimer KS, Puma JP, Power MD, Powers MA, George-Nascimento C, Stephans JC, Levy JA, Sanchez-Pescador R, Luciw PA, Barr PJ, Hallewell RA (1986). Differential antibody responses of individuals infected with AIDS-associated retroviruses surveyed using the viral core antigen p25gag expressed in bacteria. Virology, in press
13. Hallewell RA, Masiarz FR, Najarian RC, Puma JP, Quiroga MR, Randolph A, Sanchez-Pescador R, Scandella CJ, Smith B, Steimer KS, Mullenbach GT (1985). Human Cu/Zn superoxide dismutase cDNA: isolation of clones synthesizing high levels of active or inactive enzyme from an expression library. Nucleic Acids Research 13:2017

14. Travis J, Owen M, George P, Carrell R, Rosenberg S, Hallewell RA, Barr PJ (1985). Isolation and properties of recombinant DNA produced variants of human α_1-proteinase inhibitor. J. Biol. Chem. 260:4384
15. Shuster JR (1986). Regulated expression of heterologous gene products in the yeast Saccharoymes cerevisiae. "Uniscience Series", CRC press.
16. Dowbenko DJ, Bell JR, Benton CV, Groopman JE, Nguyen H, Vetterlein D, Capon DJ, Lasky LA (1985). Bacterial expression of the acquired immunodeficiency syndrome retrovirus p24 gag protein and its use as a diagnostic reagent. Proc. Natl. Acad. Sci. USA. 82:7748
17. Robey WG, Safai B, Oroszlan S, Arthur LO, Gonda MA, Gallo RC, Fischinger PJ (1985). Characterization of envelope and core structural gene products of HTLV-III with sera from AIDS patients. Science 228:593
18. Schwartz DE, Tizard R, Gilbert W (1983). Nucleotide sequence of Rous Sarcoma Virus. J. Virol. 48:667
19. Jacks T, Varmus HE (1985). Expression of the Rous Sarcoma Virus pol gene by ribosomal frameshifting. Science 230:1237
20. Jacks T, Varmus HE, these authors. Unpublished observations.
21. Power MD, Marx PA, Bryant ML, Gardner MB, Barr PJ, Luciw PA (1986). The nucleotide sequence of a type D retrovirus, SRV-1, etiologically-linked with the simian acquired immunodeficiency syndrome. Science, in press.
22. Yoshinaka Y, Katoh I, Copeland TD, Oroszlan S (1985). Murine leukemia virus protease is encoded by the gag-pol gene and is synthesized through suppression of an amber termination codon. Proc. Natl. Acad. Sci. USA. 82:1618
23. Brake AJ, Merryweather JP, Coit DG, Heberlein UA, Masiarz FR, Mullenbach GT, Urdea MS, Valenzuela P, Barr PJ (1984). α-Factor-directed synthesis and secretion of mature foreign proteins in Saccharomyces cerevisiae. Proc. Natl. Acad. Sci. USA. 81:4642
24. Barr PJ, Bleackley RC, Brake AJ, Merryweather JP (1984). Yeast α-factor directed secretion of human interleukin-2 from a chemically synthesized gene. J. Cell Biochem. 8A:23
25. Gardell SJ, Craik CS, Hilvert D, Urdea MS, Rutter WJ (1985). Site-directed mutagenesis shows that tyrosine 248 of carboxypeptidase A does not play a crucial role in catalysis. Nature 317:551
26. Chang NT, Huang J, Ghrayeb J, McKinney S, Chanda PK,

Chang TW, Putney S, Sarnagadharan MG, Wong-Staal F, Gallo RC (1985). An HTLV-III peptide produced by recombinant DNA is immunoreactive with sera from patients AIDS. Nature 315:151

27. Steimer KS, Higgins KW, Powers MA, Stephan JC, Gyenes A, George-Nascimento C, Luciw PA, Barr PJ, Hallewell RA, Sanchez-Pescador R (1986). Recombinant polypeptide from the endonuclease region of the acquired immune deficiency syndrome retrovirus polymerase (pol) gene detects serum antibodies in most infected individuals. J. Virol., in press
28. Allan JS, Coligan JE, Barin F, McLane MF, Sodroski JG, Rosen CA, Haseltine WA, Lee TH, Essex M (1985). Major glycoprotein antigens that induce antibodies in AIDS patients are encoded by HTLV-III. Science 228:1091
29. Veronese FD, DeVico AL, Copeland TD, Oroszlan S, Gallo RC, Samgadharan MG (1985). Characterization of gp41 as the transmembrane protein coded by the HTLV-III/LAV envelope gene. Science 229:1402
30. Luciw PA, Parkes D, Van Nest G, Dina D, Gardner MB (1986). "Genetic Engineering of Animals: An Agricultural Perspective", New York: Plenum Press.
31. Scolnick EM, Mclean AA, West DJ, McAleer WJ, Miller WJ, Buynak EG (1984). Clinical evaluation in healthy adults of a Hepatitis B vaccine made by recombinant DNA. J. Am. Med. Assoc. 251:2812

Viruses and Human Cancer, pages 221–235

EXPRESSION OF THE HUMAN T-CELL LEUKEMIA VIRUS x GENE IN VITRO

Alan J. Cann, Joseph D. Rosenblatt, Neil P. Shah, Jan Williams, William Wachsman, and Irvin S.Y. Chen

Division of Hematology-Oncology, Department of Medicine, UCLA School of Medicine, Los Angeles, CA 90024

ABSTRACT Both known human T-cell leukemia viruses (HTLV) are associated with specific human T-cell malignancies. HTLV-I is the etiologic agent of adult T-cell leukemia (ATL), an aggressive leukemia-lymphoma endemic in several parts of the world (1). HLTV-II has been found associated with a rare T-cell variant of hairy-cell leukemia (2, unpublished data). Both viruses are able to transform T cells in culture, which then continue to proliferate without exogenous interleukin-2 (3). However, the precise mechanism of cellular transformation induced by HTLV is, at present, unknown. The relation between cellular transformation *in vitro* and leukemogenesis *in vivo* is also unclear, and despite the differences in the diseases associated with the two viruses, the properties and growth of the transformed cells *in vitro* appear to be similar for both viruses (3). However, the studies of T-cell transformation *in vitro* which have been performed are imprecise, largely because of the lack of a quantitative assay for HTLV-induced transformation. Partly for this reason, and partly because of the difficulties involved in growing and working with HTLV, we have chosen to investigate the transforming and other biological

This work was supported by NCI grants CA30388, CA32737, CA38597, CA09279, and CA16042; grants PF-2182 and JFRA-99 from the American Cancer Society; and grants from the California Institute for Cancer Research and California University-wide Task Force on AIDS.

properties of these viruses using molecular biological techniques. Information obtained from these molecular studies will then be applied to the problems of the mechanisms of cellular transformation and leukemogenesis.

Expression of the HTLV x Gene

HTLV and the related bovine leukemia virus (BLV) differ from other retroviruses in that they possess a unique fourth gene in addition to the usual retrovirus genome structure of 5'-gag-pol-env-3'. Although HTLV-I and -II share only approximately 65% nucleotide sequence homology overall, the region between the env gene and the 3' LTR which was originally identified by nucleic acid sequencing (4) has a much greater homology (82% amino acid sequence homology). This region of the HTLV genome, called X because of its unknown function, has been shown to encode a protein of 40 kd in HTLV-I and 37 kd in HTLV-II (5,6), variously referred to as the x, lor, or x-lor protein. These proteins have been shown to be located primarily in the nucleus of HTLV-infected cells, and have short half-lives of approximately 120 minutes (7). Both proteins are expressed with an unusual double-splicing mechanism (8,9). The three nucleotides encoding the env initiation codon plus one additional nucleotide are joined to the major X region open reading frame to form a mature mRNA with an open reading frame of 1125 nucleotides in HTLV-I (8) and 1017 nucleotides in HTLV-II (9). This mechanism of expression may afford the virus opportunities for closely regulating x gene expression, and thus, for control of its replication (see below).

Early studies on the function of this region of the HTLV genome suggested an involvement with transcriptional activation of the promoter in the virus long terminal repeat (LTR). Transfection of recombinant HTLV LTR-promoted chloramphenicol acetyl transferase (CAT) constructions demonstrated that much higher levels of CAT activity were produced in HTLV-infected cells than in non-infected cells (9), implying activation of transcription from the LTR in-*trans*. We have shown that this *trans*-regulatory function of the χ gene is an integral and essential part of the HTLV life cycle. Using an infectious, molecularly cloned HTLV-II genome, we have developed a unique transfection system. Transfection of HTLV-II DNA into an Epstein-Barr virus-transformed B-cell line, 729-6, gives rise to infectious HTLV-II (10). To examine χ gene function, HTLV-II subclones bearing large deletions in the χ gene were stably transfected into 729-6 cells (11). The resulting cell lines were assayed for virus transcription and for their ability to transform normal T lymphocytes in co-culture experiments. Cells containing χ gene-deleted proviruses produced at least 100-fold less HTLV RNA than cells containing the entire HTLV genome and did not transform T lymphocytes, indicating that no infectious virus was being produced. However, the defective virus could be rescued by infection with replication-competent HTLV-II, showing that wild-type virus could complement the χ-deleted provirus in-*trans*.

The kinetics of HTLV infection *in vitro* are unusual for a retrovirus. Unlike typical retrovirus infections where stable levels of RNA production merely require passage of the infected cell through S phase of the cell cycle, initial slow rates of HTLV transcription are probably due to low intracellular levels of χ protein (unpublished data). Maximal rates of viral RNA transcription are reached only after several months' cultivation following transfection of 729-6 cells with the infectious HTLV-II construction pH6-neo or after co-cultivation with lethally irradiated HTLV-infected cells. Transcription and χ protein gradually increase due to a positive feedback system as the infection progresses. This process probably has a bearing on the long latent period which seems to follow HTLV infection and the development of leukemia (12). However, it is also likely that multiple events in addition to virus replication are necessary for the generation of HTLV-associated neoplasia.

Although much valuable information was provided by the above studies, the lengthy time course of HTLV infections and the difficulties of quantifying HTLV infection have limited the application of this approach to studying the χ gene. The lack of an infectious HTLV-I clone has also precluded the extension of these studies to HTLV-I infections. Expression of the χ gene *in vitro* offers a much more rapid and convenient experimental system. In addition, the availability of the necessary molecular reagents for both HTLV-I and -II offers the possibility of accurate quantitative comparisons between the two viruses. We have therefore developed a number of systems for efficient *in vitro* expression of the χ gene of both HTLV-I and -II.

Construction of χ Gene Expression Vectors

The logical starting point for the development of expression systems are constructions in which transcription of the χ gene is promoted by its native promoter, the HTLV LTR. One such construct contains the entire infectious genome of HTLV-II (referred to above), pH6-neo (10). A similar construction, cH6-H11, differs from pH6-neo in the vector portion of the construction, pSVod (13) in place of pSV2-neo (14) (see Figure 1). Both constructions contain an infectious HTLV-II provirus plus some flanking cellular DNA. Transcription of the HTLV insert is promoted by the provirus LTR. A major problem with this type of construction is that the basal level of HTLV LTR transcription is extremely low unless stimulated by the χ protein (see earlier). Although the intracellular concentration of χ protein gradually rises during the course of a natural infection as discussed above, e.g., on stable transfection of pH6-neo into 729-6 cells, the low basal rate of transcription effectively prevents this type of construction from being of use in transient transfection experiments, where LTR function is assayed after only 48 hours. Moreover, the low activity of the LTR in fibroblast cells, especially the HTLV-II LTR, is an additional problem, since transfection experiments are more conveniently performed with adherent cells rather than non-adherent lymphoid cells.

To facilitate further analysis, we constructed vectors which express the χ gene at high efficiency after transfection into fibroblast cell lines. This in turn permitted the development of rapid assays for the χ protein by transient transfection of these constructions into any cell line in which the promoter is active. For

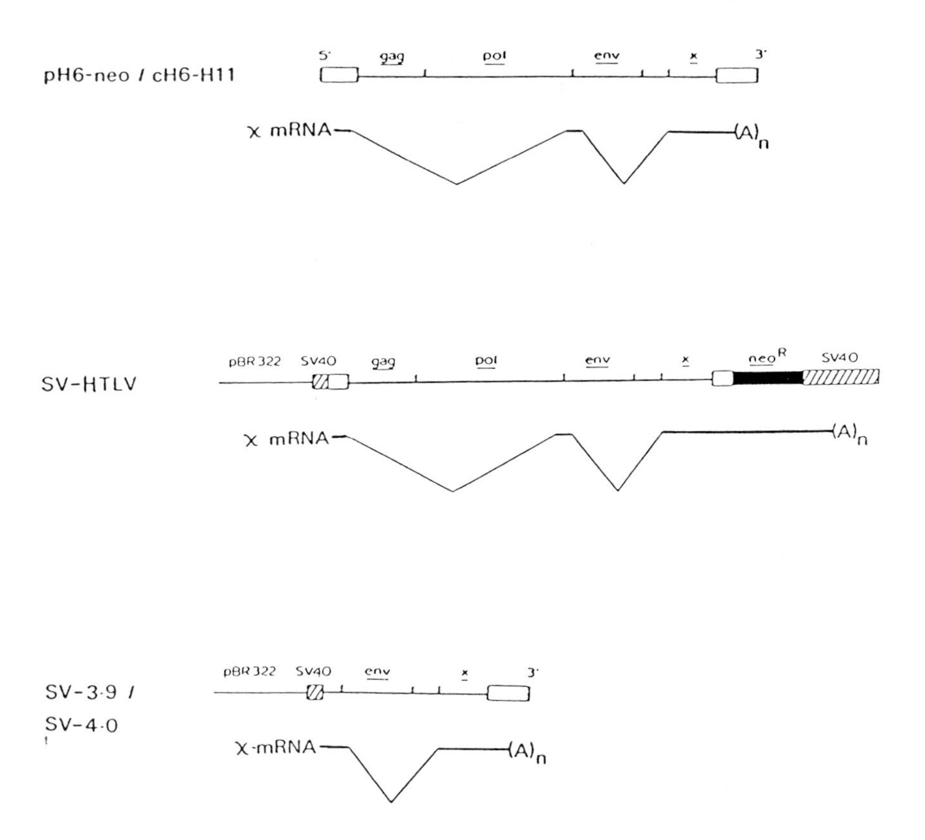

Figure 1. Schematic structure of recombinant χ gene constructions used in transfection experiments. Upper line of each section shows structure of plasmid; lower line, the structure of the (spliced) mRNA produced.

our initial studies, we used vectors which directed expression of the χ gene by SV-40 initiation-termination signals. The vector pSV2-neo (14) contains the strong SV-40 early promoter and upstream enhancer elements, neomycin resistance gene, which confers resistance to the synthetic aminoglycoside compound G418 and provides termination/polyadenylation signals from SV-40. A unique *Hind*III cloning site is located between the SV-40 early promoter and the *neo*^{R} gene. The first construction in this series, SV-HTLV (Figure 1), was constructed in pSV2-neo as follows. Partial *Bam*HI restriction endonuclease digestion of HTLV-II provirus DNA results in an 8.2 kbp *Bam*HI fragment bounded by sites in the repeat region of the 5' and 3' LTRs, removing the promoter elements located in the U3 region of the 5' LTR. This fragment was treated with the large fragment of *E. coli* DNA polymerase I to produce blunt ends. *Hind*III linkers were added, and the fragment cloned in the *Hind*III site of pSV2-neo. Transcription of the HTLV-II χ gene in this construction is promoted by the strong SV-40 promoter upstream of the HTLV-II insert and terminates in the termination-polyadenylation sequences downstream from the *neo*^{R} gene.

This construction was used to develop a transient co-transfection assay for χ gene function (15). Two DNA clones, one an χ gene expression construction and the other an LTR-CAT construction, are transfected into COS (or other) cells by the calcium phosphate precipitation method (16). Recombinant CAT constructions for both HTLV-I and -II LTRs were used in these studies (15). The cells are harvested 40-48 hours after transfection, and intracellular levels of CAT tested. This system is a rapid, sensitive and convenient procedure, and has the two major advantages of permitting analysis of large numbers of samples and also providing a quantitative assessment of transcription from the LTR. Transfection of SV-HTLV plasmid DNA into the COS fibroblast cell line results in activation of the SV-40 early promoter due to endogenous T-antigen in the COS cells (17). The χ protein is efficiently expressed by this construction, as shown by *trans*-activation of HTLV LTR-CAT constructions (Table 1), and makes a correctly spliced χ gene transcript (15). By sequential introduction of frameshift and deletion

TABLE 1. QUANTITATIVE ASSESSMENT OF χ GENE FUNCTION IN TRANSFECTED COS CELLS

Construction	LTR-I-CAT	LTR-II-CAT
pH6-neo/cH6-H11	<1	<1
SV-HTLV	78	77
SV-3.5/SV-4.0	~70	~70
SV-3.9	29	2
91023B-χI	40	<1
91023B-χII	100	56

Results of co-transfection of constructions introduced into COS cells as previously described (15). Relative χ gene activity measured by conversion of ^{14}C-chloramphenicol into acetylated forms. Results are expressed as a percentage of the conversion seen with 91023B-χII and LTR-I-CAT (highest values observed) and represent the average of several assays.

mutations into the _gag_, _pol_, and _env_ genes, we have shown that only mutations which affect expression of the χ gene (i.e., mutations affecting the _env_ methionine region or the X open reading frame) prevent _trans_-activation of the HTLV-LTR (15). One of these mutant genomes, SV-HTLV-EXB, has deletions of _gag_, _pol_, _env_, and the untranslated region, which total 5.1 kbp of the 8.2 kbp HTLV insert in SV-HTLV, yet this construction efficiently _trans_-activates both LTR-CAT constructions, indicating that the χ gene is solely responsible for _trans_-activation of transcription and that no other virus genes are required. Similar results have also been obtained by other investigators who expressed a fragment of the HTLV genome and assayed for χ gene function in co-transfection experiments (18,19). In the case of Rous sarcoma virus, a transcriptional activator locus has recently been identified in a different reading frame of the gag gene (20). It is not known whether the gag trans-activation locus is essential for the replication of Rous sarcoma virus. Our results demonstrate that there is no comparable locus in HTLV, as constructions which completely lack the gag gene still activate the LTR efficiently. These studies also demonstrate directly that the requirement for the χ gene for viral replication is due to the transcriptional regulatory function of the χ gene product on the LTR.

With knowledge of the sequences in the HTLV genome required for expression of the χ gene, we made a number of other constructions in the vector pSV2-neo which also express the χ gene. A 3.5 kbp BamHI fragment from the 3' end of the HTLV-II genome (positions 5090-8550) encodes the 3' terminal portion of the pol gene, env, the untranslated region of the genome, and the X open reading frame, but lacks the usual polyadenylation signals from the 3' LTR. This fragment was made blunt-ended by treatment with the large fragment of E. coli DNA polymerase I and ligated to similarly treated HindIII-digested pSV2-neo DNA. As with SV-HTLV, this construction, SV-3.5, uses the SV-40 early promoter and SV-40 termination/polyadenylation sequences of the pSV2-neo vector for expression of the χ gene. As expected, SV-3.5 is comparable to SV-HTLV in its ability to activate transcription from the HTLV LTR, since gag-pol sequences are not required for trans-activation and the χ gene would be expected to be expressed from SV-3.5 by a single spliced mRNA, compared with the double-spliced χ mRNA of SV-HTLV (see Table 1).

To investigate the effect of HTLV regulatory sequences on χ gene expression, a construct was made which utilizes the intact 3' LTR for termination/polyadenylation rather than the SV-40 sequences in pSV2-neo. This construct, SV-4.0, was made as follows. SV-HTLV was doubly digested with BamHI and EcoRI, and a 3086 bp fragment lacking the neomycin resistance and SV-40 termination sequences of the vector isolated. A 4.0 kbp fragment of HTLV-II from a BamHI site at genomic position 5090 to an EcoRI site in cellular DNA flanking the 3' LTR was ligated into the pSV2-neo vector fragment. A similar construct, SV-3.9, was made for HTLV-I as follows. A clone containing the 3' half of the HTLV-I genome plus some flanking DNA was digested with Bal 31 nuclease to remove the flanking DNA and an EcoRI linker added to the 3' LTR. A 3.9 kbp HindIII (position 4990) to EcoRI fragment was ligated to HindIII, EcoRI doubly-digested pSV2-neo DNA, producing an HTLV-I construct which expresses the χ gene in a comparable way to SV-4.0 (Figure 1).

Co-transfection of these various constructions with LTR-CAT DNAs into COS cells reveals several features of χ gene expression and function. All of the pSV2-neo-based constructions function in a variety of cells which do not contain SV-40 T-antigen, although in these cases, expression of the χ gene occurs at lower levels because of less transcription from the SV-40 promoter. Cells tested include fibroblast cell lines such as HeLa, CV-1, and NIH 3T3, as well as lymphoid cells such as 729-6 B cells and the T-cell line, Jurkat. The production of functional χ protein in these cell lines indicates that the χ gene, per se, is not responsible for determining the tissue tropism of HTLV infection. The χ protein does not appear to act directly on the HTLV LTR, and it is likely that it must interact with cellular factors to stimulate transcription. This possible interaction may be involved in the tissue tropism seen with these viruses, as many cellular transcription factors are likely altered or missing in different cell types or species (see below).

Altering the structure of the vector so that SV-40 termination/polyadenylation sequences 3' of the neo^R gene are used (e.g., SV-HTLV, SV-3.5) rather than the native sequences in the 3' LTR (e.g. SV-4.0, SV-3.9) (Figure 1) does not appear to affect the efficiency of χ gene expression. This is surprising, since SV-HTLV and SV-3.5 have a distance of approximately 2.0 kbp between the termination codon and polyadenylation site. However, this distance is also relatively long in the intact LTR when compared with most mRNAs. It has been proposed that a hairpin loop structure forms, bringing the two sites close together for polyadenylation, allowing efficient expression of the mRNA to occur (8,9). Possibly there is sufficient structural flexibility to enable a similar event to occur in SV-HTLV and SV-3.5, and this allows efficient expression of the mRNA.

We have used these constructions to examine the untranslated region of the HTLV genome, (542 nucleotides in HTLV-II) between env and the X open reading frame. Mutants of SV-HTLV bearing deletions of 200-300 bp at the 5' end of this region do not show impaired expression of the χ gene, as assayed by co-transfections and S_1 nuclease analysis (15). Similarly, deletion of a 300 base pair PstI fragment from the untranslated region of SV-HTLV does not affect the ability of these mutants to produce a fully functional χ protein. Therefore, it is likely that the untranslated region is not involved with χ gene expression with regard to either the quantity or accuracy of the mRNA produced. Although the double splicing mechanism used for χ gene expression does not seem to require sequences in the untranslated region, the conservation of this region argues strongly that it is functionally important, perhaps for the assembly or stability of mature virions, although these studies do not investigate these possible functions.

The HTLV-I construction SV-3.9 and its HTLV-II equivalent, SV-4.0, show a significant difference in specificity of χ gene function. As had been suggested by earlier studies in HTLV-infected cells (21), we observed that the HTLV-II χ protein produced by SV-4.0 efficiently activates both LTR-I-CAT and LTR-II-CAT DNAs when co-transfected into COS cells, but the HTLV-I χ protein produced by SV-3.9 only activates LTR-I-CAT and does not increase the basal rate of transcription from LTR-II-CAT DNA (15) (Table 1).

This observation led us to make constructions which make it possible to compare directly the function of the two χ proteins, without considerations such as possible differences in splicing which might introduce quantitative or qualitative differences between χ proteins. These recombinant χ gene constructions, 91023B-χI and -χII, contain only the native coding sequences of the HTLV-I and -II χ genes (i.e., do not express CAT-χ or metallothionein-χ fusion products (18,19)), and do not contain intervening sequences which must be spliced out to form a mature mRNA (submitted for publication). The χ proteins expressed by these recombinant constructions behave similarly to the proteins from the pSV2-neo constructions described above (Table 1).

Functional Similarity Between HTLV χ Protein and Other Transcriptional Regulatory Proteins

We have recently demonstrated that functional similarities exist between the HTLV χ gene and the transcriptional activator ElA gene of adenoviruses (22). The HTLV-II χ protein is able to activate transcription from the adenovirus ElA-dependent early promoters, E2 and E3, with comparable or greater efficiency than ElA itself, suggesting a possible common mechanism of action. Since there is no obvious amino acid sequence homology between the HTLV-I and -II χ proteins and ElA, this observation does not help to explain the actual mechanism of *trans*-activation. Similarly, the fact that there is little or no nucleotide sequence homology between promoters activated by χ and ElA does not illuminate the mechanism of action of the χ protein. Our observations on the different promoter specificities of the HTLV-I and -II χ proteins suggest at least two possible mechanisms of action. The χ proteins may act by inducing different modifications of a common cellular protein (co-factor) which acts directly by binding to the LTR. Alternatively, the HTLV-I and -II χ proteins may act through completely distinct mechanisms within the cell. For the adenovirus ElA gene product, it is known that specific upstream promoter elements are not essential for the activation of ElA-dependent genes, since large deletions can be made upstream of the transcription start site without a reduction in the amount of transcription observed (for a review, see ref. 23). For this reason, regulatory sequences specific for ElA activation have not been identified, and probably do not exist.

To examine the effect of sequences in the HTLV LTR which are specifically required for χ gene activation, we have taken the approach of chemically synthesizing DNA oligonucleotides corresponding to regions of interest in the U3 region of the LTR, upstream of the start site of transcription. These fragments have been inserted into plasmid vectors upstream of the CAT gene to examine their effect on transcription. We have centered our attention on the 21 base pair repeated sequences in the U3 region of the LTR (24). These studies are still in progress, but preliminary results indicate that these sequences do not function as strong enhancer elements, nor do they confer responsiveness to the χ protein on the recombinant CAT constructs. Although part of the reason for these findings may be the inability of these sequences to function in isolation from their correct locations within the LTR, the situation is also reminiscent of the lack of specific regulatory sequences in E1A-dependent promoters, once again suggesting a similar mechanism of action for the two proteins.

While the precise mechanism of χ gene activation remains unknown, the similar function of the χ gene and transcriptional regulatory proteins of DNA viruses implies that the HTLV χ gene induces malignant transformation of T cells by causing aberrant transcriptional regulation of specific cellular genes. This model of carcinogenesis may be applicable to transcriptional regulatory genes of DNA tumor viruses such as adenoviruses, herpesviruses, papova viruses, and papilloma viruses, as well as cellular oncogenes such as myc and myb (23). Our studies on the expression of the χ gene and its functional abilities have contributed towards an understanding of altered transcriptional regulation as a cause of oncogenesis.

ACKNOWLEDGMENTS

We thank J. Fujii and C. Nishikubo for technical assistance, and W. Aft for preparation of the manuscript.

REFERENCES

1. Gallo RC, Kalyanaraman VS, Sarngadharan MG, Sliski A, Vonderheid EC, Maeda M, Nakao Y, Yamada K, Ito Y, Gutensohn N, Murphy S, Bunn PA Jr, Catovsky D, Greaves MF, Blayney DW, Blattner W, Jarrett WFH, zur Hausen H, Seligmann M, Brouet JC, Haynes BF, Jegasothy BV, Jaffe E, Cossman J, Broder S, Fisher RI, Golde DW, Robert-Guroff M (1983). Association of the human type C retrovirus with a subset of adult T-cell cancers. Cancer Res 43:3892-3899.

2. Kalyanaraman VS, Sarngadharan MG, Robert-Guroff M, Miyoshi I, Blayney D, Golde D, Gallo RC (1982). A new subtype of human T-cell leukemia virus (HTLV-II) associated with a T-cell variant of hairy cell leukemia. Science 218:571-573.
3. Chen ISY, Quan SG, Golde DW (1983). Human T-cell leukemia virus type II transforms normal human lymphocytes. Proc Natl Acad Sci USA 80:7006-7009.
4. Seiki M, Hattori S, Hirayama Y, Yoshida M (1983). Human adult T-cell leukemia virus: complete nucleotide sequence of the provirus genome integrated in leukemia cell DNA. Proc Natl Acad Sci USA 80:3618-3622.
5. Slamon DJ, Press MF, Souza LM, Cline MJ, Golde DW, Gasson JC, Chen ISY (1985). Studies of the putative transforming protein of the type I human T-cell leukemia virus. Science 228:1427-1430.
6. Lee TH, Coligan JE, Sodroski JG, Haseltine WA, Salahuddin SZ, Wong-Staal F, Gallo RC, Essex M (1984). Antigens encoded by the 3'-terminal region of human T-cell leukemia virus: evidence for a functional gene. Science 226:57-61.
7. Slamon DJ, Shimotohno K, Cline MJ, Golde DW, Chen ISY (1984). Identification of the putative transforming protein of the human T-cell leukemia viruses HTLV-I and HTLV-II. Science 226:61-65.
8. Seiki M, Hikikoshi A, Taniquchi T, Yoshida M (1985). Expression of the pχ gene of HTLV-I: general splicing mechanism in the HTLV family. Science 228:1532-1535.
9. Wachsman W, Golde DW, Temple PA, Orr EC, Clark SC, Chen ISY (1985). HTLV χ gene product: requirement for the _env_ methionine initiation codon. Science 228:1534-1537.
10. Shimotohno K, Takahashi Y, Shimizu N, Gojobori T, Chen ISY, Golde DW, Miwa M, Sugimura T (1985). Complete nucleotide sequence of an infectious clone of human T-cell leukemia virus type II: a new open reading frame for the protease gene. Proc Natl Acad Sci USA 82:3101-3105.
11. Chen ISY, Slamon DJ, Rosenblatt JD, Shah NP, Quan SG, Wachsman W (1985). The χ gene is essential for HTLV replication. Science 229:54-58.

12. Kinoshita K, Amagasaki T, Ikeda S, Suzuyama J, Toriya K, Nishino K, Tagawa M, Ichimaru M, Kamihira S, Yamada Y, Momita S, Kusano M, Morikawa T, Fujita S, Ueda Y, Ito N, Yoshida M (1985). Preleukemic state of adult T-cell leukemia. Blood 66:120-127.
13. Mellon P, Parker V, Gluzman Y, Maniatis T (1981). Identification of DNA sequences required for transcription of the human α1-globin gene in a new SV40 host-vector system. Cell 27:279-288.
14. Southern PJ, Berg P (1982). Transformation of mammalian cells to antibiotic resistance with a bacterial gene under control of the SV40 early region promoter. J Mol Appl Genet 1:327-341.
15. Cann AJ, Rosenblatt JD, Wachsman W, Shah NP, Chen ISY (1985). Identification of the gene responsible for human T-cell leukemia virus transcriptional regulation. Nature (London) 318:571-574.
16. Chatis PA, Holland CA, Hartley JW, Rowe WP, Hopkins N (1983). Role for the 3' end of the genome in determining disease specificity of Friend and Moloney murine leukemic viruses. Proc Natl Acad Sci USA 80:4408-4411.
17. Gluzman Y (1981). SV-40 transformed simian cells support the replication of early SV-40 mutants. Cell 23:175-182.
18. Felber BK, Paskalis H, Kleinman-Ewing C, Wong-Staal F, Pavlakis GN (1985). The pX protein of HTLV-I is a transcriptional activator of its long terminal repeats. Science 229:675-679.
19. Sodroski J, Rosen C, Goh WC, Haseltine W (1985). A transcriptional activator protein encoded by the χ-lor region of the human T-cell leukemia virus. Science 228:1430-1434.
20. Broome S, Gilbert W (1985). Rous sarcoma virus encodes a transcriptional activator. Cell 40:537-546.
21. Sodroski JG, Rosen CA, Haseltine WA (1984). Trans-acting transcriptional activation of the long terminal repeat of human T lymphotropic viruses in infected cells. Science 225:381-385.
22. Chen ISY, Cann AJ, Shah NP, Gaynor RB (1985). Functional relationship of HTLV-II χ and adenovirus E1A proteins in transcriptional activation. Science 230:570-573.

23. Kingston RE, Baldwin AS, Sharp PA (1985). Transcription control by oncogenes. Cell 41:3-5.
24. Shimotohno K, Golde DW, Miwa M, Sugimura T, Chen ISY (1984). Nucleotide sequence analysis of the long terminal repeat of human T-cell leukemia virus type II. Proc Natl Acad Sci USA 81:1079-1083.

Viruses and Human Cancer, pages 237–256

EXPRESSION OF HTLV-III GENES USING MAMMALIAN CELL EXPRESSION VECTORS

Nancy T. Chang, Chip Shearman[1] and Ruey Liou[2]

Center for Biotechnology
Baylor College of Medicine
Houston, Texas 77030

ABSTRACT We have developed two systems for producing HTLV-III proteins in cultured mammalian cells. In one approach, subgenomic segments of HTLV-III have been inserted into a eukaryotic expression vector containing the bovine papilloma virus (BPV) genome and the neomycin (NEO) resistance gene. Mouse fibroblast cell line C127 transformed by these plasmids produced substantial amounts of HTLV-III peptides as demonstrated by Western immunoblot analyses using sera of patients with AIDS or AID-related complex (ARC). Transformants containing the *gag* gene of HTLV-III produce immunoreactive proteins with molecular weights of 55 kd, 48 kd, 44 kd, 41 kd and 39 kd. Transformants containing the entire HTLV-III coding region produce immuno-reactive proteins with molecular weights the same as the *gag* transformants as well as those of 12 kd, 17 kd, 24 kd, 65 kd and 110 kd. Competition studies with *E. coli* extracts expressing the *gag* gene confirmed the complex pattern of the *gag* related proteins. The amount of HTLV-III peptides expressed in mouse C127 cells has been estimated to be on the order of 5 mg per 10^8 cells. In a separate approach, we have inserted a 5.4 kb HTLV-III DNA encoding the *gag*-*pol* region into a pSV-dhfr vector. This plasmid was then transfected into dhfr-deficient Chinese hamster ovary (CHO) cells.

[1]Present address: Centocor, Malvern, Pennsylvania 19355
[2]Present address: Biotech Research Laboratory, Inc. Rockville, Maryland 20850

The expression of HTLV-III gag and pol proteins in the dhfr+ cells were also analyzed by Western blot techniques using sera from AIDS patients. A polypeptide of molecular weight around 53 kd was detected in HTLV-III DNA transfected dhfr+ cells. Further analysis of using mouse monoclonal antibodies specific for gag or pol proteins showed that this polypeptide is likely to be the gag precursor protein. The lack of any expression of pol related peptides in the HTLV-III-dhfr DNA transfected CHO cell suggested that the ribosomal frame shift mechanism which allow retroviral pol gene to be translated is probably not operative in CHO cells. This cell line and the expressed 55 kd gag precursor peptide may provide an ideal model system for the analysis of HTLV-III encoded protease function.

INTRODUCTION

Human T cell lymphotrophic virus type III (HTLV-III) also referred to as lymphadenopathy-associated virus (LAV) or AIDS related virus (ARV) is a human cytopathic exogenous retrovirus which has been identified as the etiological agent of acquired immune deficiency syndrome (AIDS) (1-8). Since its discovery, the structure and the biology of HTLV-III has been the subject of intensive investigation of many research laboratories around the world. More than 100 viral strains have been isolated (9). The proviral genome of several virus isolates have been molecularly cloned and sequenced (10, 11, 12, 13). Many viral structure proteins which include the major envelope proteins gp120 and gp41 (14, 15) and the ploymerase (16) the gag protein p24 (13), etc., have been isolated and characterized and expressed in *E. coli* (17-22). A picture emerging from these analyses in that HTLV-III constitute a family of viruses whose genome can vary up to 10% while the most divergent part lies within the major exterior glycoprotein of the env gene (8, 23, 24). Although it is not clear that the genomic drift in the env gene was due to immune selection, it dampens somewhat the hope an effective vaccine can be developed for the prevention of the devastating AIDS and HTLV-III infection. Recently, it has been demonstrated that patients with AIDS or ARC and healthy HTLV-III infected individuals do contain neutralizing antibodies that can prevent virus infection *in vitro* on susceptible culture cells (25, 26). Moreover, the titer of these neutralizing antibodies present in patients from diverse geographic areas were shown to be similar despite the genomic variations present among the various virus isolates (26). These findings have alleviated some of the concerns. It has been proposed that a constant region of the HTLV-III envelope antigen is likely to be the target for the neutralizing antibodies (8).

One of the modern approaches toward the development of an effective vaccine against HTLV-III infection is to immunize with viral surface antigens that can elicit neutralizing antibodies. This approach required that large quantities of viral surface protein be produced and that the antigenic site that elicit neutralizing antibodies is preserved.

Since the HTLV-III viral envelope protein is heavily glycosylated and since the carbohydrate moieties and other possible modifications of the envelope protein may play an integral role in the induction of neutralizing antibodies against HTLV-III, we are studying the expression of HTLV-III proteins in mammalian cells. Our goal is to develop an efficient expression system that would stably produce HTLV-III peptides at high levels. For this purpose we have inserted several HTLV-III proviral DNA segments into two mammalian cell expression vectors which have been used successfully in producing large quantities viral glycoproteins (27, 28, 29). In one system, the cloning vector contains a recombinant bovine papilloma virus in which the HTLV-III DNA are inserted 3' to the mouse metallothioein-I (Mt-I) promoter. This system has been used for high level production of human growth hormone and hepatitis B virus surface antigen (27, 29). Using this system we report here the expression of high levels, 1 to 5 mg HTLV-III specific proteins per 10^8 cells. In the second system, the cloning vehicle contains the murine dihydrofolate reductase (dhfr) gene under the transcriptional control of a SV40 early promoter. The HTLV-III DNA was inserted 3' to a second SV40 early promoter. This system has been used to produce in large quantities of the gD glycoprotein of Herpes Simplex Virus (28). Here we report the expression of an unprocessed 53 kd *gag* precusor protein in the transfected dhfr-deficient CHO cells.

RESULT

Expression of HTLV-III Envelope Proteins in E. coli.

Before we attempt to study the synthesis of HTLV-III protein in mammalian culture cells we have constructed many E. coli expression clones which produce various HTLV-III envelope peptides in E. coli. The 2.7 kb env gene fragment of HTLV-III used in these constructions was isolated from λBH-10 (a recombinant phage containing a 9.0 kb segment of HTLV-III DNA) by KpnI digestion (30). We took advantage of the KpnI StuI, BglII, HindIII and BamHI restriction sites spanning the env gene encoding the envelope prescursor protein of 856 amino acids. Various restriction fragments generated from the HTLV-III env gene were inserted into two high level expression vectors ompA (31) or REV (32) (Figure 1). The construction was made in such a way that the

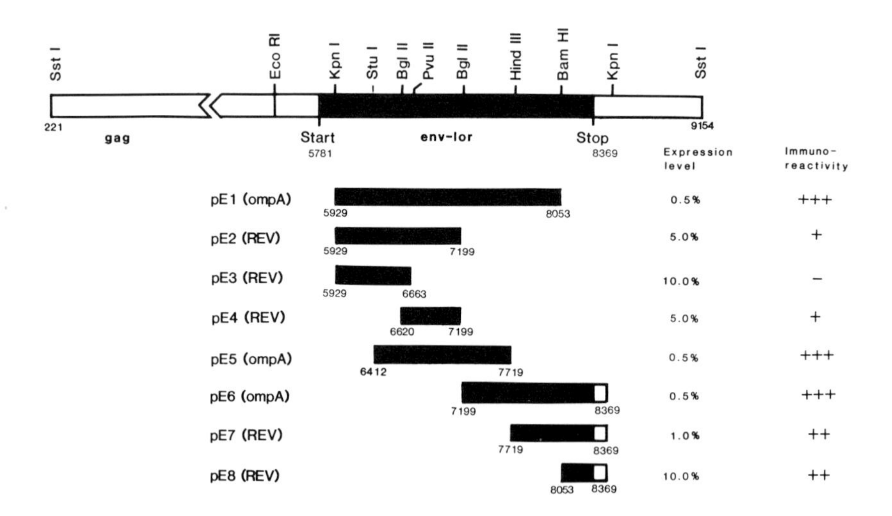

FIGURE 1. Summary of the recombinant HTLV-III env peptides produced in E. coli.

reading frame of the inserted HTLV-III env fragment was in phase with the leader peptide encoded by the vector. The env peptides produced by the recombinant E. coli clones were analyzed with sodium dodecyl sulfate polyacrylamide gel electrophoresis (SDS-PAGE). The immunoreactivity of the expressed env peptides were determined by Western blot technique using a panel of sera from patients of AIDS, ARC or seropositive healthy homosexual individuals. We were surprised to learn that the recombinant env peptides derived from the major exterior glycoprotein portion of the env gene (pENV 2, 3, 4) were not reactive or only weakly reactive with most sera from the AIDS, ARC or healthy homosexuals. In contrast, recombinant HTLV-III env peptide derived from clones containing gene segments encoding partial or entire membrane envelope protein gp41 reacted strongly with the antibodies present in the test sera. It is also apparent that the most immune dominant region of the gp41 protein residing at the N-terminus between the processed site to the HindIII site located at position 7718. This is consistent with our earlier results in identifying HTLV-III encoded peptides that reacted immunologically strongly with antibodies in sera from AIDS patients employing open reading frame cloning and expression strategies (17). Similar finding has also been obtained from Petteway et.al. (33). The lack of immunoreactivity of the E. coli produced env peptides derived from gP120 env region could be due to more than one of the following reasons. It is known that in most cases the heterologous proteins synthesized by the recombinant E. coli were in denatured forms. The expressed protein aggregated and formed inclusion bodies in the bacteria. These inclusion bodies segregated with the membrane fraction when the cells were broken. Moreover, E. coli are not capable of forming disulfide linkages which in most cases play an integral role in maintaining the structural stability and functional activity of the native proteins. In addition, the E. coli produced env antigens are devoid of any glycosylations which may influence the immunoreactivity of the HTLV-III proteins in mammalian cell systems.

Construction of pNM9-BPV-HTLV Plasmids and Transfer into Mouse Cells

The construction of the BPV expression vector was prepared by inserting the BamHI-SalI fragments, containing the entire BPV genome, from NSI-8 into BamHI-SalI cleaved pNM9 (Figure 2). pNM9-BPV consists of three parts: (1) a portion of pML2 (a "poison-minus" derivative of pBR322) containing the origin of replication and amp^r gene, (2) the complete BPV-1 genome linearized at the BamHI site, and (3) a transcriptional cassette composed of the mouse Mt-I

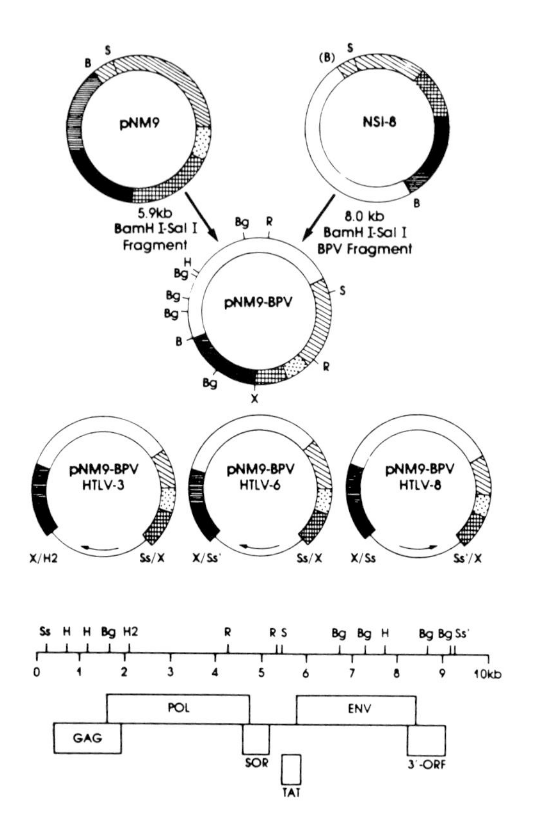

FIGURE 2. Construction BPV-metallothionein - HTLV-III recombinant plasmids.

promoter linked to the murine sarcoma virus (MSV) enhancer sequence, the neo^r gene of Tn5 and the SV40 early-region transcriptional processing signals. The unique XhoI site downstream from the Mt-I promoter was used to insert HTVL-III sequences. Two fragments of HTLV-III used in these studies were isolated from λBH-10 (30). The 9.0 kb HTLV-III fragment was released from the vector by digestion BH-10 with SstI. This fragment was end-repaired and ligated into pNM9-BPV at the XhoI site. Clones in both orientations relative to the Mt-I promoter were isolated and are designated pNM9-BPV-HTLV-6 for the "+" orientation and pNM9-BPV-HTLV-8 for the "-" orientation. The other HTLV-III segment used was the 1.86 kb SstI-HindII (between nucelotide position

221 and 2079 on HTLV-III genome) fragment which contains only the gag gene. The fragment was inserted into pNM9-BPV as above in the "+" orientation and is designated pNM9-BPV-HTLV-3 (Figure 2). The pNM9-BPV clones were used to tranfect mouse C-127 cells using the DNA-calcium phosphate precipitation technique (35). Transformants were selected for the resistance to G-418, a neomycin derivative. The transfer of G-418 resistance was relatively efficient averaging 5000 transformants per pmol DNA per 10^6 cells. Several G-418 resistance colonies were isolated and subcloned for further study. All clones exhibited a transformed morphology indicative of BPV transformation.

State of the BPV-HTLV-III DNA in Transformed Cells

Southern blot analysis was performed on representative clones transformed with pNM9-BPV, pNM9-BPV-HTLV-3 and pNM9-BPV-HTLV-6. In the absence of restriction endonuclease digestion, DNA homologous to the input plasmid do not associate with the high-molecular-weight cellular DNA suggesting that the plasmids were maintained extrachromosomally. By comparing the band intensities of various hybridized band with the known standard we estimated most clones analyzed contained low copies of the input plasmid DNA averaging around ten copies per cell. To further test this result, circular DNA was extracted from the cell by Hirts extraction procedure (36) and subjected to Southern blot analysis (data not shown). Using HTLV-III DNA as probe, no hybridizable DNA is detected with extracts from C-127 control cells or pNM9-BPV transformants. Extracts from pNM9-BPV-HTLV-3 and pNM9-BPV-HTLV-6 on the other hand show prominent bands with intensities corresponding to copy numbers of approximately 10 and 20 per cell respectively. There appears to be no gross rearrangements or deletions within the plasmids. Restriction enzyme patterns with six restriction enzymes, SstI, SmaI, SalI, EcoRI, HindIII and BglII, are identical to the restriction maps of the original plasmids.

Transcription of the HTLV-III genes

RNA blotting experiments indicated that the transfected HTLV-III genes were transcribed in mouse cells. Total cytoplasmic RNAs were purified from pNM9-BPV-HTLV-3 and pNM9-BPV-HTLV-6 transformed cell lines as well as from control

C127 cells and pNM9-BPV transformed cells. After electrophoresis and transfer to nitrocellulose filters, HTLV-III RNAs were detected by hybridization to 32p-labeled HTLV-III DNA probes. Two probes were used, one contains the entire

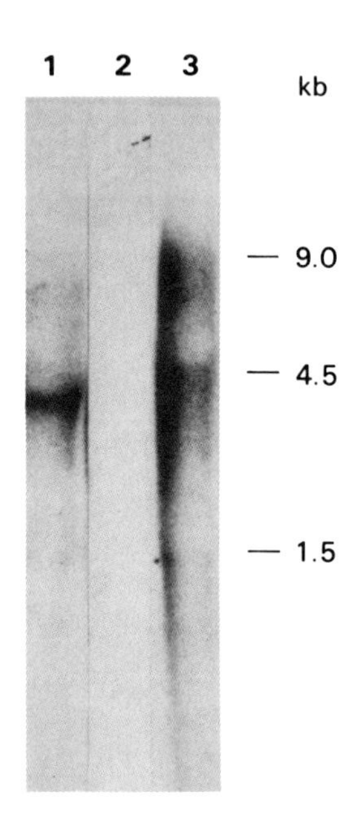

FIGURE 3. Analysis of HTLV-III RNA in pNM9-BPV-HTLV-3 and pNM9-BPV-HTLV-6 transformed cell lines. Cytoplasmic RNA was isolated from the transformants, resolved by agarose gel electrophoresis, blotted onto nitrocellulose and probed for HTLV-III specific sequences using nick-translated HTLV-III DNA (9.0 kb SstI-SstI fragment from λBH-10). The apparent size (kb) of each band was calculated from internal standards, 18S and 28S ribosomal RNAs. The samples in each lane are as follows: 1, pNM9-BPV-HTLV-3; 2, pNM9-BPV; and 3, pNM9-HTLV-6.

HTLV-III genome (figure 3) and the other contains the gag specific sequences (data not shown). No HTLV-III transcripts were detected in pNM9-BPV transformants or in control C-127 cells. pNM9-BPV-HTLV-3 transformants produce a 3.2 kb transcript and a minor 1.5 kb transcript corresponding most likely to a gag-neo precursor transcript and a processed gag transcript. The pNM9-BPV-HTLV-6 transformants synthesize three transcripts of 9.0 kb, 4.5 kb and 1.5 kb. Similar sized transcripts have been observed in cells infected with HTLV-III (11).

HTLV-III Peptides Production in Transformed Mouse Cells

The expression of HTLV-III peptides was determined by Western blot analysis. Total cytoplasmic protein from the G-418 resistant transformants were resolved by SDS-PAGE and electroblotted onto nitrocellulose filters. The immunoreactivities of the expressed HTLV-III peptides were analyzed using sera from patients with AIDS or ARC (Figure 4).

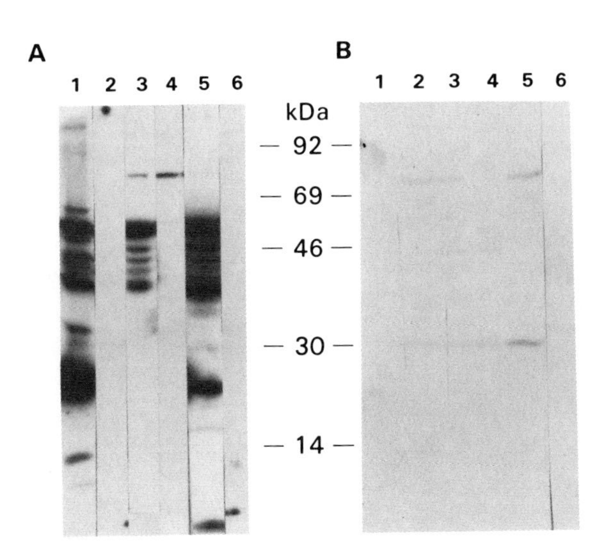

FIGURE 4. Expression of HTLV-III peptides in transformed cells. Cytoplasmic protein lysates were prepared and analyzed by Western immunoblot technique as described (17). Samples in panel A were tested with pooled sera from AIDS patients and include: 1, HTLV-III virus; 2, C-127 cells; 3, pNM9-BPV-HTLV-3; 4, pNM9-BPV-HTLV-8; 5, pNM9-BPV-HTLV-6; and 6, pNM9-BPV. Samples in panel B were tested with a normal control serum and include: 1, HTLV-III virus; 2, pNM9-BPV-HTLV-6; 3, pNM9-BPV-HTLV-3; 4, pNM9-BPV; 5, pNM9-BPV-HTLV-8; and 6, C-127 cells.

Several immunoreactive proteins are observed in pNM9-PBV-HTLV-3 transformants with apparent molecular weights of 55, 48, 44, 41 and 39 kd. Transformants containing the entire HTLV-III coding sequence (pNM9-BPV-HTLV-6) produced immunoreactive proteins with apparent molecular weights the same as those produced by the gag transformants as well as proteins with molecular weights of 12, 17, 24, 30, 65 and 110 kd. Proteins with identical molecular weights are observed in purified HTLV-III preparations (Figure 4A). Lysates from control C127 cells, pNM9-BPV transformants and a pNM9-BPV-HTLV-8 (minus orientation) transformants do not produce proteins that are reactive with the test sera. The immunoreactive bands are HTLV-III specific since they cannot be detected using sera from normal individuals (Figure 4B).

The gag region of HTLV-III encodes the internal structural proteins of virion. In HTLV-III infected cell a 55 kd precursor polypeptide is synthesized and subsequently processed into mature 12, 17 and 24 kd gag proteins. To more clearly assign the 48, 44, 41 and 39 kd proteins to the gag gene, lysates were tested with AIDS serum deficient in gag P24 reactivity (Figure 5A) and subjected to competition studies using E. coli lysates containing the gag 55 kd protein (Figure 5B) (21). Immunoblot analysis using serum deficient in anti P24 gag reactivity shows no reactivity with the proteins (48, 44, 41 and 39 kd) produced by the transformants as compared to the usual AIDS serum. Competition with E. coli produced p55 also eliminates reactivity with the proteins produced by the transformants. Control E. coli lysate has no effect on the immunoreactivity. Clearly, the 48, 44, 41 and 39 kd proteins are gag encoded and may represent processing intermediates of the gag precursor protein in C127 cells. The additional proteins produced by the pNM9-BPV-HTLV-6 transformants most likely correspond to the mature gag proteins (12 kd, 17 kd, 24 kd), the exterior env presursor (120 kd) and the reverse transcriptase (p65). The origin of the 30 kd protein is unclear. It may be encoded by the SOR or 3'-orf region of HTLV-III or it may be a nuclease encoded in the HTLV-III or even be of cellular origin. It is of interest to note the presence of gag related proteins whose sizes comparable to the mature gag proteins in clones containing the entire HTLV-III coding regions while transformants with only the gag gene appear to lack them. This may indicate the presence of a virally encoded protease necessary for gag protein processing. The amount of HTLV-III peptides produced by the transformants was estimated by comparing the immunoreactivity of the

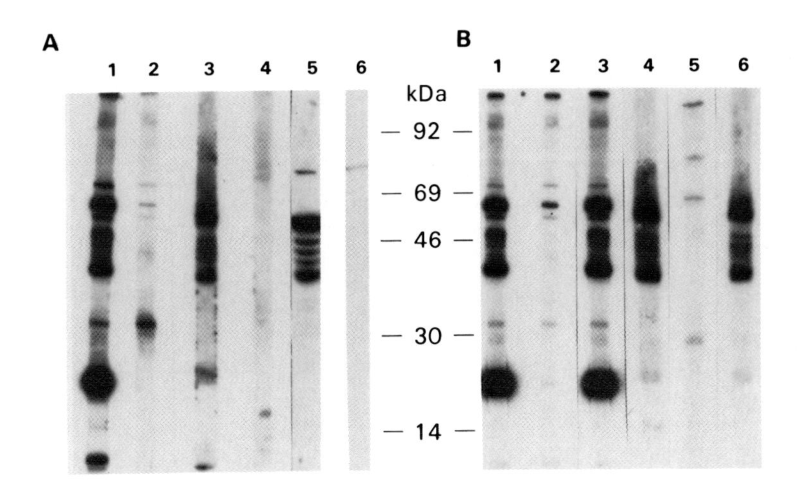

FIGURE 5. Reactivity of HTLV-III peptides with anti-gag deficient AIDS serum. Cytoplasmic protein lysates were prepared and analyzed by Western immunoblot technique. Samples in panel A were tested with either pooled sera from AIDS patients (lanes 1, 3, 5) or serum from an AIDS patient which was deficient in antibodies to gag sepcific proteins (lanes 2, 4, 6) and include: 1 and 2, disrupted HTLV-III; 3 and 4, pNM9-BPV-HTLV-6; and 5 and 6, pNM9-BPV-HTLV-3. Samples in panel B were tested with pooled sera from AIDS patients either alone (lanes 1 and 4) or in the presence of an E. coli lysate expressing (lanes 2 and 5) or not expressing (lanes 3 and 6) the gag gene and include: 1, 2, 3, HTLV-III; and 4, 5, 6, pNM9-BPV-HTLV-6.

transformant produced peptides to the immunoreactivity of a purified, recombinant HTLV-III 55 kd peptide encoded by the gag gene. Graded amounts of the peptide were separated by SDS-PAGE and analyzed by the Western immunoblot technique. Intensities of the immunoreactive bands were subsequently quantitated using a soft laser densitometer. For both pNM9-BPV-HTLV-3 and pNM-BPV-HTLV-6, expression levels ranged from 1 to 5 mg per 10^8 cells. The Mt-I promoter can be induced by treatment with heavy metals such as cadmium (29). Growth of the transformants in the presence of cadmium enhances the expression of HTLV-III proteins approximately 3-5 fold (data not shown).

Expression of HTLV-III *gag* Protein in Chinese Hamster Ovary Cells

In a separate approach we have inserted a 5.4 kb HTLV-III *gag-pol* region DNA fragment released by SstI cleavage from a biologically active HTLV-III proviral clone HXB-2 (30) into another mammalian cell vector, pSV-dhfr. Figure 6 shows a diagram of pSV·GP·dhfr. This plasmid contained the following features: (1) The 5.4 kb *gag-pol* gene of HTLV-III under the transcriptional control of SV40 early promoter and utilizing the 3' termination and polyadenylation signals of the SV40 early region genes. (2) A cDNA encoding the murine dhfr gene under the transcriptional control of a second SV40 early promoter that allows for the selection of dhfr-deficient cell lines, and (3) a bacterial vector containing the ampicillin resistant gene and *E. coli* DNA replication origin derived from the plasmid pBR322. pSV·Gp·dhfr was used to transfect dhfr-deficient CHO cells using calcium phosphate madiated gene transfer techniques (35). Cell lines that can grow in medium lacking hypoxanthine, glycine and thymidine were selected. The expression of HTLV-III *gag* or *pol* proteins in the selected cells were studied by SDS-PAGE and Western blot techniques as described above using sera from

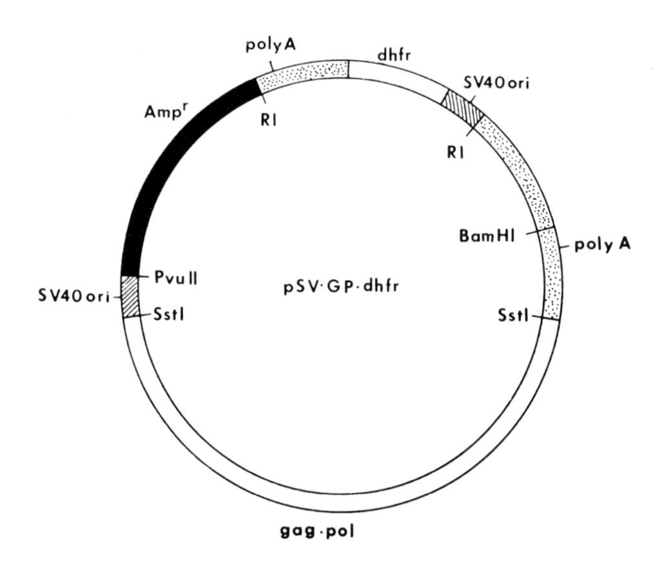

FIGURE 6. Diagram of pSV·Gp·dhfr.

AIDS patients. Figure 7 illustrates the result of the Western blot analysis. A single band of molecular weight approximately 53 kd was detected in cells transformed with pSV·Gp·dhfr using serum from AIDS patients. Control CHO cells or cells transformed by PSV·dhfr do not display this immunoreactive band. The immunoreactive 53 kd protein is specific for HTLV-III since it cannot be detected using serum from normal individuals. To further assign this 53 kd polypeptide to the _gag_ gene, protein from pSV·Gp·dhfr transformed cells were tested by Western Blot analyses using rabbit polyclonal or mouse monoclonal antibodies specific for the purified HTLV-III viral _gag_ P24 protein (data not shown). In both cases the P24 _gag_ specific antibody reacted with the 53 kd protein in CHO cells transformed with pSV·Gp·dhfr and not in normal CHO cell or cells transformed with the control plasmid pSV·dhfr. Immunofluorescence staining of actone/methanol fixed cells using antibodies specific for p24 _gag_ protein also confirmed the results and revealed the accumulation of _gag_ related 53 kd protein in the transformed cells.

DISCUSSION

The development of reagents useful in the treatment and prevention of AIDS and ARC are actively being pursued. One of our approaches is to use recombinant HTLV-III envelope peptides as immunogen to elicit a neutralizing antibody response.

We have cloned and expressed various HTLV-III _env_ peptides in _E. coli_. Significant quantities of recombinant _env_ antigens have been produced and purified. We were disappointed to learn that the _E. coli_ produced _env_ antigens derived from the _env_ gene encoding the major exterior envelope glycoprotein gp120 were not reactive with sera from most patients with AIDS or ARC. On the contrary the native HTLV-III _env_ protein gp120 is the major target antigen for serum antibodies in AIDS patients (37).

Because modifications such as carbohydrate and phosphate moieties on the _env_ protein may be important in elicing antibody production and protective immunity, we have concentrated our efforts in producing HTLV-III peptides in eukaryotic cells. The eukaryotic expression vectors, pNM9-BPV and pSV·dhfr, were chosen for several reasons. Firstly, they have been used successfully and reliably in a number of systems (27-29). Secondly, the vectors encodes the neomycin resistence gene or dhfr gene which can be used as a dominant

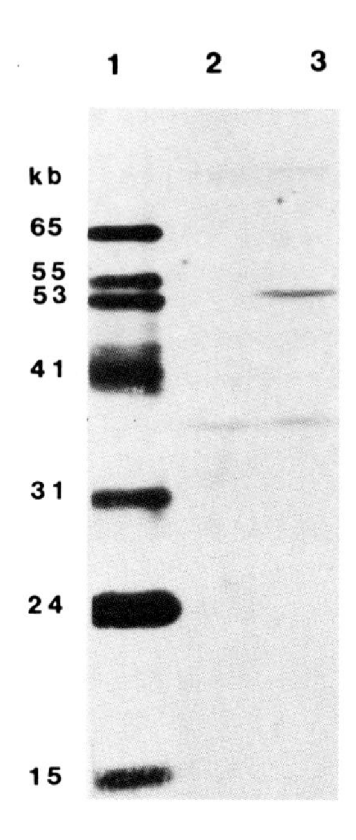

FIGURE 7. Expression of HTLV-III peptide in pSV·Gp·dhfr transfected CHO cells. Cytoplasmic protein lysates were prepared and analyzed by SDS-PAGE and Western immunoblot technique as described (17). Samples were tested with pooled sera from AID patients and included: 1, HTLV-III virus; 2, CHO cells transfected by pSV·dhfr; 3, CHO cells transfected by pSV·Gp·dhfr.

selection marker in eukaryotic cells. Moreover, the presence of the bovine papilloma virus genome in pNM9-BPV allow transformation of recipient cells for enhanced growth characteristics. Since BPV is maintained in an extrachromosomal state, linked foreign genes remain intact. Also, the extrachromosomal state of BPV-HTLV-III plasmids provide a homogenous sequence environment for each copy of the plasmid. These plasmids can be readily rescued from the mammalian cells.

Expression of the HTLV-III *gag* was initially chosen for the model system for several reasons. 1) HTLV-III viral preparations contain an abundance of *gag* proteins as a source of comparison. 2) Patient antisera deficient in reactivity to *gag* protein were readily available. 3) The *gag* gene has been recently expressed in *E. coli* and the purified protein was available in competition studies. 4) The *gag* gene is processed naturally so its ability to be processed in our transformants could be studied.

Transfection of the BPV-HTLV-clones into C-127 cells was fairly efficient at 5000 neomycin resistant transformants per pmol DNA per 10^6 cells. Southern blot analysis of the DNA from the transformants consistently showed that the BPV-HTLV-III clones are maintained extrachromosomally and that no gross rearrangements or deletions have occurred. Copy number appears to be low, averaging 10 copies per cell. Analysis of transcription of HTLV-III specific sequences in the transformants revealed several important features. The level of transcription was fairly low with abundances averaging only 1000 copies per cell. This may be due to a lack of stability of the transcripts since the metallothinine gene promoter is proported to be very strong. Also, the HTLV-III inserts lack the HTLV-III "cap" site which may affect mRNA stability. Because of the low abundancy of HTLV-III transcripts, only the primary transcript is readily detected. This would be the 3.2 kb transcript with pNM9-BPV-HTLV-6. Processed transcripts are more difficult to detect including the 1.5 kb transcript of pNM9-HTLV-3 and the 4.5 kb and 1.5 kb transcripts of pNM9-BPV-HTLV-6. Nevertheless, these are the same sized transcripts identified in HTLV-III infected cells (11).

The transformants containing the *gag* gene produce immunoreactive peptides with molecular weights of 55, 48, 44, 41 and 39 kd. The source of this heterogeneity is at present unclear but similar sized proteins are present in HTLV-III viral lysates (Figure 4). Transformants containing the entire HTLV-III coding sequence produce similar *gag* proteins as well as those with sizes of 12, 17 and 24 kd; the sizes of the mature *gag* proteins. The fact that processed *gag* proteins are found in clones containing the entire HTLV-III coding sequence but not in clones containing only the *gag* gene may indicate the presence of a virally encoded protease located in the *pol* region (38). Transformants containing the entire HTLV-III coding sequence also produce immunoreactive proteins with molecular weights of 30, 65 and 120 kd. The 120 kd protein may correspond to the exterior envelope protein and the 65 kd protein may be the reverse transcriptase (16). The 30 kd protein may be virally encoded nuclease (18), the product from the *SOR* or 3'-*orf* regions or possible of cellular origin. It is not clear at the present time, why there is no detection of gp41 in these cells. Substantial amounts of HTLV-III specific peptides are produced by the transformants. Expression levels range from 1 to 5 mg per

10^8 cells. The metallothionine gene promoter used in these constructs can be induced by the presence of heavy metals. Growth of the transformants in the presence of cadmium increases the expression levels approximately 3 to 5 fold. The results of these studies show the feasibility of expressing biologically interesting proteins in mammalian cells. The high expression levels seen with the BPV system should make it possible to produce proteins in sufficient quantities for purification and subsequent studies. We are currently devising experiments to express the env gene of HTLV-III using this system.

The production of only the 53 kd gag protein in CHO cells transformed with pSV·Gp·dhfr was unexpected. Despite the presence of the entire coding sequence of the pol and the putative protease genes in pSV·Gp·dhfr, there is little expression of these two encoded proteins. Recently Jacks and Varmus (39) demonstrated that the expression of the Rous sarcoma virus pol gene is facilitated by a ribosomal framshift mechanism of the host cell. It has been proposed that either ribosomal frame shifting or RNA splicing is required for the expression of the pol gene which resides at a different reading frame down stream from the gag polypeptide (10, 11). The lack of any detectable synthesis of the pol gene polypeptide in transformed CHO cells suggest these cells may lack both cellular mechanisms required for pol gene expression. The lack of any functional protease activity in the transformed cell was again illustrated by the lack of processing of the 53 kd gag precursor polypeptide. The size difference of the gag precursor protein 53 kd compared to the native HTLV-III gag 55 kd precursor is not clear. Nevertheless, we suggest that the pSV·Gp·dhfr transformed CHO cells which produce the 53 kd protein would provide a novel system to characterize the protease of HTLV-III.

REFERENCES

1. Gallo, R. C., Salahuddin, S. Z., Popovic, M., Shearer, G. M. Kaplan, M., Haynbes, B. F., Palther, T. J., Redfield, R., Oleske, J., Safai, B., White, G., Foster, P., and Markham, P. D. (1984). Frequent detection and isolation of cytopathic retroviruses (HTLV-III) from patients with AIDS and at risk for AIDS. Science 224:500.

2. Popovic, M., Sarngadharan, M. G., Read, E., and Gallo, R. C. (1984). Detection, isolation and continuous production of cytopathic retroviruses (HTLV-III) from patients with AIDS and pre-AIDS. Science 224:497.
3. Sarngadharan, M. G., Popovic, M., Bruch, L., Schupbach, J., and Gallo, R. C. (1984). Antibodies reactive with human T-lymphotropic retroviruses (HTLV-III) in serum of patients with AIDS. Science 224:506.
4. Schupbach, J., Popovic, M., Gilden, R. V., Gonda, M. A., Sarngadharan, M. G., and Gallo, R. C. (1984). Serological analysis of a subgroup of human T-lymphotropic retroviruses (HTLV-III) associated with AIDS. Science 224:503.
5. Barre-Sinoussi, F., Chermann, J. C., Reyn, F., Nugeyre, M. T., Chamaret, S., Gruest, J., Dauguet, C., Axler-Blin, C., Vezinet-Brun, F., Rouzioux, C., Rozenbaum, W., and Montagnier, L. (1983). Isolation of a T-lymphotropic retrovirus from a patient at risk for acquired immune deficiency syndrome (AIDS). Science 220:868.
6. Levy, J. A., Hoffman, A. D., Kramer, S. M., Landis, J. A., Shimabukuro, J. M., and Oshiro, L. S. (1984). Isolation of lymphocytopathic retroviruses from San Francisco patients with AIDS. Science 225:840.
7. Broder, S., and Gallo, R. C. (1984). A Pathogenic Retrovirus (HTLV-III) Linked to AIDS. New Eng. J. Med. 311:1292.
8. Wong-Staal, F. and Gallo, R. C. (1985). Human T-lymphotropic retroviruses. Science 317:395.
9. Salahuddin, S. Z., Markham, P. D., Popovic, M., Sarngadharan, M. G., Orndorff, S., Fladagar, A., Patel, A., Gold, J., and Gallo, R. C. (1985). Isolation of infectious human T-cell leukemia/lymphotropic virus type III (HTLV-III) from patients with acquired immunodeficiency syndrome (AIDS) or AIDS-related complex (ARC) and from healthy carriers: A study of risk groups and tissue sources. Proc. Natl. Acad. Science 82:5530.
10. Ratner, L., Haseltine, W., Patarca, R., Livak, K. J., Starcich, B., Josephs, S. F., Doran, E. R., Rafalski, J. A., Whitehorn, E. A., Baumeister, K., Ivanoff, L., Petteway, S. R., Pearson, M. L., Lautenberger, J. A., Papas, T. S., Ghrayeb, J., Chang, N. T., Gallo, R., C., and Wong-Staal, F. (1985). Complete nucleotide sequence of AIDS virus, HTLV-III. Science 313:277.

11. Muesing, M. A., Smith, D. H., Cabradilla, C. D., Benton, C. V., Lasky, L. A., and Capon, D. J. (1985). Nucleic acid structure and expression of the human AIDS/lymphadenopathy retrovirus. Science 313:450.
12. Wain-Hobson, S., Sonigo, P., Danos, O., Cole, S. and Alizon, M. (1985). Nucleotide sequence of the AIDS virus, LAV. Cell 40:9.
13. Sanchez-Pescador R., Power, M. D., Barr, P. J., Steimer, K. S., Stempien, M. M., Brown-Shimer, S. L., Gee, W. W., Renard, A., Randloph, A., Levy, J. A., Dina, D., and Luciw, P. A. (1985). Nucleotide sequence and expression of an AIDS-associated retrovirus (ARV-2). Science 227:484.
14. Allan, J. S., Coligan, J. E., Barin, F., McLane, M. F., Sodroski, J. G., Rosen, C. A., Haseltine, W. A., Lee, T. H. and Essex, M. (1985). Major glycoprotein antigens that induce antibodies in AIDS patients are encoded by HTLV-III. Science 228:1091.
15. Veronese, F. D., DeVico, A. L., Copeland, T. D., Oroszlan, S., Gallo, R. C., and Sarngadharan, M. G. (1985). Characterization of gp41 as the transmembrane protein coded by the HTLV-III/LAV envelope gene. Science 229:1402.
16. Veronese, F. D., Sarngadharan, M. G., DeVico, A. L., Rahman, R., Joseph, B., Copland, T. D., Oroszlan, S. and Gallo, R. C. (1986). Identification of p65 and p51 as the HTLV-III/LAV reverse transcriptase. J. Cell Biol. 10A:205
17. Chang, N. T., Chanda, P. K., Barone, A. D., McKinney, S., Rhodes, D. P., Tam, S. H., and Shearman, C. W., Huang, J., Chang, T. W., Gallo, R. C., and Wong-Staal, F. (1985). Expression of *Escherichia coli* of open reading frame gene segments of HTLV-III. Science 228:93.
18. Chang, N. T., Huang, J., Ghrayeb, J., McKinney, S., Chanda, P. K., Chang, T. W., Putney, S., Sarngadharan, M. G., Wong-Staal, F., and Gallo, R. C. (1985). An HTLV-III peptide produced by recombinant DNA is immunoreactive with sera from patients with AIDS. Nature 315:151.
19. Barone, A. D., Silva, J. J., Ho, D. D., Gallo, R. C., Wong-Staal, F., and Chang, N. T. (1986). Reactivity of *E. coli*-derived trans-activating protein of HTLV-III with AIDS sera. J. Immunology (in press).

20. Franchini, G., Robert-Guroff, M., Wong-Staal, F., Ghrayeb, J., Kato, I., Chang, T. W., and Chang, N. T. (1986). Expression of the protein encoded by the 3' open reading frame of HTLV-III in bacteria: demonstration of its immunoreactivity with human sera. Proc. Natl. Acad. Sci. USA (in press).
21. Ghrayeb, J., Kato, I., McKinney, S., Huang, J. J., Chanda, P. K., Sarangadharan, M. G., Chang, T. W., and Chang, N. T. (1986). Human T cell lymphotropic virus type III (HTLV-III) core antigens: synthesis in _E. coli_ and immunoreactivity with human sera. DNA (in press).
22. Crowl, R., Ganguly, K., Gordon, M., Conroy, R., Schaber, M., Kramer, R., Shaw, G., Wong-Staal, F., and Reddy, E. P. (1985). HTLV-III _env_ gene products synthesized in _E. coli_ are recognized by antibodies present in the sera of AIDS patients. Cell 41:979.
23. Hahn, B. H., Gonda, M. A., Shaw, G. M., Popovic, M., Hoxie, J. A., Gallo, R. C., and Wong-Staal, F. (1985). Genomic diversity of the acquired immune deficiency syndrome virus HTLV-III: different viruses exhibit greatest divergence in their envelope genes. Proc. Natl. Acad. Sci. 82:4813.
24. Starcich, B. R., Hahn, B. H., Shaw, G. M., Modrow, S., Josephs, S. F., Wolf, H., Gallo, R. C., and Wong-Staal, F. (1986). Identification and characterization of conserved and divergent regions in the envelope gene of AIDS viruses. Cell (in press).
25. Weiss, R. A., Clapham, P. R., Cheingsong-Popov, R., Dalgleish, A. G., Carne, C. A., Weller, I. V. D. and Tedder, R. S. (1985). Neutralization of human T-lymphotropic-virus type III by sera of AIDS and AIDS-risk patients. Nature 316:69.
26. Robert-Guroff, M., Brown, M., and Gallo, R. C. (1985). HTLV-III-neutralizing antibodies in patients with AIDS and AIDS-related complex. Nature 316:72.
27. Hsiung, N., Fitts, R., Wilson, S., Milne, A., and Hamer, D. (1984). Efficient production of hepatitis B surface antigen using a bovine papilloma virus - Metallothionein vector. J. Mol. Appl. Genet. 2:497.
28. Laskey, L. A., Dowbenko, D., Simonsen, C., and Berman, P. W. (1984). Production of an HSV subunit vaccine by genetically engineered mammalian cell lines. In Chanock, R. M., Lerner, R. A. (eds): "Modern Approaches to Vaccines: Cold Spring Harbor Laboratory, p. 189.

29. Pavlakis, G. N., and Hamer, D. H. (1983). Regulation of a metallothionein-growth hormone hybrid gene is bovine papilloma virus. Proc. Natl. Acad. Sci USA 80:397.
30. Hahn, B. H., Shaw, G. M., Arya, S. K., Popovic, M., Gallo, R. C., and Wong-Staal, F. (1984). Molecular cloning and characterization of the HTLV-III virus associated with AIDS. Nature (London) 312:166.
31. Masui, Y., Mizuno, T., and Inouye, M. (1984). Novel high-level expression cloning vehicles: 10^4-fold amplification of _Escherichia coli_ minor protein. Biotechnology 2:81.
32. An expression vector developed by Repligen, Inc., Cambridge, Massachusetts.
33. Petteway, S. R., Tritch, R. J., Whitehorn, E. A., Ivanoff, L. A., Reed, D. L., and Kenealy, W. R. (1986). Proc. Natl. Acad. Sci. USA (in press).
34. Alexander, S. and Elder, J. H. (1984). Carbohydrate dramatically influences immune reactivity of antisera to viral glycoprotein antigens. Science 226:1328.
35. Graham, F. L., and Van der Eb, A. J. (1973). A new technique for the assay of infectivity of human adenovirus 5 DNA. Virology 52:456.
36. Hirt, B. 1969. Replicating Molecules of Polyoma Virus DNA. J. Mol. Biol. 40:141.
37. Barin, F., McLane, M. F., Allan, J. S., Leu, T. H., Groopman, J. E., and Essex, M. (1985). Virus envelope protein of HTLV-III represents major target antigen for antibodies in AIDS patients.
38. Shimotohno, K. Takahashi, Y., Shumizu, N., Gojobori, T., Golde, D. W., Chen, I. S. Y., Miwa, M., and Sugimura T. (1985). Complete nucleotide sequence of an infectious clone of human T-cell leukemia virus type II: An open reading frame for the protease gene. Proc. Natl. Acad. Sci. USA 82:3101.
39. Jacks, T., and Varmun, H. E. (1985). Expression of the Rous sarcoma virus _pol_ gene by ribosomal frame shifting. Science 230:1237.

ACKNOWLEDGMENTS

We wish to thank Dr. Jame Huang and Gia-Fan Chan for technical help, Dianne Marcotte for preparing the manuscript. This research was supported by Centocor, Inc. and Biotech Research Laboratory, Rockville, MD.

Viruses and Human Cancer, pages 257–267

ANTI-IDIOTYPES AND IMMUNOREGULATION: QUESTIONS AND ANSWERS

Hilary Koprowski and Dorothee Herlyn

The Wistar Institute, 3601 Spruce Street
Philadelphia, Pennsylvania 19104

ABSTRACT In spite of the enormous progress in the field of molecular biology, concepts of vaccine production today are not very different than those envisaged by Pasteur 100 years ago. There are, however, developments in research to produce more modern vaccines which may lead to drastic changes in our approaches to deriving more efficient vaccines. Better presentation of antigen is provided by the use of micelle preparations (ISCOM) to which the antigen in question is adsorbed. The use of viral components for immunization instead of complete virions has been tried in many viral infections but has not been very successful. Expression of viral antigen by either bacteria or yeast carrying the viral genome may lead in isolated cases to the production of a vaccine. Insertion of a viral genome into a vaccinia virus has led to the successful immunization of animal hosts with the vaccinia virus against a variety of viral diseases, and this method holds good prospects for the future. Once the modulation of an immune response to peptide antigens is better understood, synthetic peptides may find their place in vaccination against some viral infections.

Vaccination with anti-idiotype is based on the premise that it carries an internal image of the antigen and induces immunity similar to that obtained using an inactivated virion preparation or a live attenuated virus. One of the more important aspects of these studies is the ability of anti-idiotype to induce an antibody (Ab3) after immunization of different animal species. This Ab3 must inhibit the binding of anti-idiotype to the animal antibody (Ab1), inhibit binding of Ab1 to the viral antigen, and express characteristics similar to, if not identical with, those of Ab1. Only then can one assume that the anti-idiotype carries the internal image of the antigen and is potentially useful for vaccination purposes.

The obvious advantages in using anti-idiotype for vaccination are its complete inocuousness and the ability to immunize neonates who are often unable to respond directly to viral or bacterial antigen.

This paper is written in a form of question and answer accompanied by pertinent illustrations. A selected bibliography is appended. Readers who are curious enough to ask for more detailed information concerning subject of this presentation are kindly requested to contact directly the authors.

ANTI-IDIOTYPES (DEFINITION AND PRODUCTION):

Q. What is an anti-idiotype?

A. An anti-idiotype (α-Id) is a component of immunoregulatory system induced as a response of the host to immunization with an antibody (Ab1). It is essentially an anti-antibody and is referred to also as Ab2.

Q. How do you produce Ab2?

A. Through immunization of mice or rats with Ab1 and production of hybridomas secreting monoclonal Ab2 or through immunization of larger animals such as rabbits or goats and production of polyclonal Ab2.

Q. How do you purify a polyclonal Ab2?

A. By repeated passages of Ab2 serum globulin through columns of unrelated Ab1 preferably of the same isotype as Ab1 used for immunization and final adsorption to and elution from a column of Ab1 used for immunization.

ANTI-IDIOTYPES (CHARACTERIZATION):

Q. How do you characterize preliminary on Ab2?

A. By its binding to the Ab1 used for its induction and <u>not</u> binding to the antigen defined by Ab1.

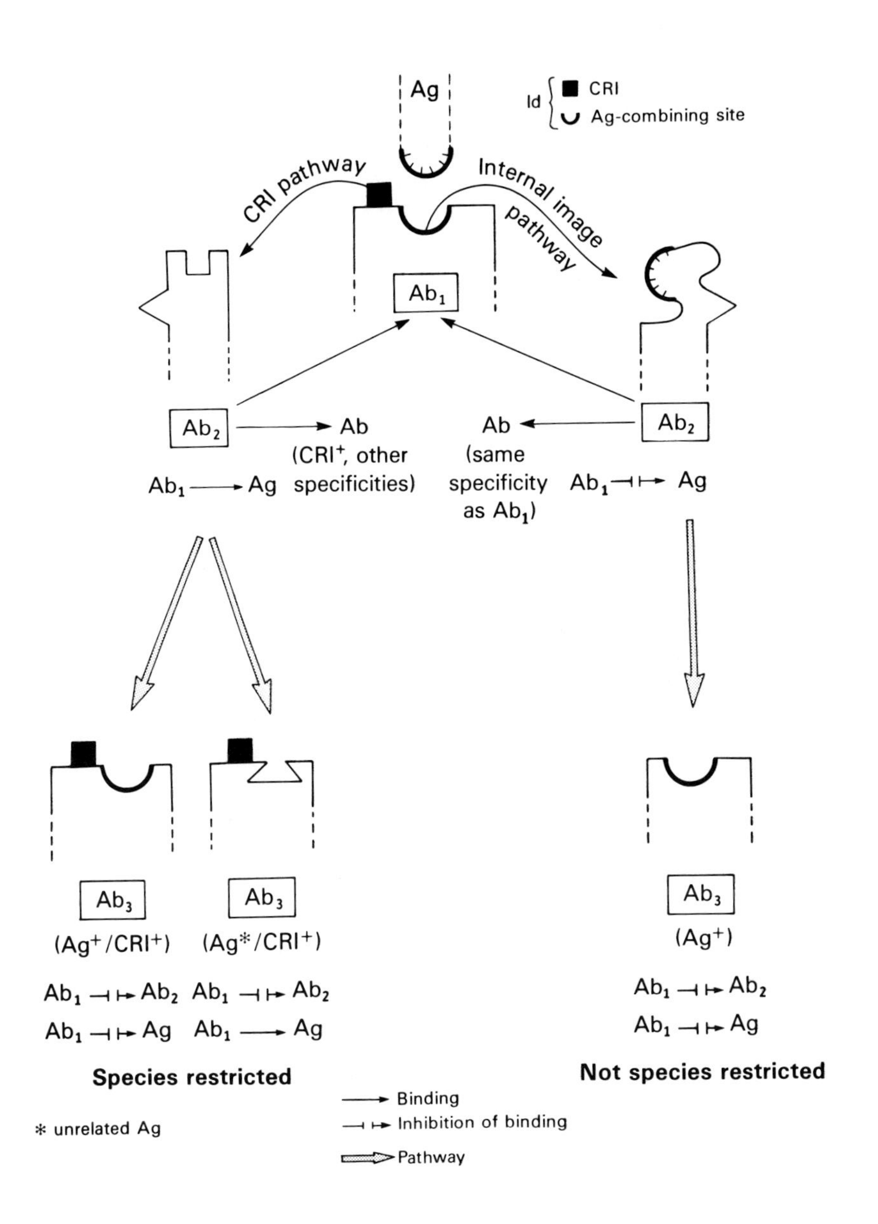
Ag
Id
CRI
Ag-combining site
CRI pathway
Internal image pathway
Ab1
Ab2
Ab
(CRI+, other specificities)
Ab1 → Ag
Ab
(same specificity as Ab1)
Ab2
Ab1 ⊣⊢ Ag
Ab3
(Ag+/CRI+)
Ab1 ⊣⊢ Ab2
Ab1 ⊣⊢ Ag
Ab3
(Ag*/CRI+)
Ab1 ⊣⊢ Ab2
Ab1 → Ag
Ab3
(Ag+)
Ab1 ⊣⊢ Ab2
Ab1 ⊣⊢ Ag
Species restricted
Not species restricted
* unrelated Ag
Binding
Inhibition of binding
Pathway

Fig. 1

Q. Are there different types of Ab2?

A. Yes, essentially two types: one carrying the internal image of the antigen (int. img. α-Id) and the other reacting with crossreacting idiotypes (anti-CRI Ab2).

Q. How do you distinguish between the two types?

A. The int. img. α-Id binding to the combining site will inhibit binding of Ab1 to the antigen and in turn will bind to various Ab1s of different origin and derived from different species provided they express identical binding specificities. The anti-CRI Ab2 will not inhibit binding of Ab1 to the antigen but will bind in turn to various Ab1s which share the same CRI but are crossreactive either with variants of the agent used for immunization or with different proteins (antigen) expressed by the same agent.

Q. Is characterization of the α-Ids along the above criteria sufficient to distinguish unequivocally int. img. α-Id from anti-CRI Ab2?

A. No. The most conclusive evidence of the type of Ab2 produced can be obtained by immunization of animals with Ab2 and induction of Ab3. This is particularly applicable to cases when sufficient number of sera of different origins containing Ab1 necessary for characterization of Ab2 is not available.

ANTIBODY 3 (Ab3) (DEFINITION):

Q. What is Ab3?

A. It is a component of the immunoregulatory system induced as a response of the host to immunization with Ab2. It is essentially an anti-anti-antibody.

Q. Can you induce an Ab3 in any host species?

A. The int. img. α-Id induced-immunity is not species restricted, whereas the anti-CRI Ab2 induced immunity may be genetically restricted to a given strain or species of the host.

Ab3 (CHARACTERIZATION):

Q. How do you characterize Ab3 and how do you distinguish those which were induced by int. img. α-Id from those induced by anti-CRI Ab2?

A. All Ab3s bind to the inducing Ab2 and inhibit its binding (see Fig. 1) to Ab1. Ab3 induced by int. img. α-Id will bind specifically to the antigen used for induction of Ab1 and will inhibit binding of Ab1 to this antigen. Ab3 induced by anti-CRI Ab2 will not bind to a specific antigen but will bind to variants of the same agents or to different proteins (antigens) expressed by the same agent.

VACCINATION WITH Ab2:

Q. What are the advantages of Ab2 immunization instead of conventional immunization with antigen preparations?

A. There are several advantages.

1) Safety: In cases of vaccination against pathogenic agents there is no need to search Ab2 for traces of non-inactivated pathogens or information conveying nucleic acids.

2) Immunization of newborns who are incapable of mounting an immune response to the antigen but are capable of being primed by its internal image on globulin molecule (Ab2).

3) High specificity for the antigen in question without danger of induction of antibodies crossreacting with normal components of host tissue (molecular mimicry).

4) Protection against variants of a pathogen particularly those escaping immunologic surveillance during infection.

5) Economical production: Ab2 vaccines, particularly those derived from hybridomas and administered in doses of a few milligrams of globulin per animal or human host are much more economical and less cumbersome to produce than large doses of purified antigen.

IMMUNIZATION WITH Ab2:

Q. Does immunization with Ab2 protect the host against challenge with lethal pathogen?

A. Yes, this has been shown in cases of several viral infections such as hepatitis, Sendai and reoviruses.

Table 1. Induction of Immunity Against Viruses by Anti-Id Antibodies

	RABIES	HEPATITIS B	SENDAI	REO	POLIO
Ab1					
Species of Origin	Balb/c Mouse	human	B10.D2	Balb/c	Balb/c
Clonality	monoclonal	polyclonal	monoclonal[b]	monoclonal	monoclonal
Binding Specificity	various nucleocapsid glycoproteins	Hepatitis B surface antigen group a determinant	Sendai virus (not further specified)	reovirus type 3 hemagglutinin (HA)[c]	polio virus type II
Ab2					
Species of Origin	rabbit	rabbit	B10.D2	Balb/c	Balb/c
Clonality	polyclonal	polyclonal	monoclonal	monoclonal	monoclonal
Binding Specificity	bind to homologous Ab_1 only	Crossreactive ID on human chimpanzee, swine, goat, rabbit, guinea pig, mouse; but not chicken anti-hepatitis virus Ab	Bind to 37% of the various Sendai virus-specific T helper cell clones	binds to homologous Ab_1 only	crossreactive Id on anti-polio virus Ab from mice, rabbits, guinea pigs and humans
IMMUNITY INDUCED BY Ab2					
Mouse Strain	ICR	Balb/c	B10.D2 and other strains of different haplotype	Balb/c and mice of various other Igh allotypes CTL:[c] Balb/c	Balb/c
Humoral/Cell Mediated	neutralizing polyclonal Ab_3	polyclonal Ab_3	Cytotoxic T cells	DTH,CTL[c]	neutralizing polyclonal Ab_3
Induction by Ab_2 alone or Ab_2 + antigen	Ab_2 alone; but enhancement by subsequent antigen exposure	Ab_2 alone; but enhancement by subsequent antigen exposure	Ab_2 alone	Ab_2 (DTH) or Ab_2 hybridomas (CTL)	Ab_2 alone
Specificity	identical or different from Ab_1 depending on source of Ab_2	Hepatitis B surface antigen group a only	Syngeneic and allogeneic Sendai virus infected but not uninfected targets	DTH: reovirus type 3HA only. CTL: reovirus type 3HA (variant viruses not tested)	poliovirus type II (other type not tested)
Effect on Challenge with Virus	protection	protection[a]	protection	protection	none

[a] In chimpanzees

[b] Instead of Ab_1 a Sendai virus-specific T helper cell clone was used for generation of Ab_2.

[c] Abbreviations: HA, hemagglutinin; DTH, delayed-typed hypersensitivity; CTL, cytotoxic T lymphocytes.

Ab2 IN HUMAN CANCER:

Q. Is there a possibility of immunization against human cancer with Ab2 even though there is no evidence that cancer patients develop Ab1 against antigen(s) of their tumor?

A. A monoclonal antibody (Ab1) reactive with a defined antigen of human gastrointestinal cancer induced a polyclonal Ab2 in immunized goat. Following purification, the Ab2 was found to inhibit binding of Ab1 to cancerous cells and the binding of this Ab2 to Ab1 was inhibited by the specific tumor antigen. This Ab2 induced in mice and rabbits an Ab3 which was found to immunoprecipitate the tumor antigen of the same molecular weight as that precipitated by Ab1; the Ab3 showed the same binding specificity to various target cells as Ab1. These results indicate that the Ab2 carried the internal image of the tumor antigen and was capable to induce an immune response mimicking a response to the tumor antigen.

COMPONENTS OF HUMAN IMMUNOREGULATORY SYSTEM:

Q. Was Ab2 detected in humans immunized with heterologous Ab1?

A. Yes: The vast majority of patients injected for immunotherapeutic purposes with murine monoclonal antibody (MAb) reacting with tumor antigen developed anti-mouse protein antibody; more than 80 percent of these patients also developed an anti-Id reacting specifically with the MAb only.

Table 2. Summary of Characteristics of Human Ab2 to Anti-Colon Carcinoma (CRC) MAb 17-1A

Parameter Investigated	Results[a]	Total Number of Patients
Number of patients with Ab2 responses	35	41
μg/ml Ab2 isolated from patients' sera (range in various patients)	2.8-42.0	4
Ab2 in % of total anti-mouse IgG (range)	21.0-80.2	4
Anti-combining site Ab2:		
a) Maximal % inhibition of ^{125}I-17-1A binding to CRC cells by Ab2 (range)	32.7-70.5	4
b) Maximal % inhibition of ^{125}I-17-1A binding to human Ab2 by extracts from CRC cells (range)	20.0-68.5	3
Crossreactivities between various human Ab2. Maximal % inhibition of ^{125}I-Ab2 (from patient no. 8) binding to MAb 17-1A by Ab2 derived from various other patients (range)	17.0-91.2	4

Q. What fraction of the anti-mouse response is the α-Id response?

A. Depending on individual responses 2.8 to 42.0 mgs or 21-80% of anti-mouse response was the (Ab2) α-Id fraction.

Q. Did the Ab2 carry the internal image of the tumor antigen?

A. Binding of MAb to cancer cells was inhibited by the purified Ab2s and binding of several Ab2s to MAb was inhibited by tumor antigens.

Q. Were the Ab2s isolated from different individuals crossreactive?

A. Yes: A significant crossreactivity was found between Ab2 isolated from four subjects, suggesting a common determinant of the MAb.

Q. Was there an Ab3 induced in human subjects?

A. Ab3 was not detected in serum but lymphocytes obtained from patients immunized four times with homologous Ab2 and then exposed to Ab2 in vitro secreted in culture an Ab3 binding to cancer cells.

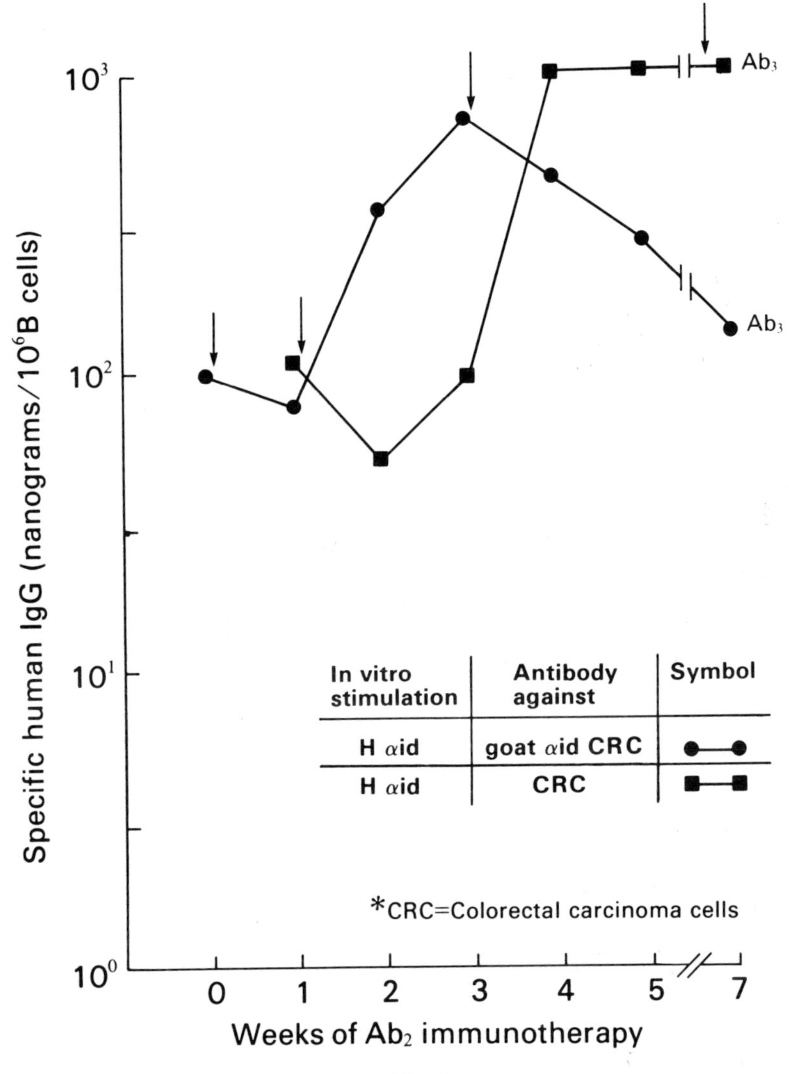

Fig. 2

Ab2 AS A DIAGNOSTIC TOOL:

Q. Will Ab2 play another role in addition to that of "unconventional vaccines"?

A. Yes: Inhibition of binding of MAb to its int. img. α-Id by serum will indicate presence of antibodies reacting with the same antigen as MAb. In addition, binding of int. img. α-Id to antigen primed B and T cells may detect immunoglobulin receptors for antigen in the absence of antibody secreted into circulation.

CONCLUSION:

It is feasible to produce an Ab2 response to any desired Ab1. It is easy and economical to produce Ab2 in large quantities. Among the population of Ab2s it is possible to distinguish between int. img. α-Id and anti-CRI Ab2. The internal image Ab2s can be used successfully for engendering an immune response of the host to a variety of disease conditions such as cancer, infectious diseases, etc. α-Ids may also be useful in diagnostic procedures for detection of an immune response of the host to an antigen.

SELECTED BIBLIOGRAPHY:

Bona, C.A., Heber-Katz, E., and Paul, N.E. J. Exp. Med. 153, 951 (1981).

Bona, C.A. Fed. Proc. 43, 2558, (1984).

Cazenave, P.A., and Roland, J. Immunol. Rev. 79, 157, (1984).

DeFreitas, E., Suzuki, H., Herlyn, D., Lubeck, M., Sears, H., Herlyn, M., and Koprowski, H. Curr. Top. Microbiol. Immunol. 119, 76 (1985).

Ertl, H.C.J., and Finberg, R.W. Proc. Natl. Acad. Sci. USA, 81, 2850 (1984).

Gaulton, G.N. Abstract, Symposium, 2nd S-W Foundation for Biomedical Research Virology/Immunology, San Antonio, TX. Dec. 4-6, 1985.

Herlyn, D., Ross, A.H., and Koprowski, H. Science, accepted for publication.

Herlyn, D., Lubeck, M., Sears, H.F., and Koprowski, H. J. Immunol. Meth., in press.

Jerne, N.K. Ann. Immunol. (Inst. Pasteur) 125C, 373, (1974).

Kelsoe, G., Reth, M., and Rajewski, K. Immunol. Rev. 52, 75 (1980).

Kennedy, R.C., and Dreesman, G.R. J. Exp. Med. 159, 655 (1984).
Kennedy, R.C., Adler-Storthz, K., Henkel, R.D., Sanchez, Y., Melnick, J.L., and Dreesman, G.R. Science 221, 853, (1983).
Kennedy, R.C., Melnick, J.L., and Dreesman, G.R. Science 223, 930 (1984).
Kennedy, R.C. Abstract, Symposium 2nd S-W Foundation for Biomedical Research Virology/Immunology, San Antonio, TX Dec. 4-6, 1985.
Koprowski, H., and Herlyn, D. In: "Concepts in Viral Pathogenesis", Notkins, A.L., and Oldstone, M.B.A. (eds) Springer-Verlag, New York, N.Y., in press.
Koprowski, H. Cancer Res. Suppl. 45, 4689s (1985).
Koprowski, H., Herlyn, D., Lubeck, M., DeFreitas, E., and Sears, H.F. Proc. Natl. Acad. Sci. USA, 81, 216 (1984).
Lindemann, J. Ann. Immunol. (Inst. Pasteur) 124C, 171 (1985).
Moran, T., Liu, Y-N.C., Schulman, T.L., and Bona, C. Proc. Natl. Acad. Sci. USA 81, 1809 (1984).
Reagan, K.J., Wunner, W.H., Wiktor, T.J., and Koprowski, H. Virol. 48, 660, (1983).
Reagan, K.J., Wunner, W.H., and Koprowski, H. In: "High Technology Route to Virus Vaccines," Dreesman, G.R., Bronson, T.G., Kennedy, R.C. (eds.) Am. Soc. J. Microbiol. Publ. pp. 117 (1985).
Sharpe, A.H., Gaulton, G.N., McDade, K.K., Fields, B.N. and Greene, M.I. J. Exp. Med. 160, 1195 (1984).
Stein, K., and Soderstrom, J.J. Exp. Med. 160, 1001, (1984).
Urbain, J., Slaoui, M., Mariame, B., and Leo, O. In: "Idiotypie in Biology and Medicine", Kohler, H., Urbain, J., Cazenave, P.A. (eds.) Academic Press, N.Y., p. 15 (1984).
Uytdehaag, IGCM, Osterhaus, A.D.M.E.J. Immunol. 134, 1225 (1985).

Viruses and Human Cancer, pages 269–281

ANALYSIS OF THE MECHANISM OF TRANS-ACTIVATION MEDIATED BY SIMIAN VIRUS 40 LARGE T ANTIGEN AND OTHER VIRAL TRANS-ACTING PROTEINS[1]

James C. Alwine[2], Susan Carswell, Chris Dabrowski, Gregory Gallo, Janis M. Keller, Jane Picardi, and John Whitbeck

Department of Microbiology, School of Medicine, University of Pennsylvania, Philadelphia, PA 19104-6076

ABSTRACT

The simian virus 40 (SV40) late promoter is activated in trans by a mechanism mediated by the viral early gene product, large T antigen. There appears to be two separate activatable elements in the late promoter. We describe these elements and outline a model for their mechanism of activation. A salient feature of this model is that one of the promoter elements (the tau element) is not acted upon directly by T antigen, but appears to be activated by a cellular trans-acting factor which is modified, activated or induced by T antigen. We provide data which support this model by demonstrating that the tau element is the target element for trans-activation of the late promoter by the pseudorabies virus immediate early protein. Overall, the tau element appears to be a trans-activatable promoter element acted upon by a cellular trans-acting factor. Since T antigen also mediates the activation of a number of non-SV40

[1]This research was supported by Public Health Service Grants CA28379 and CA33656 awarded by the N.I.H. to J.C.A.

[2]To whom correspondence should be addressed.

genes it is postulated that the cellular trans-acting factor affected by T antigen also activates its normal cellular target promoters. Thus the tau element and the late promoter offer an excellent model system for the understanding of the mechanism of trans-activation and its role in viral diseases and transformation.

INTRODUCTION

The transcriptional activation of the Simian virus 40 (SV40) late promoter has been shown to occur by a trans-activation mechanism mediated by the viral early protein, large T antigen (2,4,5,14,15). As we will discuss below, the trans-activation mechanism may not be mediated directly by T antigen but by a cellular transactivation mechanism which is utilized by T antigen. This is supported by the observation that the presence of T antigen causes the activation of a number of very different, non-SV40 promoters. The utilization of a cellular mechanism to trans-activate the late promoter would necessitate that some feature(s) of the late promoter must mimic features of cellular promoters subject to trans-activation.

In order to determine the mechanism of trans-activation, several laboratories, including ours, have undertaken studies to determine the elements of the late promoter which are necessary for trans-activation to occur (5,7,9,15). Figure 1 shows elements of the late promoter previously defined in this laboratory by mutational analysis (3,15); they are denoted omega, rho, tau and delta. It should be noted that these elements overlap the defined elements of the early viral promoter, the GC rich 21 base pair (bp) repeats, the 72 bp repeat enhancer region and the T antigen binding sites (I, II and III). However, the late promoter elements must, in large part, be considered unique and not analogous to the early promoter elements with which they share sequences. For example, the enhancers (72 bp repeats) of the early promoter do not function as enhancers for the late promoter, although elements of the late promoter are located within this region.

In the following text we describe the late promoter elements as we have defined them (3,15). In addition, we present a model for T antigen mediated trans-activation of the late promoter which is consistent with our

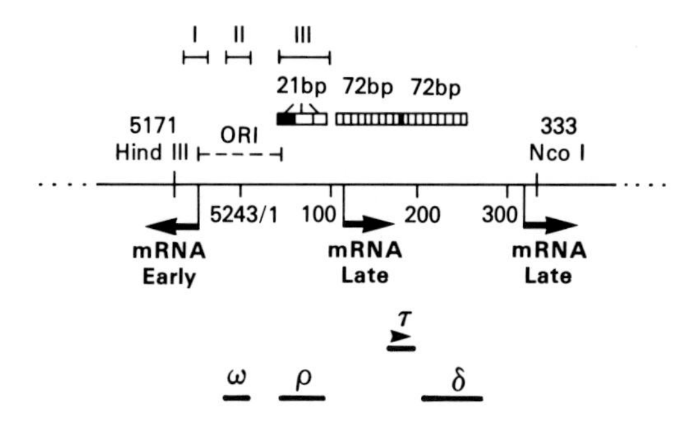

FIGURE 1. Diagram of the SV40 early and late promoter elements. The sequences between SV40 nucleotides 5171 and 333 (through the origin of replication, 5243/1) are shown; they represent the promoter sequences contained in the CAT transient expression vectors used in these studies (see Figure 3). The elements of the early promoter (21 and 72 bp repeats) and the late promoter (omega, rho, tau and delta) are marked as well as the T antigen binding sites (I, II and III) and the origin of replication (Ori). The late promoter elements are described in the text.

observations of the activities of the promoter elements. Finally, we will examine data which suggest that T antigen does not mediate trans-activation directly; instead these data suggest that T antigen activates, modifies or induces a cellular trans-acting factor which directly affects the promoter. Such utilization of cellular trans-acting factors would account for the general activation of many cellular genes which is known to occur during lytic infection in permissive cells and during the process of transformation in nonpermissive cells (2,22).

THE ELEMENTS OF THE SV40 LATE PROMOTER

Figure 1 diagrams the SV40 sequence between nucleotides 5151 and 333, through the origin of replication

(5243/1; nucleotide numbering is described in reference 22), which contain the early and late promoters as well as the origin of replication (Ori) and the T antigen binding sites (I, II, and III). This segment of DNA has been used in chloramphenicol acetyltransferase vectors (see Figure 3) for transient expression analysis of the elements of the late promoter (3,14,15). The rho element, located within the 21 bp repeat region, appears to be the site of repressor interaction which inhibits late promoter activity early in the lytic cycle (3). The significance of this element and late promoter repression will be discussed below. The delta element, within SV40 nucleotides 200-270, is active in the presence of T antigen only when the origin region, the omega element, is intact. The delta/omega combination accounts for approximately 35% of the late promoter activity under wild type conditions (i.e. in the presence of T antigen and under replicative conditions; 15). The tau element (SV40 nucleotides 150-200) does not require an intact origin region, nor the T antigen binding sites, to cause late promoter activation in the presence of T antigen. Under wild type conditions the tau element contributes approximately 65% of the late promoter activity (15). The tau element also has been shown to demonstrate an orientation dependence indicated by the arrow in Figure 1 (15). Overall these data indicate two means of activating the late promoter which appear to function somewhat independently.

Because the delta element appears to require the specific T antigen binding sites at the origin (and possibly the act of replication) we feel that the activation mechanism mediated by the delta element may be specific to the SV40 system. Conversely, the tau element's independence of the specific T antigen binding sites and the origin suggests that it may be representative of a more general trans-activatable element which can be studied as a model for understanding the mechanism of trans-activation.

A MODEL FOR THE ACTIVITY OF THE LATE PROMOTER DURING LYTIC INFECTION

Based on our studies (3,15) of the activity of the late promoter in the presence and absence of T antigen, i.e. basal activity and trans-activated activity, we postulate the following model for <u>in vivo</u> late promoter activity during lytic infection. Upon entry of the viral

DNA into the nucleus a factor binds to the rho element, repressing transcription in the late direction but allowing transcription to proceed in the early direction. Such repression at this point in the lytic cycle would be a reasonable strategy, for it is initially very important that transcription be dedicated to the early genes in order to establish the infection. The repression of the late promoter elements at early times may eliminate competition between the promoters so that the early promoter has the greatest advantage. Whether or not this effect relates to the binding of the transcription factor Spl within the 21 bp repeat region (6,18) is unknown at this point.

During a period of transition between early and late viral gene expression, large T antigen reaches physiologically active concentrations and binds to the origin (omega) region. We postulate that the binding may neutralize the effect of the rho element-repressor interaction, thus freeing the late promoter elements for activation. Our previous data (15) suggest that this activation is a result of: 1) activation of the omega-dependent delta element mediated either by the binding of T antigen to the origin binding sites or by the onset of DNA replication; and 2) trans-activation of the omega-independent tau element. As discussed and tested in the next sections, a variety of evidence indicates that the activation of the tau element may be mediated by a cellular trans-acting factor (TAF) which is induced, activated or modified by T antigen and other viral trans-acting proteins.

EVIDENCE THAT A CELLULAR TRANS-ACTING FACTOR IS INDUCED, ACTIVATED OR MODIFIED BY LARGE T ANTIGEN

A number of observations have been reported which indicate that trans-activation of the tau element of the late promoter may be mediated indirectly by T antigen through the utilization of a cellular trans-acting factor. These will be listed and discussed below:

1. The ability of T antigen to bind to DNA is not necessary for trans-activation to occur. The SV40 transformed simian cell line T22 (20) produces a mutant T antigen which has lost the ability to bind to viral DNA as measured by in vitro assay (16). As expected, this T antigen is also unable to initiate viral DNA synthesis (16). In data to be presented elsewhere, we have determined that the T22 T antigen is able to trans-activate

the late promoter as well as other mutant T antigens which maintain DNA binding capacity but are defective for viral DNA replication. This observation is best explained by the prediction that T antigen does not need to interact with the promoter elements directly to cause trans-activation, but instead it acts through an intermediate.

2. We have previously demonstrated (2) that both T antigen and the pseudorabies virus immediate early protein (PrV IE) can similarly activate the late promoter as well as a number of non-SV40 promoters. This activation of the late promoter is demonstrated in Figure 2 which shows a time course analysis of late promoter activation using the CAT transient expression assay. The late promoter CAT plasmid, pL16 (see Fig 3 and reference 15) was cotransfected into NIH-3T3 cells with plasmids which expressed T antigen (p6-1d1,15) PrV IE (pIE,10) or the adenovirus Ela protein (pEla,19). Both IE and Ela have previously been demonstrated to have trans-activation functions (8,11,12,13,17). It is clear that PrV IE and T antigen are equivalent in their ability to activate the late promoter. Conversely, the Ela protein represses the basal activity of the late promoter. This effect of Ela has been noted previously (2,23) and will not be discussed here. However, the similarity in activation ability by T antigen and IE protein is striking. The necessity to directly interact with the promoter for activation can be ruled out in the case of PrV IE protein since there is no evidence that PrV IE protein binds to DNA. Thus the simplest explanation for the data is again the assumption that both T antigen and the IE protein act through a similar intermediate, presumably a cellular trans-acting factor. This assumption has recently been supported by in vitro transcription data (1) which suggests that the PrV IE protein interacts with cellular factors to mediate its trans-activation.

3. Recent data (21) have shown that SV40 late transcriptional complexes contain no bound T antigen. A conclusion of this work was that cellular trans-acting factors mediate the activation of the promoter. Our model that T antigen utilizes a cellular trans-activation mechanism predicts that no T antigen would be bound to late transcriptional complexes.

Overall these data present a strong argument in favor of the model where T antigen, and PrV IE, mediate trans-activation through an intermediate, most likely a cellular trans-acting factor (TAF) which has been induced,

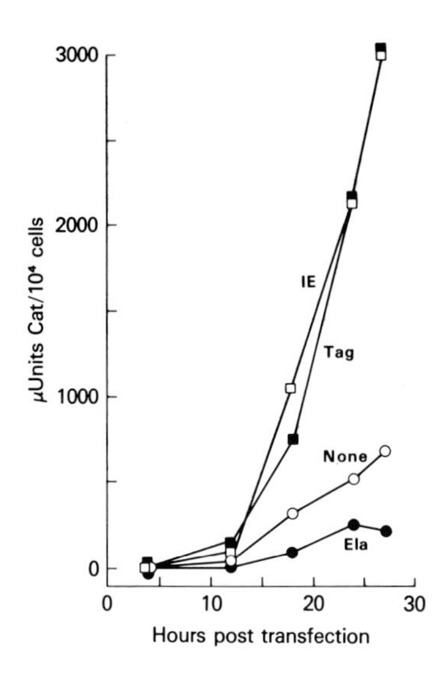

FIGURE 2. Time course analysis of trans-activation of the SV40 late promoter mediated by T antigen, pseudorabies virus IE protein and adenovirus Ela protein. The late promoter-CAT transient expression vector pL16-cat (see Fig. 3) was transfected into NIH-3T3 cells either alone or cotransfected with a plasmid which expresses one of the viral early proteins, T antigen (p6-1d1,14), PrV IE (pIE,10) or Ela (pEla,19). Transfections and CAT assays were done as described in the text.

activated or modified by these viral proteins. In the next section we will test this model.

THE TAU ELEMENT IS NECESSARY FOR TRANS-ACTIVATION OF THE SV40 LATE PROMOTER BY T ANTIGEN AND PrV IE PROTEIN

Obviously the direct study of a putative cellular trans-acting factor is not possible at this point. However, if it is assumed that T antigen and the PrV IE protein utilize the same, or similar, cellular factors to mediate

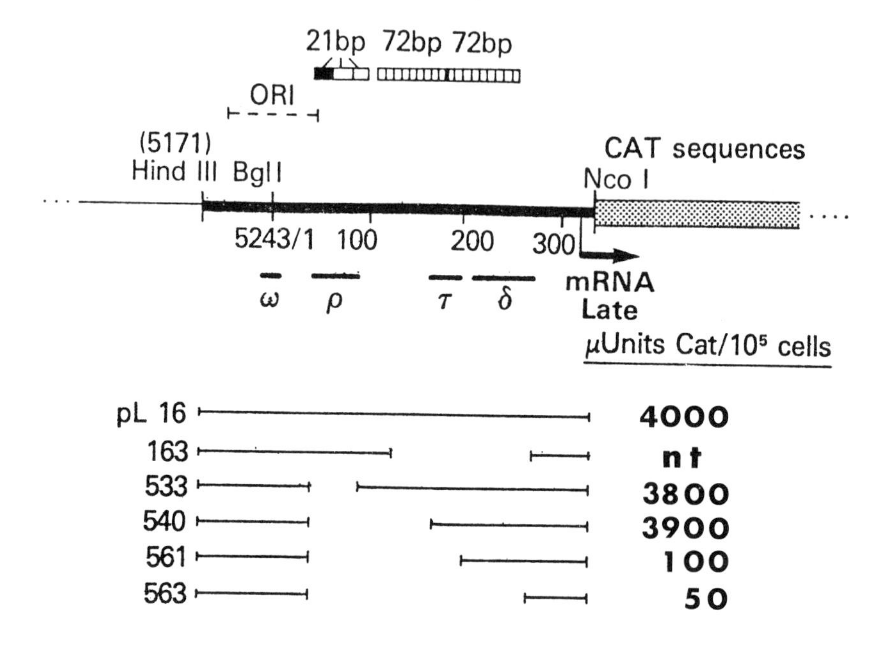

FIGURE 3. Analysis of the activity of SV40 late promoter elements in the presence of the PrV IE protein. The wild type late promoter CAT plasmid, pL16, contains the promoter region described in Figure 1. The open spaces indicate the regions deleted in the various mutants. The position of the omega, rho, tau and delta elements are indicated with respect to the deletions. Transfection and assay conditions are described in the text, the final results are expressed as microunits CAT activity per 10^5 transfected cells. These activities have been standardized for transfection efficiency variations using described procedures (2).

trans-activation then we can predict that the same promoter elements are necessary for the trans-activation mediated by either viral protein. We have previously demonstrated (15) that the tau element is necessary for T antigen mediated trans-activation, thus our model would predict that the tau element is also essential for trans-activation mediated by PrV IE protein. The deletion mutant analysis shown in Figure 3 indicates that this is the case. The specific plasmids shown are described in detail elsewhere (15). The open spaces denote deletions which successively remove the promoter elements. For the analysis of the late promoter elements 5μg of each plasmid was cotransfected with 5μg of pIE (which expresses PrV IE protein) into NIH3T3 cells. Transfected cells were harvested 40 hours after transfection and assayed for CAT activity. All CAT activities have been standardized for variations in transfection efficiencies by described procedures (2,15).

The salient feature of the data is that the deletion of the tau element eliminates the activity of the promoter in the presence of PrV IE (compare pL540 and pL561). Despite the fact that each of these plasmids contains the omega and delta element, there is no activity generated from the omega-dependent delta element. The lack of utilization of the delta element during PrV IE trans-activation is expected since our previous data suggests that the activity of delta is SV40 specific, resulting from the direct interaction of T antigen with its binding sites within the omega region. Likewise, the necessity of the tau element for trans-activation mediated by PrV IE is the predicted result from the model where tau is a trans-activatable element acted upon by a cellular TAF which has been induced, activated or modified by PrV IE protein.

CONCLUSIONS

We have presented an argument, and data to support it, which states that SV40 T antigen mediates trans-activation of the tau element of the SV40 late promoter through the utilization of a cellular trans-acting factor. This TAF is activated, induced or modified by T antigen; considering the results with the PrV IE protein, this TAF may be similarly affected by a number of viral proteins known to activate gene expression.

The activation of a cellular TAF by T antigen may

accomplish more than turning on the late genes; it may be a necessary event for the induction of expression of cellular genes whose products are needed for the progression of the viral infection. It is well documented that T antigen induces cellular replicative enzymes (and a round of cell replication) in order to insure that viral DNA will be replicated (23). Perhaps the promoters of these genes are targets for the TAF affected by T antigen.

In the case of nonpermissive cells, where viral DNA synthesis cannot occur, thus blocking the lytic infection, the presence of T antigen can still cause the trans-activation of the late promoter and of cellular genes (See Fig. 2; 2,22). If the viral DNA integrates a continual presence of T antigen will result; this, in turn, would result in constant activity of cellular TAFs, hence the continual activation of cellular target promoters. If, as is indicated, the cellular genes activated are cell cycle genes and genes which affect replication, then induction of cell growth and devision would result and be maintained as long as T antigen was present. This putative effect of TAFs would explain, in large part, the transformed state induced by T antigen and may indicate a general mechanism for viral transformation and viral diseases which alter the phenotype of cells.

REFERENCES

1. Abmayr, S.M., L.D. Feldman and R.G. Roeder. 1985. In vitro stimulation of specific RNA polymerase. 11-mediated transcription by the pseudorabies virus immediate early protein. Cell 43:821-829.

2. Alwine, J.C. 1985. Transient gene expression control: effects of transfected DNA stability and trans-activation by viral early proteins. Mol. Cell. Biol. 5:1034-1042.

3. Alwine, J.C., and J. Picardi. 1986. Activity of the Simian Virus late promoter elements in the absence of large T-antigen: Evidence for repression of late gene expression. Submitted to J. Virol.

4. Brady, J., J.B. Bolen, M. Radonovich, N.P. Salzman, and G. Khoury. 1984. Stimulation of simian virus 40 late gene expression by simian virus 40 late tumor antigen. Proc. Natl. Acad. Sci. U.S.A. **81**:2040–2044.

5. Brady, J., and G. Khoury. 1985. Trans-activation of simian virus 40 late transcription unit by T antigen. Mol. Cell. Biol. **5**:1391–1399.

6. Dynan, W.S., and R. Tjian. 1983. The promoter-specific transcription factor Spl binds to upstream sequences in the SV40 early promoter. Cell **35**:79–87.

7. Ernoult-Lange, M., and E. May. 1983. Evidence of transcription from the late region of the integrated simian virus 40 genome in transformed cells: location of the 5' ends of late transcripts in cells abortively infected and in cells transformed by simian virus 40. J. Virol. **46**:756–767.

8. Ernoult-Lange, M., P. May, P. Moreau, and E. May. 1984. Simian virus 40 late promoter region is able to initiate simian virus 40 early gene transcription in the absence of the simian virus 40 origin sequence. J. Virol. **50**:163–173.

9. Green, M.R., R. Treisman, and T. Maniatis. 1983. Transcriptional activation of cloned human beta-globin genes by viral immediate-early gene products. Cell **35**:137–148.

10. Hartzell, S.W., B.J. Byrne, and K.N. Subramanian. 1984. The SV40 minimal origin and the 72 base pair repeats are required simultaneously for efficient induction of late gene expression with large tumor antigen. Proc. Natl. Acad. Sci. U.S.A. **81**:6335–6339.

11. Ihara, S., L. Feldman, S. Watanabe, and T. Ben-Porat. 1983. Characteristics of immediate-early functions of pseudorabies virus.

Virology **131**:437-454.

12. Imperiale, M.R., L.T. Feldman, and J. R. Nevins. 1983. Activation of gene expression by adenovirus and herpesvirus regulatory genes acting in trans and by cis-acting adenovirus enhancer elements. Cell **35**:127-136.

13. Imperiale, M.J., H.-T. Kao, L.T. Feldman, J.R. Nevins, and S. Strickland. 1984. Common control of the heat shock gene and early adenovirus genes: evidence for a cellular Ela-like activity. Mol. Cell. Biol. **4**:867-874.

14. Jones, N., and T. Shenk. 1979. An adenovirus 5 early gene product function regulates expression of other early viral genes. Proc. Natl. Acad. Sci. U.S.A. **76**:3665-3669.

15. Keller, J.M., and J.C. Alwine. 1984. Activation of the SV40 late promoter: direct effects of T antigen in the absence of viral DNA replication. Cell **36**:381-389.

16. Keller, J.M., and J.C. Alwine. 1985. Analysis of an activatable promoter: sequences in the Simian Virus 40 late promoter required for T-antigen-mediated trans-activation. Mol. Cell. Biol. **5**:1859-1869.

17. Manos, M.M., and Y. Gluzman. 1985. Genetic and biochemical analysis of transformation-competent replication-defective Simian Virus 40 large T-antigen mutants. J. Virol. **52**:120-127.

18. Nevins, J.R. 1981. Mechanism of activation of early viral transcription by adenovirus Ela product. Cell **26**:213-220.

19. Rio, D.C., and R. Tjian. 1984. Multiple control elements involved in the initiation of SV40 late transcription. J. Mol. Appl. Genet. **2**:423-435.

20. Ruley, H.E. 1983. Adenovirus early region 1A enables viral and cellular transforming genes to transform primary cells in culture. Nature (London) **304**:602-606.

21. Shiroki, K., and H. Shimojo. 1971. Transformation of green monkey kidney cells by SV 40 genome: the establishment of transformed cell lines and the replication of human adenovirus and SV40 in transformed cells. Virology **45**:163-171.

22. Tack, L.C., and P. Beard. 1985. Both trans-acting factors and chromatin structure are involved in the regulation of transcription from the early and late promoters in simian virus chromatin. J. Virol. **54**:207-218.

23. Tooze, J. (ed.). 1981. DNA tumor viruses: molecular biology of tumor virus. 2nd ed. Cold Spring Harbor Laboratory, Cold Spring Harbor, N.Y. p. 799-841.

24. Velcich, A., and E. Ziff. 1985. Adenovirus Ela proteins repress transcription from the SV40 early promoter. Cell **40**:705-716.

Viruses and Human Cancer, pages 283–296

REVERSE TRANSCRIPTASE INHIBITORS PROLONG LIFE OF RETROVIRUS-INFECTED MICE[1]

Ruth M. Ruprecht and Lucia D. Rossoni

Dana-Farber Cancer Institute, Boston MA 02115

ABSTRACT Reverse transcriptase (RT) is key to retroviral pathogenesis. We developed murine models to show that: a) inhibition of RT decreases viral titers, b) RT inhibitors must be administered continuously once proviral integration has occurred in order to protect uninfected target cells, and c) continuous administration of RT inhibitors prolongs life of animals infected with pathogenic retroviruses. BALB/c mice were infected with Rauscher Murine Leukemia Virus (MuLV) and either mock-treated with saline or given suramin at 40 mg/kg twice per week i.v. Suramin treatment not only led to a significant decrease in virus-induced splenomegaly but also prolonged the median survival from 35.5 days for infected mock-treated mice to 83 days for infected suramin-treated animals (p=0.0025). Actinomycin D was also effective in decreasing Rauscher MuLV-induced splenomegaly. No synergy was seen when Rauscher MuLV-infected BALB/c mice were treated with both suramin and actinomycin D. Newborn NSF/N mice were infected with the T-cell tropic MuLV SL3-3 and either mock-treated with saline or given suramin at 40 mg/kg per week i.v. Suramin-treated animals had a median survival of 150 days as compared to 91 days for the mock treated, virus-infected mice (p=0.02). We propose that our murine systems are cost effective animal models for testing the biological effectiveness of candidate drugs directed against the RT of T-cell tropic retroviruses such as HTLV-III.

[1] This work was supported in part by a contract from the State of Massachusetts, and grants from the Elsa Pardee Foundation and the Sandoz Foundation to RMR.

INTRODUCTION

Since 1980, when the first human retrovirus was discovered (1), the following diseases have been linked to retroviruses: T-cell leukemia/lymphoma endemic in either southern Japan (2) or the Caribbean Basin (3), the T-cell variant of hairy cell leukemia (4), and the Aquired Immune Deficiency Syndrome (AIDS) (5,6). Among these illnesses, a retroviral etiology is accepted for T-cell leukemia/lymphoma, associated with the human T-cell leukemia virus type I, HTLV-I (1) as well as for AIDS, associated with the human T-lymphotropic virus type III, HTLV-III (5), also called lymphadenopathy associated virus (LAV) (6). For AIDS, the prognosis is grave (7), and to date no curative therapy exists. The spreading AIDS epidemic calls for development of therapeutic intervention.

Retroviruses of different species are alike in their life cycles. Viral propagation and successful establishment of infection in the host cell crucially depends on reverse transcriptase (RT). Since normal mammalian cells do not contain this enzymatic activity, RT represents an ideal target for development of antiretroviral therapy, and since it differs in its structure and mode of action from the mammalian DNA polymerases, compounds could theoretically be found that inhibit RT to a high degree with little effect on the DNA polymerases of the host cell. The *pol* gene encoding RT is not affected by an abnormally high mutation rate such as the one found for certain regions of the *env* gene in HTLV-III (8).

The amino acid sequence of RT has been remarkably conserved through evolution of retroviruses of different species. The RT of LAV has a 250 amino acid domain which is 38% homologous to Rous sarcoma virus (RSV), 28% to HTLV-I and 21% to Moloney murine leukemia virus (MuLV) (9). This domain is thought to represent the core amino acid sequence of RT. Thus, this enzyme appears to be under evolutionary pressure to conserve the amino acid sequences related to enzyme functions which include not only RNA-directed DNA synthesis and DNA-directed synthesis but also ribonuclease H activity. The *pol* region furthermore encodes a protease as well as a DNA endonuclease required for integration. Some of these complex functions are unique to retroviruses and represent, therefore, logical targets for the development of antiretroviral compounds. Due to the high degree of amino acid homology of retroviral RTs derived from different species, murine retroviruses could be used to study the biological effectiveness of candidate antiretroviral drugs intended for treatment of human diseases such as AIDS while ascertaining _in vitro_ activity against HTLV-III in parallel. The advantage of such an approach is the rapidity and cost-effectiveness with which candidate drugs could be evaluated (10).

The antitrypanosomal drug suramin (11) was evaluated for antiretroviral activity in our murine systems _in vitro_ and _in vivo_. De Clerq found previously that it competitively inhibited the RT activity of avian and murine viruses (12). Suramin inhibited the cytopathic effects of HTLV-III on T-helper cells cocultivated with a virus-producing cell line (13), and in a clinical trial, decreased HTLV-III titers in AIDS patients (14). Since follow-up and drug treatment were only for a limited time, no conclusions about benefit in survival of treated patients could be drawn.

Our murine retroviral systems included the erythrotropic Rauscher MuLV (15) and the T-cell tropic SL3-3 (16). Rauscher MuLV-infected mice develop splenomegaly in 2 weeks and die of leukemia in 30-40 days. The degree of splenomegaly is proportional to the viral titer (17). This short incubation period allows rapid determination of effective dosage. Biological activity of suramin against T-cell tropic retroviruses was then studied in NSF/N mice infected with SL3-3, which induces thymic lymphoma within 60-150 days after a period of viremia.

Suramin was also studied for synergy with actinomycin D, a known RT inhibitor (18,19).

MATERIALS AND METHODS

Cell lines, virus preparations and BALB/c mice were obtained from sources previously described (10). NSF/N mice, originally obtained from the National Institutes of Health, were bred in our facility. Actinomycin was obtained from Sigma Chemical Co., St. Louis, MO.

Previously published methods were used except as follows. Newborn NSF/N littermates were injected i.p. with 1.1×10^5 plaque forming units (pfu) of SL3-3 or normal saline. Four hr later, suramin at 10 or 40 mg/kg or normal saline was given and repeated weekly. All dead mice were autopsied, and 150 days post-inoculation all mice were sacrificed.

Suramin levels were determined by the method of Ruprecht _et al_. (20).

RESULTS

Inhibition of RT Activity by Suramin.

The effect of suramin on the RT activity of SL3-3 and Rauscher MuLV is shown in Fig. 1. The IC_{50} for both viruses is 6-7 µg/ml of suramin.

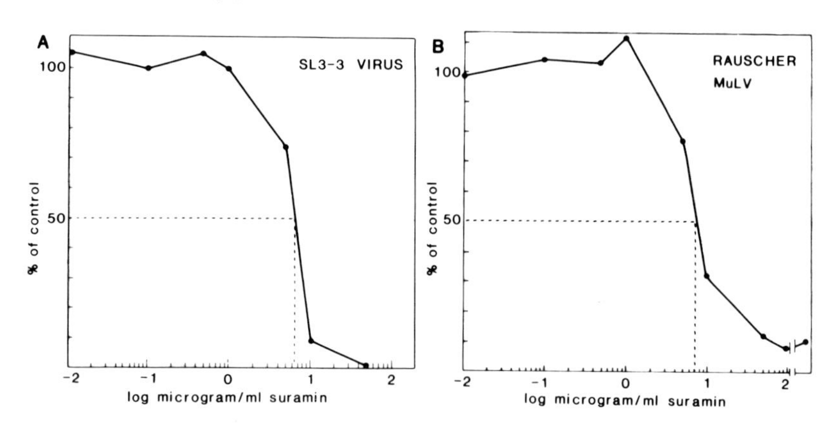

FIGURE 1. Inhibition of RT activity of crude preparations of SL3-3 (A) or Rauscher (B) MuLV as a function of suramin concentration.

Protection of Murine T Cells from SL3-3 Viral Infection.

Figure 2 shows the effect on viral titer and RT activity in tissue culture supernatants if murine L691-6 cells are infected in the presence of various concentrations of suramin. Pretreatment and continuous presence of suramin led to a 2-3 logarithmic decrease in viral titer at 100 µg/ml of suramin with a concomitant decrease in RT activity.

Suramin Acts as a Virustatic Agent.

Figure 3 demonstrates the need for continued presence of suramin in cultures exposed to virus in order to prevent spread of the infection. L691 cells were pretreated for 24 hr with 100 µg/ml of suramin, infected with SL3-3 virus and grown in the presence of the same concentration of drug. Uninfected L691 cells and L691 cells infected in the

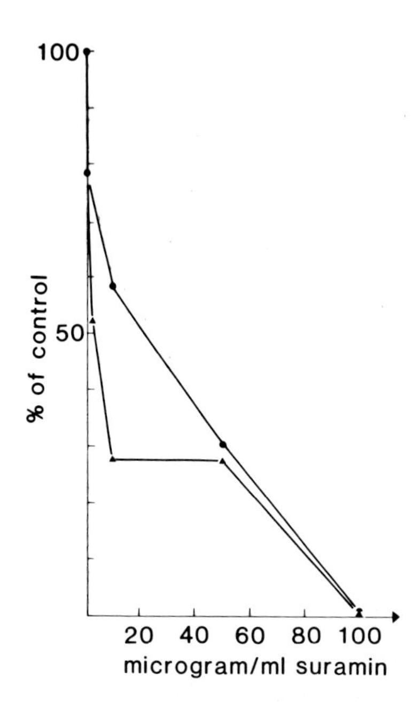

FIGURE 2. The percentage of residual RT activity (▲) and pfu (●) is shown as a function of suramin concentrations used before, during and after viral infection of L691 cells. Pfus were measured by the XC plaque assay (21).

absence of suramin served as controls. When the positive control cells showed a high level of RT activity 17 days later, the number of virus-positive cells was assayed in the infectious center assay (22). The results are shown in Fig. 3A. Sixty percent of the positive L691 control cells were virus positive, whereas infection of L691 cells in the presence of 100 µg of suramin per ml yielded 0.09% virus-positive cells. Suramin was removed from an aliquot of these latter cells, and all cultures were incubated again. Three days after removal of suramin, the supernatant of the L691 cells initially infected in the presence of 100 µg of suramin per ml showed very high RT activity. XC infectious center assays performed subsequently showed a dramatic increase in the number of virus-positive cells (Fig. 3B).

We conclude that only very few cells become infected by high titers of virus in the presence of suramin, but that suramin must be present continuously to be effective. Thus, suramin is not a virucidal but rather a virustatic drug.

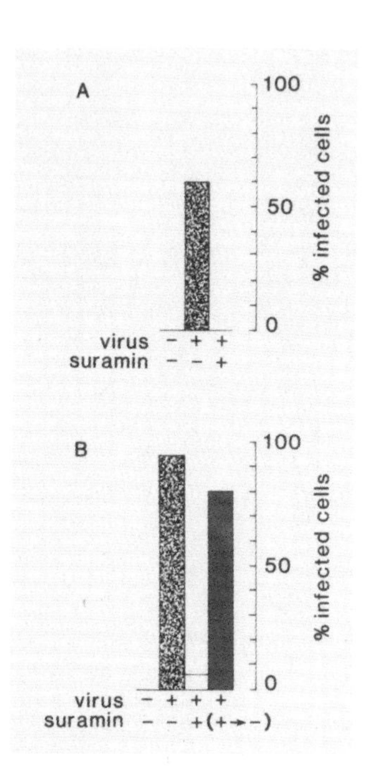

FIGURE 3. (A) Need for continuous presence of suramin. The percentage of virus-positive cells of uninfected L691 cells (left) and of L691 cells infected with SL3-3 virus without (stippled bar) or with suramin present at 100 μg/ml (right) is shown. (B) Same as (A). Suramin was removed from an aliquot of L691 cells infected in the presence of 100 μg/ml of suramin (black bar), and cultures were assayed for infectious cells 6 days later.

Suramin Does Not Inhibit Release of Virus from Infected Cells.

Next, we analyzed whether suramin has any effect on the second part of the retroviral life cycle, i.e. on the steps leading to the release of infectious virions encoded by proviral sequences already integrated into the host cell genome. L691 cells containing the SL3-3 provirus and producing high titers were grown in the presence of 100 µg/ml of suramin for 36 hr. Uninfected L691 cells and virus-producing L691 cells grown in the absence of suramin served

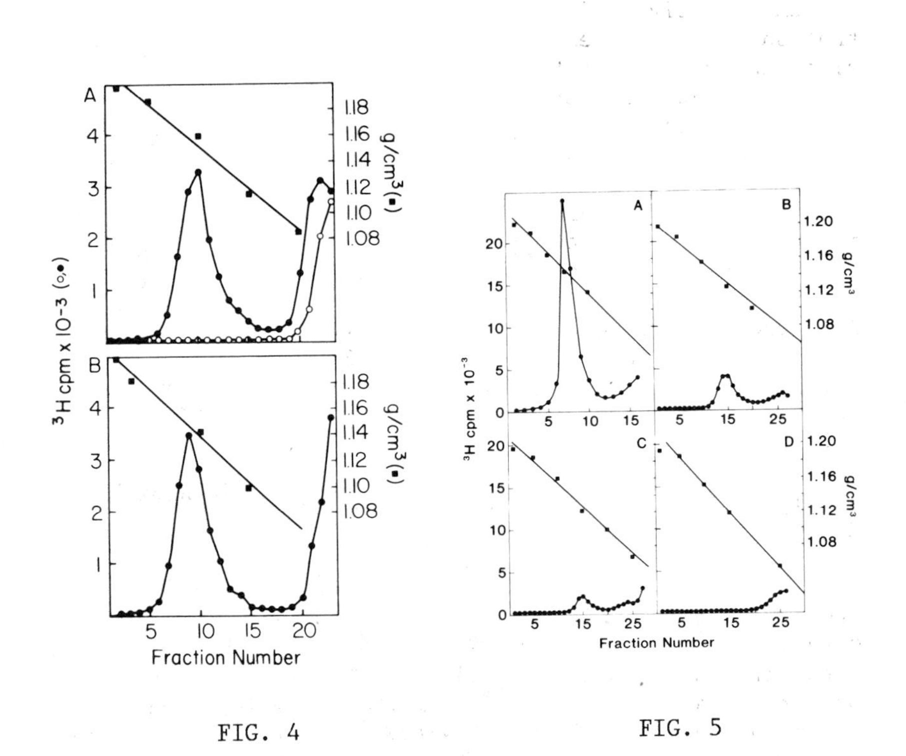

FIGURE 4. Release of [^{3}H] uridine labelled virus from an established SL3-3 producing L691 culture. Infected cells were grown in the absence (4A, ●) or in the presence of 100 µg/ml of suramin (4B, ●); uninfected L691 cells (4A, ○) served as negative control. Virus particles were labelled with [^{3}H] uridine for 12 hr, concentrated and analyzed on 10-50% sucrose gradients. The TCA precipitable radioactivity per fraction is shown.

FIGURE 5. Inhibition of release of [^{3}H] uridine-labelled virus from SL3-3 producing L691 cells by actinomycin D at various concentrations. A, no actinomycin D; B, 3 ng/ml; C, 5 ng/ml; and D, 8 ng/ml. For details, see Figure 4.

as controls. During the last 12 hr, the cells were labelled with 10 μCi/ml of [^{3}H] uridine. Virus was spun out of tissue culture supernatants and analyzed by sucrose density gradient centrifugation. Fig. 4A shows the amount of radioactivity banding a density of 1.14 to 1.16 g/cm^3 for the positive and negative controls. Fig. 4B indicates that suramin treatment of virus-producing cells did not alter the release of the virus. SL3-3 virus released from suramin-treated infected cells was as infectious as control virus (data not shown).

In contrast to suramin, actinomycin D was able to decrease the number of virions released per infected cell. Actinomycin D has well-studied antiretroviral properties: it inhibits the synthesis of DNA-minus strand in the RT reaction (19), and cellular RNA polymerase II is very sensitive to its action. As expected, actinomycin D effectively decreased the release of viral particles. Short exposure (36 hr) of uninfected L691 cells to 3 to 8 ng/ml of actinomycin D led to a reversible decrease in cell growth, whereas longterm exposure (6 days) eventually led to growth arrest.

In Vivo Testing of Suramin and Actinomycin D in Rauscher MuLV-Infected BALB/c Mice.

Groups of 8 female BALB/c mice (16 mice for group B) age 7 weeks were injected i.v. with 1.5x10^4 pfu of Rauscher MuLV or normal saline (Table I). Four hr later, treatment with suramin at 40 mg/kg, actinomycin D at 0.1 mg/kg, both, or normal saline was started. Suramin was repeated one day later and subsequently given every 3 days. Pharmacokinetic analysis showed that this regimen yields suramin peak levels of 150 μg/ml and valley levels of 30 μg/ml (20). Actinomycin D was given at the same intervals. On day 20 post-inoculation, all mice were sacrificed and their spleen weights were recorded. Untreated and mock-infected, drug-treated animals served as controls. The observed increases in spleen weights of virus-infected mice were corrected for the small degree of enlargement caused by drug treatment only. Suramin treatment alone led to a highly significant reduction of splenomegaly. Actinomycin D as a single agent inhibited development of splenomegaly even to a greater degree. A suboptimal dosage had been used for actinomycin D (18) to allow for evaluation of synergy between suramin and actinomycin D. Both agents interfere with the RT reaction, albeit by different mechanisms. Suramin presumably binds to the template-primer site of RT which can be inferred by its action as competitive inhibitor with regard to template-primer (12). Actinomycin D, on the other hand, intercalates into DNA at deoxyguanosine residues. In addition, cellular RNA polymerases are very sensitive to actinomycin D which can lead to decreased transcription of proviral sequences. Combination of the two agents could lead to greater inhibition of viral infection, resulting in a lesser degree of viremia and thus less splenomegaly. However, a small but significant interference was noted when suramin and actinomycin D were used simultaneously. One possible explanation for this finding could be inactivation of the more active agent actinomycin D by the strongly negatively charged suramin, which is known to bind avidly to a large number of proteins (11). Actinomycin D with its peptide side chains could thus possibly become inactivated.

TABLE I.

Group	Virus	Suramin 40 mg/kg	Act D 0.1 mg/kg	Spleen weight (mg)	% inhibition (p-value, two-sided)
A	-	-	-	118 ± 18	
B	+	-	-	1506 ± 334	
C	-	+	-	151 ± 20	
D	-	-	+	145 ± 16	
E	-	+	+	193 ± 31	
F	+	+	-	766 ± 185	55.7 (<0.0001)[a]
G	+	-	+	259 ± 49	91.8 (<0.0001)[b]
H	+	+	+	364 ± 99	87.7 (0.003)[c] (0.02)[d]

[a] Two-sided p value for B vs F.
[b] Two-sided p value for B vs G.
[c] Two-sided p value for F vs H.
[d] Two-sided p value for G vs H.

To assess whether decreases in splenomegaly, an indirect measure of viremia, leads to a benefit in survival, groups of 8 Rauscher MuLV-infected BALB/c mice were given normal saline or suramin at 40 mg/kg i.v. every 3 days for a total of 19 doses. Untreated mice and mice receiving drug only served as controls. No deaths occurred among these two latter groups. The median survival of virus-infected, mock-treated mice was 35.5 days, whereas that of virus-infected, suramin-treated mice was 83 days ($p=0.0025$ by log-rank test). Theoretically, suramin could prolong survival of infected mice by killing virally infected cells rather than by decreasing viral infection and spread to new target cells, i.e. by cytotoxic rather than antiretroviral action. To address this question, a third group of 8 mice was infected with Rauscher MuLV, and no treatment was given until day 19 when suramin was started at 40 mg/kg as a loading dose, followed by the same dose on the next day and every three days subsequently for a total of 6 doses. Median survival was 50.5 days ($p=0.17$, not sig.). Physical examination of these animals during treatment revealed no decrease in the splenomegaly.

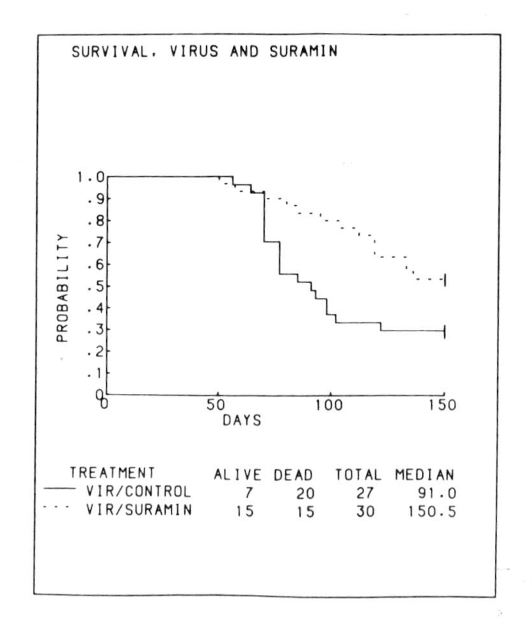

Figure 6. Kaplan-Meier survival analysis of NSF/N mice.

In Vivo Testing of Suramin in NSF/N Mice Infected with the T-Cell Tropic Virus SL3-3.

Newborn NSF/N littermates were randomly assigned to receive either SL3-3 virus or normal saline i.p. Four hrs later, saline or suramin was given at a dose of 40 mg/kg. Treatment was repeated weekly. After weaning, suramin was given i.v. On day 150 post-inoculation, all mice were sacrificed and autopsied, as well all mice that had died during the time span. Survival analysis was carried out using the Kaplan-Meier method (Fig. 6). No deaths occurred in mock-infected, mock-treated and mock-infected, suramin-treated control mice.

DISCUSSION

The AIDS epidemic calls for development of therapeutic interventions. Inhibitors of retroviral RT are logical antiviral candidate drugs for the following reasons: 1) RT is virally encoded, 2) normal host cells do not possess RT activity, and 3) without RT, retroviral infection is abortive.

We have correlated *in vitro* inhibition of RT activity and *in vitro* inhibition of retroviral propagation with biological effectiveness of candidate antiretroviral drugs in mice. Suramin, an agent undergoing clinical testing in AIDS patients, has previously been shown to be active against HTLV-III *in vitro*, but whether treatment results in a significant increase in survival has not been determined in any system yet.

We have employed our murine models to address the question of survival benefit. Since *in vitro* analysis showed a virustatic rather than virucidal mode of action for suramin, longterm treatment was administered to the mice. Suramin significantly prolonged the life of mice infected with either the erythrotropic Rauscher MuLV or the T-cell tropic SL3-3.

The mechanism of action of suramin *in vivo* raises interesting questions. The observed benefit in survival could be due to the following: 1) inhibition of RT, leading to a decrease in the number of provirus-containing host cells and consequently to a slower amplification of virus, 2) interference with virus binding to the receptors on cell membranes, or 3) cytotoxicity to virus-positive cells. A fourth possibility, namely interference with steps in the second part of the retroviral life cycle, namely transcription, translation and assembly of viral protein products followed by release of infectious virions, seems unlikely since no decrease of radioactively-labelled virus was seen _in vitro_ when virus-producing cells were grown in the presence of maximally tolerated concentrations of suramin. The other possibilities are more difficult to distinguish, and are not mutually exclusive. Studies with labelled suramin should be able to answer some of these questions.

Suramin at the doses tested was well tolerated by the animals clinically. Due to the well-known, sometimes severe side effects of suramin in humans, chronic treatment may not be possible, and the maximally tolerated length of treatment may not be long enough to lead to a benefit in survival. Less toxic, more effective antiretroviral agents may need to be developed. The murine retroviral systems described here are well-suited for preclinical evaluation of candidate RT inhibitors. Both toxicity and biological effectiveness can be evaluated cost-effectively due to the short incubation times of the viruses used.

REFERENCES

1. Poiesz BJ, Ruscetti FW, Gazdar AF, Bunn PA, Minna JD, Gallo RC (1980). Detection and isolation of type C retrovirus particles from fresh and cultured lymphocytes of a patient with cutaneous T-cell lymphoma. Proc Natl Acad Sci USA 77:7415.
2. Kalyanaraman VS, Sarngadharan MG, Nakao Y, Ito Y, Aoki T, Gallo RC (1982). Natural antibodies to the structural core protein (p24) of the human T-cell leukemia (lymphoma) retrovirus found in sera of leukemia patients in Japan. Proc Natl Acad Sci USA 79:1653.
3. Blattner WA, Takatsuki K, Gallo RC (1983). Human T-cell leukemia lymphoma virus and adult T-cell leukemia. J Am Med Assoc 250:1074.
4. Kalyanaraman VS, Sarngadharan MG, Robert-Guroff M, Miyoski I, Blayney D, Golde D, Gallo RC (1982). A new subtype of human T-cell leukemia virus (HTLV-II) associated with a T-cell variant of hairy cell leukemia. Science 218:571.
5. Popovic M, Sarngadharan MG, Read E, Gallo RC (1984). Detection, isolation and continuous production of cytopathic retroviruses (HTLV-III) from patients with AIDS and pre-AIDS. Cell 224:497.

6. Barre-Sinoussi F, Chermann JC, Rey F, Nugeyre MT, Chamaret S, Gruest J, Dauguet C, Axler-Blin C, Vezinet-Brun F, Rouzioux C, Rozenbaum W, Montagnier L (1983). Isolation of a T-lymphotropic retrovirus from a patient at risk for Aquired Immune Deficiency Syndrome (AIDS). Science 220:868.
7. Curran JW, Morgan WM, Hardy AM, Jaffe HW, Darrow WW, Dowdle WR (1985). The epidemiology of AIDS: current status and future prospects. Science 229:1352.
8. Hahn BH, Gonda MA, Shaw GM, Popovic M, Hoxie JA, Gallo RC, Wong-Staal F (1985). Genomic diversity of the acquired immune deficiency syndrome virus HTLV-III: different viruses exhibit greatest divergence in the envelope genes. Proc Natl Acad Sci USA 82:4813.
9. Wain-Hobson S, Sonigo P, Danos O, Cole S, Alizon M (1985). Nucleotide sequence of the AIDS virus, LAV. Cell 40:9.
10. Ruprecht RM, Rossoni LD, Haseltine WA, Broder S (1985). Suppression of retroviral propagation and disease by suramin in murine systems. Proc Natl Acad Sci USA 82:7733.
11. Hawking F (1978). Suramin: with special reference to onchocerciasis. Adv Pharmacol Chemother 15:289.
12. De Clerq E (1979). Suramin: a potent inhibitor of the reverse transcriptase of RNA tumor viruses. Cancer Lett 8:9.
13. Mitsuya H, Popovic M, Yarchoan R, Matsushita S, Gallo RC, Broder S (1984). Suramin protection of T cells in vitro against infectivity and cytopathic effect of HTLV-III. Science 226:172.
14. Broder S, Yarchoan R, Collins JM et al (1985). Effects of suramin on HTLV-III/LAV infection presenting as Kaposi's sarcoma or AIDS-related complex: clinical pharmacology and suppression of virus replication in vivo. The Lancet ii:627.
15. Rauscher F (1962). A virus-induced disease of mice characterized by erythrocytopoiesis and lymphoid leukemia. J Natl Cancer Inst 29:515.
16. Pedersen FS, Crowther RL, Tenney DY, Reimold AM, Haseltine WA (1981). Novel leukaemogenic retroviruses isolated from cell line derived from spontaneous AKR tumour. Nature (London) 292:167.
17. Chirigos MA (1964). Studies with the murine leukemogenic Rauscher virus. III. An in vivo assay for anti-viral agents. Cancer Res 24:1035.
18. Steeves RA, Mirand EA, Price FW (1968). Effects of

hydroxyurea (NSC-32065) and actinomycin D (NSC-3053) on several parameters of infection by Friend spleen focus-forming virus. Cancer Chemother Rep 52:557.

19. Ruprecht RM, Goodman NC, Spiegelman S (1973). Conditions for the selective synthesis of DNA complementary to template RNA. Biochim Biophys Acta 294:192.
20. Ruprecht RM, Lorsch J, Trites DH (1986). Analysis of suramin plasma levels by ion pair high performance liquid chromatography under isocratic conditions. J Chromatogr (in press).
21. Rowe WP, Pugh WE, Hartley JW (1970). Plaque assay techniques for murine leukemia viruses. Virology 42:1136.
22. Hartley JW, Rowe WP (1975). Clonal cell lines from a feral mouse embryo which lack host-range restrictions for murine leukemia viruses. Virology 65:128.

Viruses and Human Cancer, pages 297–308

ISOLATION OF A T-LYMPHOTROPIC RETROVIRUS FROM HEALTHY MANGABEY MONKEYS[1]

Patricia Fultz, Harold McClure,[2] Rita Anand, and A. Srinivasan

AIDS Program, Centers for Disease Control
Atlanta, Georgia 30333
[2]Yerkes Regional Primate Research Center, Emory University, Atlanta, Georgia 30322

ABSTRACT A T-lymphotropic retrovirus was isolated from peripheral blood mononuclear cells (PBMC) from 14 of 15 sooty mangabey monkeys (*Cercocebus atys*) that were randomly selected from among 88 healthy animals in the Yerkes Primate Center colony. Antibodies to the virus were detected by immunofluorescence and Western blot in serum samples from all virus-positive animals. Some of the mangabeys had low levels of serum neutralizing antibodies to the virus, which we call STLV-III/SMM. The new retrovirus was related to the human virus HTLV-III/LAV/ARV morphologically and serologically, was cytopathic for human $OKT4^+$ cells, and was composed of proteins with molecular weights very similar to those of the proteins of HTLV-III/LAV/ARV. However, no homology between the mangabey virus and HTLV-III/LAV/ARV was detected by RNA-DNA and DNA-DNA hybridization. Intravenous injection of STLV-III/SMM into rhesus macaques (*Macaca mulatta*) resulted in seroconversion and isolation of virus from PBMC of 5 of 6 inoculated animals. At 26 weeks after infection, one macaque was lymphopenic and had a 20% weight loss, suggesting that STLV-III/SMM may be pathogenic in rhesus macaques, although it apparently does not cause clinical immunodeficiency or disease in the host from which it was isolated.

[1]This work was supported in part by NIH grant AI 19302 and base grant RR-00165 to Yerkes Regional Primate Research Center. The Yerkes Center is fully accredited by the American Association for Accreditation of Laboratory Animal Care.

INTRODUCTION

Retroviruses cytopathic for T lymphocytes have been associated with immune deficiency diseases in both humans and nonhuman primates. There is convincing evidence that the etiologic agent of acquired immunodeficiency syndrome (AIDS) is a retrovirus called T-lymphotropic virus type III/lymphadenopathy-associated virus/AIDS-associated retrovirus (HTLV-III/LAV/ARV) (1-4). A serologically related virus, simian T-lymphotropic virus type III (STLV-III), was isolated (5) from diseased rhesus macaques (Macaca mulatta), an Asian species of Old World monkeys, and causes an AIDS-like disease in experimentally infected macaques (6). Antisera against either STLV-III or HTLV-III/LAV/ARV cross-reacted with the other virus to varying extents; in virus-infected cells the greatest reciprocal cross-reactivity was to virus proteins with molecular weights of approximately 55,000 (p55), and 24,000 (p24) daltons (7,8). The p55 and p24 proteins are encoded by the gag region of the viral genome.

Recently, a second group of STLV-III isolates was obtained from apparently healthy, wild-caught African green monkeys (Cercopithecus aethiops) and was designated STLV-III/AGM (9). Antibodies to STLV-III/AGM and HTLV-III/LAV/ARV also showed the same reciprocal cross-reactivity that was seen with the prototype STLV-III. The two simian viruses share other properties with the human isolates: they have the same virion morphology by electron microscopy, a Mg^{2+}-dependent reverse transcriptase (RT) enzyme, and a tropism for T-helper ($OKT4^{+}/Leu3^{+}$) lymphocytes.

We report here the isolation from another African primate species--the sooty mangabey monkey (Cercocebus atys)--of a retrovirus that shares all of the above properties with HTLV-III/LAV/ARV, STLV-III, and STLV-III/AGM; therefore, we will refer to the new virus as STLV-III/SMM. The mangabey monkeys from which STLV-III/SMM was isolated were selected at random from among members of a large breeding colony at the Yerkes Regional Primate Research Center. No cases of AIDS-like disease have been documented in the mangabey colony since it was established approximately 18 years ago from wild-caught animals. Therefore, STLV-III/SMM resembles STLV-III/AGM in that it apparently does not cause disease in the host from which it was isolated, whereas STLV-III and HTLV-III/LAV/ARV were isolated from species that develop immunodeficiency diseases after infection.

METHODS

Isolation and Characterization of Virus

The 15 mangabey monkeys used in this study were between 4 and 22 years old and were randomly selected from among 88 animals housed together in a compound at the Yerkes Primate Center. Peripheral blood mononuclear cells (PBMC) and serum were obtained from each animal, all of which were clinically normal. Virus was isolated from mangabey PBMC by coculture with PHA-stimulated adult human white blood cells (PHA-AWBC) and periodic assay of culture supernatants for Mg^{2+}-dependent RT activity as described (3). Characterization of STLV-III/SMM utilized published procedures for radioimmunoprecipitation (RIP) assays (10), Western blot (11), dot-blot hybridization (12), and Southern hybridization (13). The HTLV-III/LAV/ARV full-length genomic probes for DNA hybridization were prepared from molecular clones generated at the Centers for Disease Control as follows: pH9, from HTLV-III-infected H9 cells, provided by Dr. R. C. Gallo; pAS, from virus isolated from a transfusion-associated AIDS patient; and pZ6, from virus isolated from a Zairian AIDS patient. HTLV-II was cloned from 2693 cell cultures, and HTLV-I was cloned from the MT-2 cell line (14). High-molecular-weight DNA was extracted (15) from PHA-AWBC infected with STLV-III/SMM and digested with restriction enzymes before hybridization.

Neutralization Assay.

Virus was incubated with 1:10 dilutions of monkey serum for 60 min at room temperature, then was used to infect PHA-AWBC. After overnight incubation to allow for virus adsorption, the cells were washed, resuspended in fresh medium, and culture supernatants were assayed for RT activity on days 6, 9, 12, and 16. Neutralizing antibody titers are defined as the dilution of serum resulting in 80% inhibition of RT activity.

RESULTS

Virus Isolation.

PBMC from 15 randomly-selected mangabey monkeys were placed in culture with PHA-AWBC, and culture supernatants were monitored for particulate RT activity. From 6 to 23 days after establishment, 14 of 15 cultures had measurable RT activity of greater than 25,000 cpm per ml; RT activity increased to levels of from 5 x 10^5 to 3 x 10^6 cpm per ml in all but one culture. The virus-negative culture was discarded after 45 days.

Characterization of STLV-III/SMM.

Morphologic. Some of the RT-positive cultures were examined by electron microscopy. All cultures that were analyzed had retroviral particles with dense, eccentric nucleoids that appeared to be identical to both the human virus HTLV-III/LAV/ARV and the simian STLV-III viruses (16).

Biologic. Supernatants from the original cocultures that had maximum RT activity were used as virus stocks and were able to transfer infectivity to fresh PHA-AWBC. The host-range of some STLV-III/SMM isolates was tested by infecting human cell lines of different lineages. While not all isolates grew in all cell lines, some STLV-III/SMM isolates replicated in the human T-cell lines HUT-78, 6D5 (a clone of HUT-78), and HT. None of the isolates tested replicated to detectable levels in the B-cell line Raji, the myeloid cell line K-562, or SK-N-SH, a human neuroblastoma cell line. One isolate that was tested for the ability to replicate in PBMC from rhesus macaques reached maximum RT values of greater than 6 x 10^6 cpm per ml. Two other STLV-III/SMM isolates also replicated in rhesus macaques in vivo (see below). We have not been able to detect any replication of the LAV and ARV strains of the human AIDS virus in PBMC of rhesus macaques (16).

Since HTLV-III/LAV/ARV and STLV-III are tropic for T-helper cells identified by monoclonal antibodies OKT4 and Leu3 (4,5), we tested STLV-III/SMM for its ability to replicate preferentially in lymphocytes with a Leu3 phenotype. Parallel cultures of PHA-AWBC were uninfected or infected with STLV-III/SMM, and the number and percentage of Leu3 and

TABLE 1
VIRUS-SPECIFIC PROTEINS IMMUNOPRECIPITATED BY SERUM TO STLV-III/SMM and HTLV-III/LAV/ARV[a]

Cells infected with	Mangabey anti-STLV-III	Serum Normal rabbit	Rabbit anti-HTLV-III
STLV-III/SMM	p19	none	p19, p24, gp40 p55, gp120-160
HTLV-III/CDC451	p18	none	p24, gp41, p55 gp120-160
No virus	none	none	none

[a]PHA-AWBC infected with STLV-III/SMM or HTLV-III/LAV/ARV were labeled with [^{35}S]methionine and [^{35}S]cysteine for 1 hr. Proteins were immunoprecipitated and separated on a 10% acrylamide resolving gel.

Leu2 (cytotoxic/suppressor cells also identified by OKT8) were determined by double-label immunofluorescence assay (IFA) (Becton/Dickinson Test Kit). The T4:T8 ratio stayed relatively constant (greater than 1.55) in the uninfected culture, but decreased to 0.44 in the STLV-III/SMM-infected culture as RT activity in the culture supernatant increased (16).

Antigenic. Serum samples from the 15 mangabeys were screened for reactivity to HTLV-III antigens by enzyme immunoassay (EIA) and Western blot. Serum from only two of the mangabeys were positive when tested at a 1:50 dilution with the Abbott EIA Test Kit; however, sera from 11 mangabeys bound to HTLV-III p24, but not to gp41, when tested at a 1:100 dilution by Western blot, using a mini-gel system. When these same sera were tested by indirect IFA for binding to STLV-III/SMM-infected cells, all of the sera were positive except the sample from the virus-negative animal. The mangabey sera showed very little or no reactivity against LAV-infected cells by IFA. In a preliminary experiment, some of the mangabey sera were tested at a 1:100 dilution by Western blot using partially purified STLV-III/SMM as antigen. Of ten serum samples tested, nine had antibodies to a protein with an apparent molecular weight of approximately 115,000 daltons, and fewer serum samples had antibodies to proteins

TABLE 2
HYBRIDIZATION OF NUCLEIC ACID FROM STLV-III/SMM WITH PROBES DERIVED FROM HUMAN T-LYMPHOTROPIC VIRUSES

Nucleic acid	Hybridization probe[a]				
	HTLV-III/LAV			HTLV-II	HTLV-I
	pH9	pAS	pZ6		
STLV-III/SMM viral RNA	-	-	-	-	-
STLV-III/SMM proviral DNA	-	-	-	-	-
HTLV-III/LAV/ARV proviral DNA	+	+	+	-	-

[a]See Methods for description of DNA probes.

with apparent molecular weights of 50,000, 26,000, 15,000, and 11,000 daltons. Therefore, serum containing antibodies to STLV-III/SMM cross-reacted with an HTLV-III/LAV/ARV gag region protein (p24), but not to an env region gene product (gp41).

To test for reciprocal cross-reactivity by antiserum to HTLV-III/LAV/ARV, high-titered chimpanzee anti-LAV and a hyperimmune rabbit antiserum to HTLV-III/LAV/ARV isolate CDC451 were used in IFA and RIP assays. The anti-LAV serum reacted with both LAV-infected and STLV-III/SMM-infected cells to approximately the same extent in the IFA. In RIP assays, the rabbit antiserum to CDC451 immunoprecipitated from STLV-III/SMM-infected cells proteins with approximate molecular weights of 120-160,000, 55,000, 40,000, 24,000, and 19,000 daltons (Table 1). The mobilities on the SDS-polyacrylamide gels of the STLV-III/SMM proteins analogous to HTLV-III/LAV/ARV gp41 and p18 were slightly different from those of the human virus.

Molecular. Despite extensive serologic cross-reactivity between STLV-III/SMM and HTLV-III/LAV/ARV, we detected no hybridization by RNA-DNA dot blot and DNA-DNA Southern hybridization techniques (Table 2). Also, there was no hybridization detected between the simian virus and HTLV-I or HTLV-II.

TABLE 3
ISOLATION OF STLV-III/SMM FROM PBMC OF RHESUS MACAQUES

Weeks after infection[a]	Rhesus macaque number 128	129	130	131	132	134
3	+	-	+	+	+	+
6	+	-	+	+	+	+
13	-	-	+	+	+	+
19	-	-	+	+	+	-
26	-	-	+	+	+	+

[a]All 6 animals received a second iv. inoculation of virus 19 weeks after the first injection of STLV-III/SMM.

Pathogenesis of STLV-III/SMM.

Persistent infection in mangabey monkeys. Even though STLV-III/SMM replicates in and is cytopathic for OKT4/Leu3$^+$ lymphocytes, the virus is apparently nonpathogenic in mangabey monkeys. Results of a retrospective study of serum samples obtained in 1981 from mangabeys in the Yerkes colony indicated that 82% (9 of 11 tested) of the animals had been exposed to the virus; however, the colony had no unusual incidence of disease compared to other species of monkeys at the Yerkes Primate Center. One mangabey (FNa) from which virus was isolated in the present study had serum antibodies to STLV-III/SMM in 1981. This animal appears well, and when tested 3 months after virus isolation, had a T4:T8 ratio of 1.25 with 37% OKT4$^+$ and 29.7% Leu2$^+$ lymphocytes. As stated above, all of the virus-positive mangabeys had serum antibodies to STLV-III/SMM; however, neutralizing antibody titers were low or nonexistent, *i. e.*, less than or equal to 10. Low neutralizing antibody titers also have been reported (17,18) for persons infected with the human AIDS virus.

Acute infection in rhesus macaques. Because prototype STLV-III causes an immunodeficiency disease in rhesus macaques, we injected two different isolates of STLV-III/SMM by the intravenous (iv.) route into three 12 to 14 month old macaques each. Virus was isolated from PBMC of 5 of the 6 animals at various times after infection (Table 3). Macaques 128 and 134 may have eliminated the virus from their peripheral blood by 19 weeks after infection because we were unable to isolate virus after 6 and 13 weeks, respectively. Because none of the macaques exhibited any signs of illness at 13 weeks, at 19 weeks after infection all six animals were given a second iv. inoculum of STLV-III/SMM. When virus isolation attempts were made 7 weeks after the second dose of virus, no virus was recovered from macaques 128 and 129, but macaque 134 was again virus positive.

Serum samples obtained from the macaques before virus inoculation were negative for antibodies to HTLV-III/LAV/ARV by EIA and to STLV-III/SMM by IFA. However, after inoculation all animals developed antibodies to STLV-III/SMM within 6 weeks, except the one animal (129) from which virus was never isolated. Whether macaque 129 successfully cleared the virus before detectable antibody levels were reached or harbors it as a latent infection is not known.

Serum samples from sequential bleedings of macaque 132 were tested at a 1:10 dilution for neutralizing antibodies to STLV-III/SMM using two different isolates, SMM-3 and SMM-9. Essentially no neutralization of SMM-9, the isolate with which macaque 132 was infected, occurred with serum obtained 6 and 13 weeks after infection. Serum obtained at 19 weeks resulted in only 43% inhibition of virus growth compared to preinfection serum, indicating minimal neutralization. However, when these same sera were tested with SMM-3, serum obtained at 19 weeks resulted in 82% inhibition of virus growth, indicating a neutralization titer of 10. These data indicate that some macaques do not develop high neutralizing antibody titers during acute infection with STLV-III/SMM and suggest that the various virus isolates from mangabey monkeys may differ in neutralizing epitopes.

None of the STLV-III/SMM-infected macaques has shown clinical signs of illness; however, macaque 131 had a 20% weight loss and a progressive decline in numbers of lymphocytes during 26 weeks of infection (Table 4). Also, a substantial drop in platelet numbers occurred at 13 weeks after infection. These data suggest that STLV-III/SMM may be pathogenic in rhesus macaques.

DISCUSSION

Retroviruses related to the human AIDS viruses appear to be prevalent in primate species and have been isolated from both Asian and African species of monkeys. The simian viruses have the same basic characteristics as the human AIDS viruses, which include morphology, Mg^{2+}-dependent RT, and

TABLE 4
WEIGHT AND HEMATOLOGIC DATA FOR MACAQUE 131 INFECTED WITH STLV-III/SMM

Weeks after inoculation	Weight (kg)	White cells	Lymphocytes	Platelets ($x\ 10^{-3}$/ul)
0	2.45	9600	5952	471
3	2.75	6100	4087	413
6	2.45	11700	3276	403
13	2.40	6400	3072	117
19	2.36	7200	2592	314
26	2.20	5500	1320	256

cell tropism. In addition, the two groups of viruses code for proteins with similar molecular weights, and serologic tests indicate that there is substantial cross-reactivity of antibodies to viral antigens. In STLV-III/SMM-persistently infected mangabeys, the immunodominant protein (by Western blot analysis) has an apparent molecular weight of 115,000 daltons, and probably is analogous to the HTLV-III/LAV/ARV cell-surface glycoprotein, gp120, that also appears to be immunodominant in persons infected with the human virus (19).

Although the similarity between the simian and human viruses extends to shared antigenic determinants that are recognized to varying extents by antiserum to either STLV-III or HTLV-III/LAV/ARV isolates, no nucleic acid hybridization can be demonstrated (Table 2; R. Desrosiers, personal communication). These data indicate that, if the simian and human viruses diverged from a common ancestral virus, then that divergence probably did not occur within the past few decades. Isolates of HTLV-III/LAV/ARV show extensive genomic heterogeneity (20,21). It will be interesting to determine if this is true for the various isolates of STLV-III as well as for STLV-III/SMM isolates from different mangabeys.

STLV-III/SMM was isolated from more than 90% of mangabey monkeys chosen at random from among animals in the Yerkes Primate Center colony. Even though a majority of the mangabeys appear to have been infected for several years, the incidence and types of disease have not been unusually high or indicative of immunodeficiency. Therefore, the host species from which the various STLV-III isolates were initially obtained respond differently to infection: STLV-III causes immunodeficiency disease in macaques, but STLV-III/SMM does not cause disease in mangabey monkeys. The isolation of STLV-III/AGM was reported (9) to be from apparently healthy, wild-caught African green monkeys. However, no long-term study or records on incidence of disease among African green monkeys in the wild or in a colony of STLV-III/AGM-infected animals have been reported. Therefore, it is not known if STLV-III/AGM causes disease in this species or in other species of monkeys.

Our preliminary data on rhesus macaques experimentally infected with STLV-III/SMM suggest that some macaques may develop disease following infection. If disease does occur, then the different responses of mangabeys and macaques to STLV-III/SMM infection may provide a way to identify virus and host determinants of disease. Furthermore, chimpanzees are the only species other than man that reproducibly become infected after exposure to HTLV-III/LAV/ARV. Because of the cost and limited availability of chimpanzees, the STLV-III/SMM animal model system may become extremely important if development of a vaccine against the human virus proves difficult and requires extensive testing.

ACKNOWLEDGMENTS

We thank D. York, A. Brodie, and L. Wells for technical assistance, and Dr. V. Tsang for Western blots of STLV-III/SMM.

REFERENCES

1. Barre-Sinoussi F, Chermann J-C, Rey F, Nugeyre MT, Chamaret S, Gruest J, Dauguet C, Axlei-Blin C, Vezinet-Brun F, Rouzioux C, Rozenbaum W, Montagnier L (1983). Isolation of a T-lymphotropic retrovirus from a patient at risk for acquired immune deficiency syndrome (AIDS). Science 220:868.
2. Gallo RC, Salahuddin SZ, Popovic M, Shearer GM, Kaplan M, Haynes BF, Palker TJ, Redfield R, Oleske J, Safai B, White G, Foster P, Markham PD (1984). Frequent detection and isolation of cytopathic retroviruses (HTLV-III) from patients with AIDS and at risk for AIDS. Science 224:500.
3. Levy JA, Hoffman AD, Kramer SM, Landis JA, Shimabukuro JM, Oshiro LS (1984). Isolation of lymphocytopathic retro-

viruses from San Francisco patients with AIDS. Science 225:840.

4. Klatzmann D, Barre-Sinoussi F, Nugeyre MT, Dauguet C, Vilmer E, Griscelli C, Brun-Vezinet F, Rouzioux C, Gluckman JC, Chermann J-C, Montagnier L (1984). Selective tropism of lymphadenopathy associated virus (LAV) for helper-inducer T lymphocytes. Science 225:59.
5. Daniel MD, Letvin NL, King NW, Kannagi M, Sehgal PK, Hunt RD, Kanki PJ, Essex M, Desrosiers RC (1985). Isolation of T-cell tropic HTLV-III-like retrovirus from macaques. Science 228:1201.
6. Letvin NL, Daniel MD, Sehgal PK, Desrosiers RC, Hunt RD, Waldron LM, Mackey JJ, Schmidt DK, Chalifoux LV, King NW (1985). Induction of AIDS-like disease in macaque monkeys with T-cell tropic retrovirus STLV-III. Science 230:71.
7. Kanki PJ, McLane MF, King NW, Letvin NL, Hunt RD, Sehgal P, Daniel MD, Desrosiers RC, Essex M (1985). Serologic identification and characterization of a macaque T-lymphotropic retrovirus closely related to HTLV-III. Science 228:1199.
8. Kanki PJ, Kurth R, Becker W, Dreesman G, McLane MF, Essex M (1985). Antibodies to simian T-lymphotropic retrovirus type III in African green monkeys and recognition of STLV-III viral proteins by AIDS and related sera. Lancet i:1330.
9. Kanki PJ, Alroy J, Essex M (1985). Isolation of T-lymphotropic retrovirus related to HTLV-III/LAV from wild-caught African green monkeys. Science 230:951.
10. Anand R, Lilly F, Ruscetti S (1981). Viral protein expression in producer and nonproducer clones of Friend erythroleukemia. J Virol 37:654.
11. Tsang VCW, Peralta JM, Simons AR (1983). Enzyme-linked immunoelectrotransfer blot techniques (EITB) for studying the specificities of antigens and antibodies separated by gel electrophoresis. Methods Enzymol 92:377.
12. Alizon M, Sonigo P, Barre-Sinoussi F, Chermann J-C, Tiollais P, Montagnier L, Wain-Hobson S (1984). Molecular cloning of lymphadenopathy-associated virus. Nature 312:757.
13. Maniatis T, Fritsch EF, Sambrook J (1982). "Molecular Cloning: A Laboratory Manual." Cold Spring Harbor: Cold Spring Harbor Laboratory.
14. Kalyanaraman VS, Narayanan R, Feorino P, Ramsey RB, Palmer EL, Chorba T, McDougal S, Getchell JP, Holloway B, Harrison AK, Cabradilla CD, Telfer M, Evatt B (1985).

Isolation and characterization of a human T cell leukemia virus type II from a hemophilia-A patient with pancytopenia. EMBO Journal 4:1455.
15. Srinivasan A, Reddy EP, Aaronson SA (1981). Abelson murine leukemia virus: molecular cloning of infectious integrated proviral DNA. Proc Natl Acad Sci USA 78:2077.
16. Fultz PN, McClure HM, Anderson DC, Swenson RB, Anand R, Srinivasan A (1986). Isolation of a T-lymphotropic retrovirus from naturally infected sooty mangabey monkeys. Submitted.
17. Weiss RA, Clapham PR, Cheingsong-Popov R, Dalgleish AG, Carne CA, Weller IVD, Tedder RS (1985). Neutralization of human T-lymphotropic virus type III by sera of AIDS and AIDS-risk patients. Nature 316:69.
18. Robert-Guroff M, Brown M, Gallo RC (1985). HTLV-III-neutralizing antibodies in patients with AIDS and AIDS-related complex. Nature 316:72.
19. Barin F, McLane MF, Allan JS, Lee TH, Groopman JE, Essex M (1985). Virus envelope protein of HTLV-III represents major target antigen for antibodies in AIDS patients. Science 228:1094.
20. Wong-Staal F, Shaw GM, Hahn BH, Salahuddin SZ, Popovic M, Markham P, Redfield R, Gallo RC (1985). Genomic diversity of human T-lymphotropic virus type III (HTLV-III). Science 229:759.
21. Benn S, Rutledge R, Folks T, Gold J, Baker L, McCormick J, Feorino P, Piot P, Quinn T, Martin M (1985). Genomic heterogeneity of AIDS retroviral isolates from North America and Zaire. Science 230:949.

Viruses and Human Cancer, pages 309–317

A REGION OF THE HERPESVIRUS SAIMIRI GENOME REQUIRED FOR ONCOGENICITY

Ronald C. Desrosiers, Shridhara Murthy and John Trimble

New England Regional Primate Research Center
Harvard Medical School
Southboro, MA 01772

Viable experimental systems for the study of transformation and oncogenicity among the herpesviruses appear to be confined at this time to the gamma (lymphotropic) class of herpesviruses. Our research has been directed toward understanding the molecular mechanisms of oncogenic transformation by *Herpesvirus saimiri*. *H. saimiri* naturally infects squirrel monkeys, producing no signs of disease in this species. Infection of other species of New World primates, such as marmosets and owl monkeys, results in a rapidly progressing malignant T cell lymphoma.

Susceptible animals infected with *H. saimiri* die within 30-40 days of viral induced lymphoma/leukemia. Furthermore, the virus is able to transform lymphoid cells *in vitro* to continuously growing lymphoblastoid cell lines. These are properties akin to the acutely oncogenic retroviruses which carry an oncogene for transformation and the adenoviruses and papovaviruses which have a specific region of their genomes responsible for transforming functions. Thus, we might expect a region of the *H. saimiri* genome to be devoted to the function of transformation. A similar argument can be made for Epstein-Barr Virus (EBV), a virus with analogous biological properties. However, these herpesvirus genomes are much more complex than the other model tumor viruses described above and whether transformation can be ascribed to a single region of the genome has not yet been determined. Progress has been made nonetheless. A region of the *H. saimiri* genome has been identified that is not required for replication; this region *is* required for transformation and oncogenicity (1,2,3). Regions of the EBV genome that appear to contribute to transformation have also been identified (4).

Although infection of New World primate monolayer cell lines by H. saimiri is lytic yielding greater than 10^7 infectious virus particles per ml, infection of lymphoid cells is non-permissive or semi-permissive resulting in continuously growing transformed lymphoblastoid cell lines. The target cell for transformation in vitro and in vivo appears to be very specific; in all cases studied to date, the lymphoid target cell bears T cell markers, and in most instances it is the $T8^+$ cell (2,5,6,7). In a series of 17 consecutive in vitro transformations of common marmoset peripheral blood lymphocytes (PBLs) by H. saimiri, the vast majority of cells in all cultures bore the T8 phenotype of suppressor/cytotoxic cells (2). Recent results indicate that the transformed cells growing out in these cultures are a minor subpopulation of T8 cells (8). It is not clear at this time whether certain conditions might favor transformation of T4 cells or even other cell types. Nontheless, it is remarkable that the virus is able to target such a minor cell type from the complex population of peripheral blood mononuclear cells.

The region of the H. saimiri genome that we have focused on is located at the left border of unique and repetitive DNA (Fig. 1). The infectious H. saimiri genome contains an internal stretch of unique sequence DNA called L-DNA because it is low in G + C content (36%). L-DNA is approximately 110 kilobasepairs (kbp) in length. The unique L-DNA region of infectious DNA is flanked at each end by repetitive DNA called H-DNA (high in G + C content, 72%). H-DNA repeat units are 1,444 base pairs and are oriented in the same direction at both ends of the genome (9,10). In this article, we will describe: i) data demonstrating that sequences at the left end of L-DNA are required for transformation and oncogenicity; ii) evidence that these sequences are not necessary for replication of the virus; iii) RNA products from this region in lytic vs transforming infections; and iv) the possible role of these products in transformation.

The reasons for focusing on this region of the genome date back to observations made several years ago. Schaffer, Falk and their colleagues had isolated a variant of H. saimiri derived from strain 11 which had lost its oncogenic potential (11,12). This variant, called 11att, was fully replication competent in vitro and retained its ability to infect New World primates in vivo. 11att was able to persist for years in lymphocytes of inoculated marmosets and recovered virus was indistinguishable by

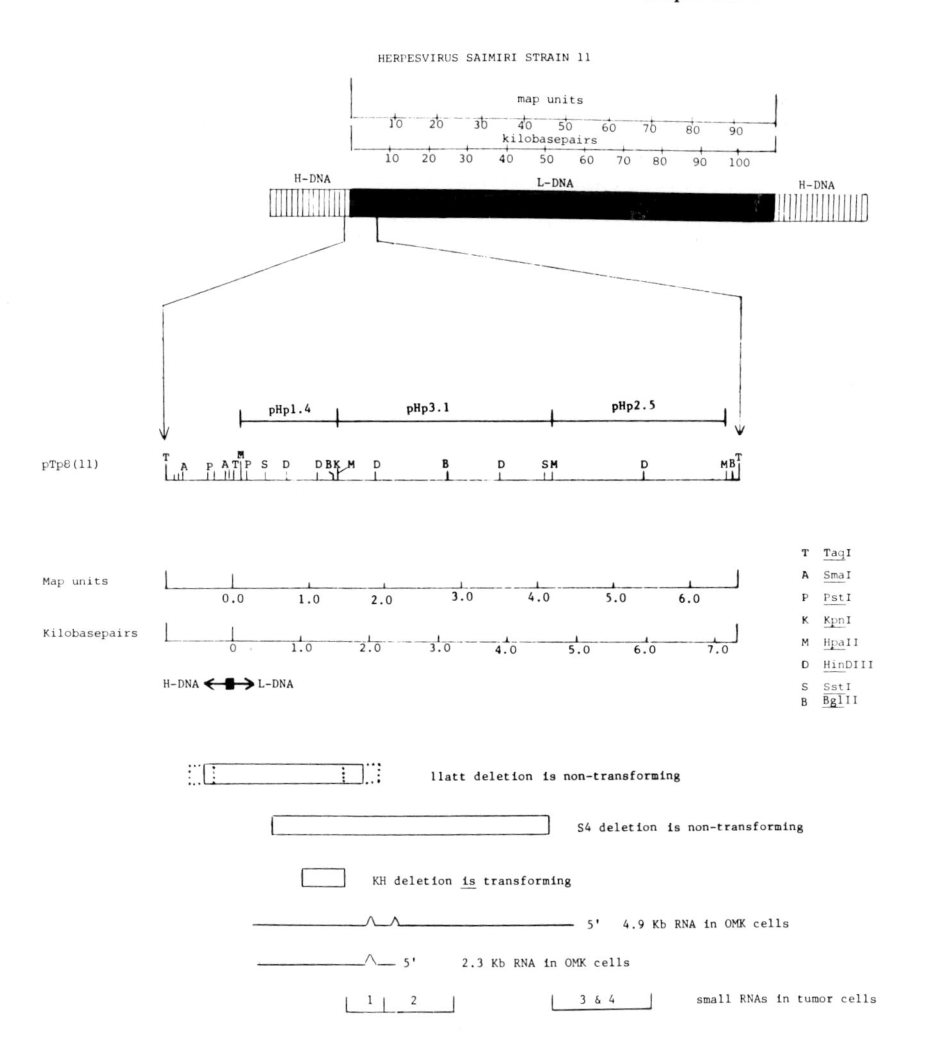

Figure 1. Summary of region of the Herpesvirus saimiri genome required for oncogenicity. Derivation of the molecular clones pT7.4, pTP8, pHp1.4, pHp3.1 and pHp2.5 from the left end of L-DNA, the spontaneous deletion in 11att, construction of the S4 and KH deletion mutants and their transforming and oncogenic potentials have been published (1,2,3,14). The 2.3 and 4.9 kb transcripts in lytically infected OMK cells were mapped previously (15). The small RNAs 1,2,3 and 4 refer to the RNAs of 135, 110, 105 and 73 nucleotides found in tumor cells.

restriction endonuclease typing from the input 11att virus (13). Restriction endonuclease analysis revealed that 11att was clearly derived from the oncogenic strain 11 and, except for one easily detectable change, 11att had not undergone other major genetic changes. Cloning and detailed restriction endonuclease mapping did reveal the deletion of 2.2 kilobasepairs from strain 11 in the generation of 11att (3).

As with the deletions in the non-transforming P3HR1 and Daudi strains of EBV, one could not confidently conclude that the deletion in 11att was responsible for the altered biological property since the strain arose via cell culture passage. It was possible that point mutations or other undetected changes contributed to or were responsible for the loss of oncogenicity. Genetic manipulations were thus performed to demonstrate the requirement for these sequences in an unambiguous manner.

A series of deletions overlapping the deletion in 11att was constructed in cloned plasmids (14). This was accomplished by simply deleting sequences between unique restriction endonuclease sites. Isolation of virus strains with these deletions was a much more arduous task. Permissive owl monkey kidney (OMK) cells growing in culture were co-transfected with infectious strain 11 virion DNA together with deletion mutant plasmid DNA. Screening of progeny virus by a limiting dilution - spot hybridization assay resulted in the isolation of _H. saimiri_ deletion mutants S4 and KH (14; Fig 1). Strains of virus in which the deleted sequences were restored to 11att, S4 and KH were isolated in a similar manner - by screening progeny virus following co-transfection of OMK cells with infectious deletion mutant virion DNA together with plasmid DNA spanning the deletion.

The results of biological testing of these constructed strains of virus are unambiguous (1,2). The S4 and KH deletions mutants, like 11att, are fully replication competent _in vitro_ and retain their ability to persistently infect New World primate lymphocytes _in vivo_. Thus a contiguous stretch of 4.5 kbp sequence information is not necessary for replication of the virus. Both the S4 and 11att deletion mutants score repeatedly negative in the reliable transformation assay using common marmoset PBLs (2) and both these mutant strains failed to induce lymphomas/leukemias in animal studies (1). Restoration of the deleted sequences to S4 and 11att restored the transforming and oncogenic capabilities (1,2). From this we can certainly conclude that the sequences defining the S4 deletion (4.0 kbp Sst 1 fragment from 0.4 to 4.0 map units) are required for transformation. From the region of overlap in the 11att and S4 deletions, these results further suggest that a region required for oncogenicity can be narrowed between the Sst 1 site at 0.4 map units and the HindIII site at 1.9 map units. Most importantly, the KH deletion mutant retains its ability to transform T cells *in vitro* and to induce lymphoma in New World primates (2). Thus, the sequences lost in the KH deletion must not be essential for transformation. These results are consistent with sequences immediately leftward from the KH deletion (0.4 to 0.8 map units) or immediately rightward from the KH deletion (1.35 to 1.9 map units) being required for transformation and oncogenicity. Additional deletion mutants are under construction to confirm and extend these observations.

During the course of lytic infection of OMK cells, two major RNAs were found to be derived from the left end of L-DNA and these RNAs were specifically altered by the deletions (15). These RNAs of 2.3 and 4.9 kilobases (kb) were overlapping and 3' co-terminal (Fig. 1). A single intron was detected in the 2.3kb RNA and two introns were detected in the larger RNA.

However, the poly A+ RNAs that were detected in lytic infection of OMK cells were not detected at all in the four viral-induced lymphomas that were examined nor in the 1670 tumor cell line. These transformed T cells did express four small RNAs of approximately 73, 105, 110 and 135 nucleotides from this region (16). These were the only RNAs detectable from this region in tumor cells. Thus, expression from this region appears to be under stringent control in a cell-type specific manner. Ribonuclease protection of ^{32}P-trancripts made in the SP6 system was used to map the small tumor cell RNAs to restriction fragments on the *H. saimiri* genome (Fig. 1). The coding sequences for two of the RNAs (135 and 110 nucleotide) fall within the S4 deletion; the coding sequence for the 135 nucleotide RNA falls within the 1.35 - 1.9 map unit region discussed above.

Two lines of reasoning make us think that synthesis of RNA(s) 1 and/or 2 (135 and/or 110 nucleotides) is/are the essential factor(s) for transformation. First, these are the only products detectable in tumor cells from this region. Second, deletion of sequences coding for one or both of these small RNAs (as in S4 and 11att) results in virus that is non-transforming while deletion of adjacent sequences that do not code for these RNAs (KH) results in virus that is transforming. These arguments will be strengthened by i) sequence information defining the ends of the small RNAs, their sequence, and their precise location relative to the deletions, introns and open reading frames ii) analysis of the biological properties of viral strains with more finely constructed deletions running through this region iii) analysis of strains where DNA fragments coding for the small RNAs are inserted into the S4 deletion mutant and iv) construction and analysis of point mutants which drastically alter the synthesis, structure or function of these small RNAs.

What is known about other viral and cellular small RNAs and what might the function of the small H. saimiri RNAs be? Epstein-Barr virus (EBV) codes for two small RNAs called EBER1 and EBER2 (17,18); however it is not known if they are dispensable for replication, if they are essential for immortalization and whether they are functionally similar to any of the H. saimiri small RNAs. Adenovirus codes for two small RNAs called VAI and VAII (19,20); these RNAs are not essential for transformation. VAI apparently prevents phosphorylation of eIF-2 after infection (21,22) and functional substitution of VAI by the EBER RNAs has been reported (23). Additionally, several examples can be cited where small RNAs have been shown or suggested to be important determinants regulating expression. 1) The small leader RNA of vesicular stomatitis virus is responsible for inhibiting cellular transcription (24). 2) Small RNAs have been found to play an essential role in the processing of RNA transcripts in the nucleus and regulation through differential processing has been proposed (25). 3) Small RNAs have been identified as tissue specific transcripts; coding sequences for these small RNAs have been described as components of controlling elements called "identifier" sequences (26,27,28,29). Coding sequences for these small RNAs occur within the introns of genes whose expression is tissue specific (29,30,31). DNA sequencing and further mapping will be needed to determine whether the 135 and 110 nucleotide RNAs are derived from the intron sequences.

If the small H. saimiri RNAs are analogous to one of the classes of small RNAs discussed above, the H. saimiri system may be ideal for unraveling how such RNAs work. One possibility that must be considered is whether the small H. saimiri RNAs activate expression of certain products important for the regulation of T-cell growth. It should be noted that the activation of transcription of specific cellular genes seems to occur during transformation by at least some DNA viruses, for example SV40 (32,33) and EBV (34).

Acknowledgements

This work was supported by PHS grant 31363 from the National Cancer Institute and by grant RR00168 to the New England Regional Primate Research Center from the Division of Research Resources, National Institutes of Health.

References

1. Desrosiers RC, Bakker A, Kamine J, Falk LA, Hunt RD, King NW (1985). A region of the Herpesvirus saimiri genome required for oncogenicity. Science 228:184.
2. Desrosiers RC, Silva DP, Waldron LM, Letvin NL (1986). Non-oncogenic deletion mutants of Herpesvirus saimiri are defective for in vitro immortalization. J Virol 57:701.
3. Koomey JM, Mulder C, Burghoff RL, Fleckenstein B, Desrosiers RC (1984). Deletion of DNA sequences in a nonocogenic variant of Herpesvirus saimiri. J Virol 50:662.
4. Wang D, Liebowitz D, Kieff E (1985). An EBV membrane protein expressed in immortalized lymphoctyes transforms established rodent cells. Cell 43:831.
5. Rabin H, Hopkins RF III, Desrosiers RC, Ortaldo JR, Djeu JY, Neubauer RH (1984). Transformation of owl monkey T cells in vitro with Herpesvirus saimiri. Proc Natl Acad Sci USA 81:4563.
6. Wright J, Falk LA, Collins D, Deinhardt F (1976). Mononuclear cell fraction carrying Herpesvirus saimiri in persistently infected squirrel monkeys. J Natl Cancer Inst 57:959.
7. Falk LA (1980). Biology of Herpesvirus saimiri and Herpesvirus ateles. In Klein G (ed): "Viral Oncology," New York: Raven Press, p. 813.
8. Kiyotaki, M, Desrosiers RC, Letvin NL, In preparation.
9. Bankier AT, Dietrich W, Baer R, Barrell BG, Colbere-Garapin F, Fleckenstein B, Bodemer W (1985). Terminal repetitive sequences in Herpesvirus saimiri virion DNA. J Virol 55:133.
10. Bornkamm GW, Delius H, Fleckenstein B, Werner FJ,

Mulder C (1976). Structure of Herpesvirus saimiri genomes: arrangement of heavy and light sequences in the M genome. J Virol 19:154.

11. Schaffer PA, Falk LA, Deinhardt F (1975). Brief communication: attenuation of Herpesvirus saimiri for marmosets after successive passage in cell culture at 39°C. J Natl Cancer Inst 55:1243.

12. Wright J, Falk LA, Wolfe LG, Ogden J, Deinhardt F (1977). Susceptibility of common marmosets (Callithrix jacchus) to oncogenic and attenuated strains of Herpesvirus saimiri. J Natl Cancer Inst 59:1475.

13. Falk LA, Desrosiers RC, Hunt RD (1980). Herpesvirus saimiri infection in squirrel and marmoset monkeys. In Essex M, Todaro G, zur Hausen H (eds): "Viruses in naturally occuring cancer," Vol. 7, Cold Spring Harbor, NY: Cold Spring Harbor Press, p. 137.

14. Desrosiers RC, Burghoff RL, Bakker A, Kamine J (1984). Construction of replication competent Herpesvirus saimiri deletion mutants. J Virol 49:343.

15. Kamine J, Bakker A, Desrosiers RC (1984). Mapping of RNA transcribed from a region of the Herpesvirus saimiri genome required for oncogenicity. J Virol 52:532.

16. Murthy S, Kamine J, Desrosiers RC (1986). Viral-encoded small RNAs in Herpesvirus saimiri induced tumors. Submitted for publication.

17. Lerner MR, Andrews NC, Miller G, Steitz JA (1981). Two small RNAs encoded by Epstein-Barr virus and complexed with proteins are precipitated by antibodies from patients with systemic lupus erythematosus. Proc Natl Acad Sci USA 78:805.

18. Arrand JR, Rymo L (1982). Characterization of the major Epstein-Barr virus-specific RNA in Burkitt lymphoma-derived cells. J Virol 41:376.

19. Reich PR, Forget B, Weissman SM (1966). RNA of low molecular weight in KB cells infected with Adenovirus type 2. J Mol Biol 17:428.

20. Soderlund H, Pettersson U, Vennstrom B, Philipson L, Mathews MB (1976). A new species of virus-coded low molecular weight RNA from cells infected with adenovirus type 2. Cell 7:585.

21. Schneider RJ, Safer B, Munemitsu SM, Samuel CE, Shenk T (1985) Adenovirus VAI RNA prevents phosphorylation of the eukaryotic initiation factor 2 alpha-subunit subsequent to infection. Proc Natl Acad Sci USA

82:4321.

22. O'Malley RP, Mariano TM, Siekierka J, Mathews MB (1986) A mechanism for the control of protein synthesis by adenovirus VA RNA. Cell 44:391.
23. Bhat RA, Thimmappaya B (1985). Construction and analysis of additional adenovirus substitution mutants confirm the complementation of VAI RNA function by two small RNAs encoded by Epstein-Barr virus. J Virol 56:750.
24. Grinnell BW, Wagner RR (1985). Inhibition of DNA-dependent transcription by the leader RNA of vesicular stomatitis virus: Role of specific nucleotide sequences and cell protein binding. Mol Cell Biol 5:2502.
25. Turner P (1985). Controlling roles for snurps. Nature 316:105.
26. Sutcliffe JG, Milner RJ, Bloom FE, Lerner RA (1982). Common 82- nucleotide sequence unique to brain RNA. Proc Natl Acad Sci USA 79:4942.
27. Sutcliffe JG, Milner RJ, Gottesfeld JM, Reynolds W (1984). Control of neuronal gene expression. Science 225:1308.
28. Sutcliffe JG, Milner RJ, Gottesfeld JM, Lerner RA (1984). Identifier sequences are transcribed specifically in brain. Nature 308:237.
29. Milner RJ, Bloom FE, Lai C, Lerner RA, Sutcliffe JG (1984). Brain-specific genes have identifier sequences in their introns. Proc Natl Acad Sci USA 81:713.
30. Barta A, Richards RI, Baxter JD, Shine J (1981). Primary structure and evolution of rat growth hormone gene. Proc Natl Acad Sci USA 78:4867.
31. Gutierrez-Hartmann A, Lieberberg I, Gardner D, Baxter JD, Cathala GG (1984). Transcription of two classes of rat growth hormone gene-associated repetitive DNA: differences in activity and effects of tandem repeat structure. Nuc Acids Res 12:7153.
32. Schutzbank T, Robinson R, Oren M, Levine AJ (1982). SV40 large tumor antigen can regulate some cellular transcripts in a positive fashion. Cell 30:481.
33. Singh K, Carey M, Saragosti S, Botchan M (1985). Expression of enhanced levels of small RNA polymerase III transcripts encoded by the B2 repeats in simian virus 40-transformed mouse cells. Nature 314:553.
34. Cheah MSC, Ley TJ, Tronick SR, Robbins KC (1986). fgr proto-oncogene mRNA induced in B lymphocytes by Epstein-Barr virus infection. Nature 319:238.

Viruses and Human Cancer, pages 319–332

STRUCTURAL, BIOCHEMICAL AND SEROLOGICAL COMPARISON OF LAV/HTLV-III AND STLV-IIIMAC TO PRIMATE LENTIVIRUSES

Schneider[1], J., Jurkiewicz[1], E., Wendler[1], I., Jentsch[1], K.D., Bayer[1], H., Desrosiers[2], R.C., Gelderblom[3], H., and Hunsmann[1], G.

[1]Deutsches Primatenzentrum, Kellnerweg 4, D-3400 Göttingen, FRG
[2]New England Regional Primate Research Center, Southborough, Massachusetts 01772, USA
[3]Bundesgesundheitsamt, Am Nordufer 20, D-1000 Berlin 65, FRG

ABSTRACT In sera of 855 non-human primates from 32 species antibodies crossreacting with LAV/HTLV-III were only found in African green monkeys as well as a red-tailed macaque. Sera of rhesus monkeys experimentally infected with STLV-IIImac likewise crossreacted with LAV/HTLV-III. Structural and serological data indicate that both the human and the cercopithecan viruses belong to the lentivirus family. However, these monkey viruses are quite distinct from human LAV/HTLV-III and therefore it seems unlikely that they play a role for the etiology of AIDS.

INTRODUCTION

Two pathogenic retroviruses have been isolated from man, the human T-lymphotropic virus type-I (HTLV-I) and the lymphadenopathy associated virus/human T-lymphotropic virus type III (LAV/HTLV-III). The oncovirus HTLV-I causes adult T-cell leukemia/lymphoma (ATLL) endemic in certain areas of the world predominantly southwest Japan, subsaharan Africa and the Caribbean basin (1). The lentivirus LAV/HTLV-III causes the acquired immunodeficiency syndrome (AIDS) as well as central nervous affections. Obviously LAV/HTLV-III is more infectious than HTLV-I since it spreads exponentially in various parts of the world. Both ATLL and AIDS are lethal and neither an etiological treatment nor a vaccine could yet be developed. To study such measures animal models are urgently needed.

Lentiviruses and oncoviruses related to the human viruses cause arthritis, pneumonia, anemia, central nervous affections, and B-cell leukemia/lymphoma in ungulates (2). However, these viral diseases are of limited relevance as models for human ATLL and AIDS. On the other hand, these human lymphotropic retroviruses were experimentally transmissable to certain non-human primate species but infected animals did not develop symptoms similar to the human diseases (3).

More appropriate models became availabe when viruses related to human ATLL and AIDS viruses were discovered in non-human primates. Serological cross reactivity with HTLV-I-related viruses were found in 13 species of cercopithecinae as well as in chimpanzees and gibbon (2). Structurally these simian T-cell tropic viruses (STLV-I) are closely related but distinguishable from each other and HTLV-I. Peptide mapping analysis of the major core polypeptides p24 and p19 allowed us to divide these viruses into three subgroups (4). Furthermore, STLV-I isolates of baboon and Japanese monkey may cause leukemia/lymphomas similar to ATLL in humans (5, 6).

Recently two lentiviruses related to LAV/HTLV-III were discovered in cercopithecinae. STLV-IIImac was found to be the etiological agent of simian AIDS (SAIDS) clinically remarkably similar to AIDS (7). The other virus STLV-IIIagm was found in green monkeys (8). It is serologically related to both LAV/HTLV-III and STLV-IIImac but obviously non-pathogenic to its natural host. It was suggested that STLV-III has recently crossed the species barrier to man perhaps in Africa, thereby causing new diseases (9).

In order to evaluate this hypothesis as well as the suitability of the STLV-IIImac infection as a model for LAV/HTLV-III we have examined the relationship of both viruses. To detect additional primate lentiviruses we have screened numerous monkey species for antibodies cross-reacting.

MATERIALS AND METHODS

Cells and Culture Conditions and Electron Microscopy

LAV/HTLV-III and STLV-IIImac were harvested (10) from LAV/HTLV-III infected Jurkat (I. Wendler, unpublished) and STLV-III-producing HUT-78 (11) cell lines. Electron microscropy on these cells was performed according to standard procedures.

Sera and ELISA Assay

Monospecific antisera were prepared against p19 and p24 of LAV/HTLV-III (12). The serum of a German AIDS patient served as prototype LAV/HTLV-III specific human serum (HS1). The other human serum (HS2) was obtained from a Caucasian female with AIDS reporting sexual contact with a Black African. Horse sera were pretested for antibodies to equine infectious anemia virus (EIAV) by a virus dependant ELISA. The monkey sera were used earlier to screen for antibodies crossreacting with HTLV-I (13). All monkeys except one red tail monkey (Cercopithecus ascanius) free-ranging in Uganda, were living in various European zoos, kept by scientific institutions or pharmaceutical companies.

In the ELISA test for LAV/HTLV-III antibodies (14) positivity was recorded when samples attained at least half of the absorbance of a positive reference serum. Borderline sera (+/-) attained optical densities between one third and one half of the positive reference serum.

Immunoprecipitation (IP), Polyacrylamide Gel Electrophoresis (PAGE) and Peptide Mapping

Cleared lysates of ^{35}S-cysteine labelled cells (15) or iodinated virus were incubated with tenfold or higher dilutions of sera at 4°C overnight. Immunocomplexes were adsorbed to protein A and analysed by PAGE on 9 to 16% gradient gels. For detection of radiolabelled polypeptides polyacrylamide gels were treated with Amplify (Amersham, UK), dried and exposed to X-ray films. To estimate the quantity of immunoprecipitated polypeptides bands on X-ray films were scanned (LKB Ultroscan Laser Densiometer, Sweden).

Gradient purified LAV/HTLV-III or STLV-IIImac was labelled with ^{125}I. For preparative immunoprecipitation the iodinated viral polypeptides were incubated with rabbit hyperimmunesera against LAV/HTLV-III p19 and p24. Further details are described by Jurkiewicz et al. (4).

Reverse Transcriptase (RT) Assays

RT assays were performed according to standard procedures which were slightly modified (16). The final concentration of Triton X-100 in test samples was adjusted to 0.6%. The influence of divalent cations was tested in ranges from 0.001 mM to 10 mM for Mn^{2+} and up tp 30 mM for Mg^{2+}. One ug of template-primers (Pharmacia) was added to 50ul of reaction mixture. ^{32}P-dATP or ^{32}P-dGTP was obtained from Amersham. The specific activity in the reaction mixture was about $1x10^5$ cpm/pmole dATP or dGTP. To test for contamination with cellular polymerases poly(dA)oligo(dT15)was also used as template-primer. The RT reaction was run at 42°C for 60 min.

RESULTS

Screening for Serum Antibodies to Lentiviruses in Non-Human Primates

We examined 855 sera from 32 species of non-human primates for antibodies cross-reacting with LAV/HTLV-III (tab. 1). All ELISA positive or questionable sera except 18 sera from crab-eating macaques and a serum of a Japanese macaque (M. fuscata) were reexamined by IP.

Elevated ELISA reactions of prosimians and New World monkeys could not be confirmed by IP. Among the 653 sera of Old World monkeys 64 reacted in the ELISA. Two sera of green monkey, one of redtail monkey and 1 of 5 rhesus monkeys experimentally infected with STLV-IIImac recognized the envelope glycoprotein of LAV/HTLV-III in IP. However, the reactivity of sera from crab-eating macaques (M. fascicularis) could not be confirmed. Antibodies to LAV/HTLV-III were found in only 2 species of the genus cercopithecus.

In addition 80 Old World monkey sera from the same collection were tested for antibodies to STLV-IIImac by IP. Only green and redtailed monkeys had antibodies. Of 29 green monkey sera 19 recognized the putative env-glycoprotein gp130 of STLV-IIImac. Among these were the two sera cross-reacting with LAV/HTLV-III. Five positive sera from Rhesus monkey originated from experimentally infected animals (Desrosier et al. 1985). In none of 106 great apes had antibodies to LAV/HTLV-III or STLV-IIImac (11 animals tested) been detectable.

Thus viruses closely related to LAV/HTLV-III are obviously rare in prosimians and New World monkeys. Natural infections with STLV-III occur often in green monkeys while other members of cercopithecinae if at all are less frequently infected.

Table 1 Seroactivity of Non-Human Primates with Lentiviruses

Species[1]		HTLV-III				STLV-III
		ELISA			IP[2]	IP
	No	+	+/-	-		
Prosimiae						
Lemuridae						
Lemur catta	7	0	1	6	0/0	1/0
Lorisidae						
Nycticebus coucang	6	0	1	5	0/0	1/0
Perodictus potto	6	0	0	6	0/0	0/0
Galagidae						
Galago crassicaudatus	6	0	0	6	0/0	0/0
Simiae						
Platyrrhina						
Callithricidae						
Leontocebus oedipus	3	0	0	3	0/0	0/0
Cebidae						
Aotes trivirgatus	9	0	0	9	0/0	0/0
Pithecia monacha	1	0	0	1	0/0	0/0
Cacajao rubicundus	1	0	0	1	0/0	0/0
Saimiri sciureus	20	0	0	20	0/0	0/0
Cebus apella	4	0	0	4	0/0	0/0
Cebus spec.	19	0	0	19	0/0	0/0
Ateles geoffroy	1	0	0	1	0/0	0/0
Lagothrix spec.	13	0	0	13	0/0	0/0
Catarrhina						
Colobinae						
Colobus polykomus	2	0	0	2	0/0	0/0
Presbytis obscurus	3	0	0	3	0/0	0/0
Presbytis entellus	2	0	0	2	0/0	0/0
Cercopithecinae						
Cerc. aetiops	58	1	3	54	7/2	29/18
Cerc. ascanius	1	1	0	0	1/1	0/0
Cerc. petaurista	2	0	0	2	2/0	0/0
Theropithecus gelada	2	0	1	1	0/0	0/0
Papio papio	31	0	0	31	0/0	0/0
Papio cynocephalus	23	0	0	23	0/0	0/0
Macaca sylvana	97	0	0	97	0/0	0/0
Macaca irus	263	4	47	212	33/0	33/0
Macaca mulatta[3]	138	1	5	132	12/1	12/5
Macaca silenus	1	0	0	1	0/0	0/0
Macaca fuscata	17	0	1	16	0/0	0/0
Hylobatidae						
Hylobates lar	13	0	3	10	0/0	4/0
Pongidae						
Pongo pygmaeus	12	0	0	12	0/0	0/0
Gorilla gorilla	23	0	1	22	1/0	1/0
Pan paniscus	5	0	0	5	0/0	0/0
Pan troglodytes	66	8	10	48	11/0	10/0

[1]Fiedler (1956) [2]No of tested / no of positives
[3]including 5 experimentally infected animals

Electron Microscopy

Both LAV/HTLV-III and STLV-IIImac cluster on small defined regions of the cell perimeter and show identical diameters and interior organization. Depending on the angle of the section plane, the cores of both viruses appear centro-symmetric, triangular or tubular in shape, indistinguishable from each other and very similar to other lentiviruses.

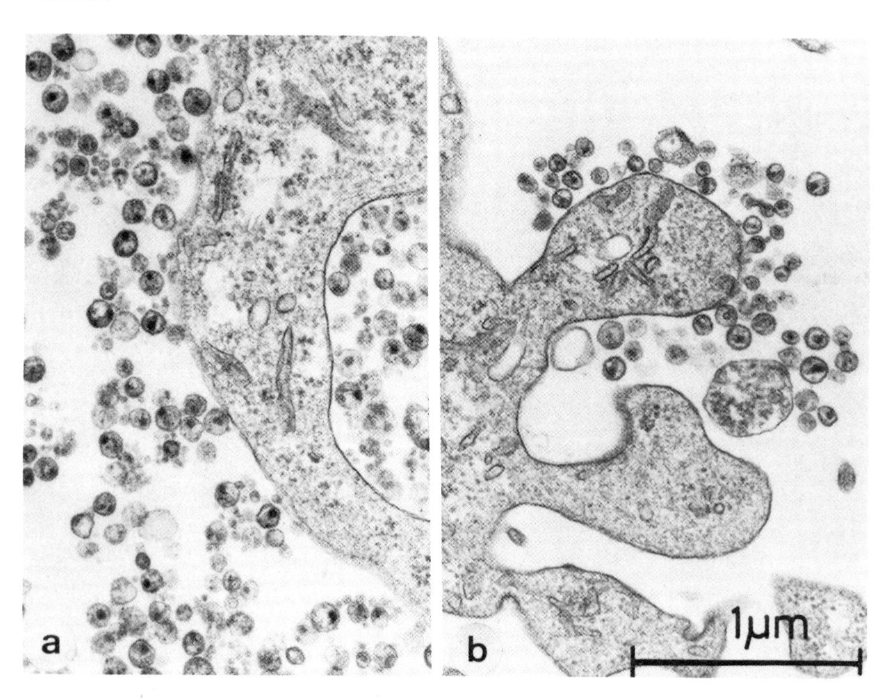

FIGURE 1. Electronmicrographs of LAV/HTLV-III (left) and STLV-IIImac (right).

Template Preferences of Primate Lymphotropic Retrovirus RT

We compared the RT reaction of HTLV-I with those HTLV-III and STLV-IIImac using as template primers the two most active homopolymers poly(rA)oligo(dT)15 and poly(rC)oligo(dG)12-18 (tab. 2). The cation dependency of LAV/HTLV-III and STLV-IIImac were very similar, however, distinct from HTLV-I. LAV/HTLV-III and STLV-IIImac showed a preference to poly(rA)oligo(dT)15 with both cations. Mg^{2+} increased the specific activity at least twice. In contrast HTLV-I RT prefers poly(rC)oligo(dG)12-18 with Mg^{2+}. Thus the RT of STLV-IIImac is closely related to that of LAV/HTLV-III but distinct from RT of HTLV-I.

TABLE 2
TEMPLATE AND DIVALENT CATION PREFERENCES OF PRIMATE LAYMPHOTROPIC RETROVIRUSES

Virus	Template Primer	Cation	Optimum Conc[1]	Specific Activity[2]	Mg/Mn Ratio
HTLV-I	poly(rA):oligo(dT)	Mg	5	0.048	0.46
		Mn	1	0.103	
	poly(rC):oligo(dG)	Mg	27	0.60	6.6
		Mn	7	0.090	
HTLV-III	poly(rA):oligo(dT)	Mg	7	58	2.4
		Mn	0.07	24	
	poly(rC):oligo(dG)	Mg	25	8.6	1.4
		Mn	0.015	6.14	
STLV-III	poly(rA):oligo(dT)	Mg	6	39	2.6
		Mn	0.1	15	
	poly(rC):oligo(dG)	Mg	20	6.1	2.1
		Mn	0.01	2.8	

[1] in mM
[2] pmoles incorporated/ug viral protein

Titration of Homologous and Heterologous Sera for Antibodies against LAV/HTLV-III and STLV-IIImac Polypeptides

To study the serological relationships between lentiviruses in more detail antisera to LAV/HTLV-III, STLV-III, and EIAV were titrated in IP against individual viral polypeptides. Immunoprecipitation of polypeptides from extracts of ^{35}S-cysteine-labelled cells was used to identify and quantitate envelope glycopolypeptides (Fig. 2, tab.3) while iodinated virus served to detect antibodies to viral core polypeptides (tab. 3).
The serum of an EIAV-infected horse (Equus caballus, EC) recognized both the external env-glycoprotein gp120 of LAV/HTLV-III, and gp130 of STLV-III with a titer of 100.

This horse serum reacted also with p54, the putative precursor of STLV-IIImac gag polypeptides. A strong band in the range of 20kd is found with positive and negative sera, and therefore represents an unspecific precipitation.

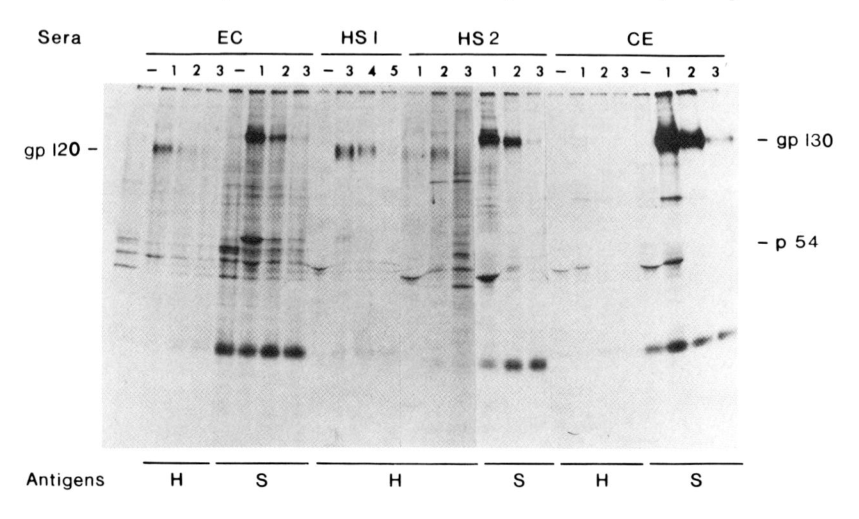

FIGURE 2. Interspecies specific determinants of LAV/HTLV-III and STLV-IIImac glycoproteins. Sera of an EIAV-infected horse (EC), two AIDS patients (HS 1, HS 2) and a serum of a green monkey (CE) cross-reacting with STLV-III were tested by immunoprecipitation for their reactivity with antigens from H9/HTLV-III cells (H), or Hut-78/STLV-III cells (S). The negative logarithm of the serum dilution is given on top of individual lanes. Negative sera (-) of respective species are included.

Sera of additional 20 AIDS patients like HS1 serum frequently attained titers against gp120 up to 10000, whereas they recognize gp130 of STLV-III only faintly or not at all (not shown). However one serum of a Caucasian female AIDS patient (HS2) reporting sexual contact with a Black African showed a titer of 100 with respect to both glycopolypeptides.

The serum of a naturally infected green monkey (CE) recognized only gp130 (titer 1000) of STLV-III, and failed to detect gp120 of LAV/HTLV-III.

STLV-IIImac seems to share major epitopes of the core polypeptides with LAV/HTLV-III, since the serum of an AIDS patient attained a titer of 10000 and 1000 against STLV-

IIImac p24 and p19 respectively. Vice versa p24 is the only polypeptide recognized by the serum of a rhesus monkey (M. mulatta) experimentally infected with STLV-IIImac. A similar reactivity was observed with the serum of a naturally infected green monkey (C. aethiops). It recognized the core polypeptide of LAV/HTLV-III, but envelope and core polypeptides of STLV-IIImac.

EIAV-specific serum of a horse (E. caballus) showed titers of 100 with the glycoproteins of LAV/HTLV-III and STLV-IIImac. The additional reaction of this serum with the probable core polypeptide precursor indicates that the reactivity is not restricted to carbohydrate epitopes of STLV-IIImac.

Out of four horse sera against EIAV three recognized the same pattern of polypeptides, one was negative (not shown).

This results demonstrate that broadly reacting epitopes exist on envelope- as well as on core-polypeptides of the lentiviruses. STLV-IIIagm is probably more closely related to STLV-IIImac than to LAV/HTLV-III.

TABLE 3
TITRATION OF ANTIBODIES TO HTLV-III AND STLV-III IN HOMOLOGOUS AND HETEROLOGOUS SERA

Sera[1]	Antigens[2]						
	HTLV-III			STLV-IIImac			
	gp120	p24	p19	gp130	p56	p24	p19
H. sapiens 1	4[3]	4	4	1	2	4	3
H. sapiens 2	3	nt	nt	3	-	nt	nt
M. mulatta	-	1.5	-	1.5	nt	2.5	nt
C. aethiops	-	3	-	3	-	2.5	-
E. caballus	2	-	-	2	2	-	-

[1]The origin of sera is described in materials and methods.

[2]Serum titers against core polypeptides p24 and p19 were determined with radioiodinated virus. Reaction against gp120/130 and p54 were assayed with ^{35}S-cysteine-labelled H9/HTLV-III or HUT-78/STLV-III cells.

[3]Titers of sera are expressed as negative logarithms of the dilution precipitating detectable amounts of an antigen. nt, not tested; -, no reaction observed.

Structural Differences between Core Polypeptides of LAV/HTLV-III and STLV-III.

Rabbit hyperimmunesera against LAV/HTLV-III structural polypeptides p19 and p24 crossreacted with respective polypeptides of STLV-III. The molecular weight of LAV/HTLV-III and STLV-IIImac polypeptides are very similar. Therefore, we compared these viral polypeptides by two-dimensional tryptic peptide mapping (fig. 2). Experiments with mixed iodinated polypeptides indicate similarities between peptide maps. In contrast to the highly conserved p24 of HTLV-I related viruses which have 54-77% of spots in common (4) the p24

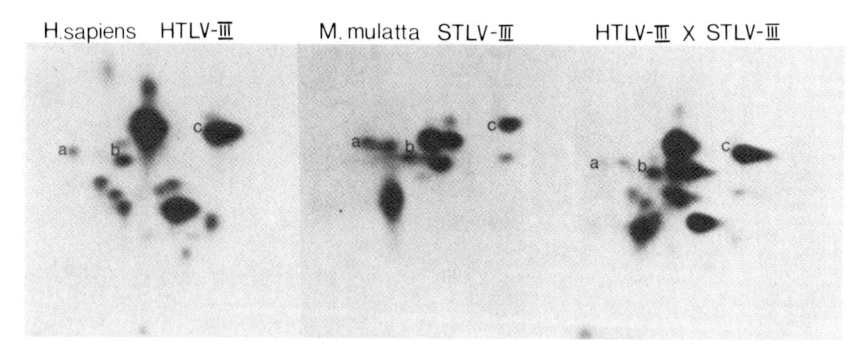

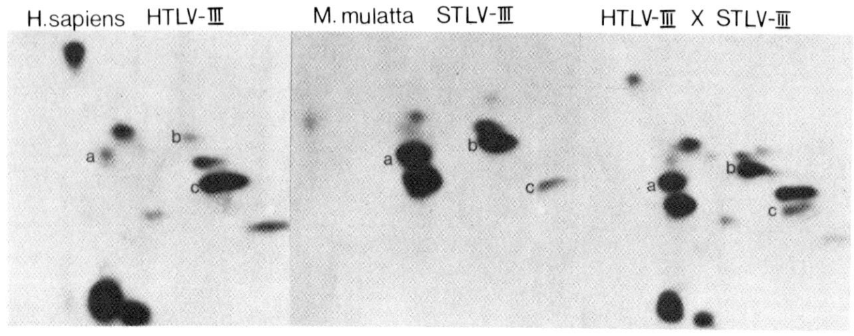

FIGURE 3. Peptide maps of viral polypeptides p24 (upper panel) and p19 (lower panel) of LAV/HTLV-III and STLV-IIImac. Iodinated polypeptides were digested with TPCK-trypsin, applied in the lower left corner and separated in two dimensions. Letters indicate common fragments.

maps of LAV/HTLV-III and STLV-IIImac have only 21% common

tyrosine containing fragments. P19s of HTLV-I-related primate oncoviruses have 40-100% homologous spots but the 2 primate lentiviruses have only 27% homology of p19. This indicates that lentiviruses of primates are more heterogeneous to each other than HTLV-related primate oncoviruses.

DISCUSSION

Our serological survey of LAV/HTLV-III-related lentiviruses in primates indicates that such viruses are found only in the species green monkey, redtail monkey and rhesus monkey. One isolate STLV-IIImac (11) was compared to the human isolate LAV/HTLV-III (17). Ultrastructurally both viruses are indistinguishable. They display structural features of the core of lentiviruses, however, they lack surface projections. Likewise the specificity for synthetic templates and divalent cations of the RT of LAV/HTLV-III and STLV-IIImac is very similar. On polyacrylamide gels core polypeptides of both viruses migrate to the same positions but the major viral glycoprotein of STLV-IIImac is about 10kd larger than gp120 of LAV/HTLV-III. This difference could be explained by variation in the carbohydrate content or by a shift of the protease cleavage site in the precursor polypeptide.

More pronounced structural differences between the human and macaque lentivirus were discovered on peptide maps of their core polypeptides p24 and p19. In this respect the group of HTLV-I-related primate oncoviruses is much more conserved (4) than the two primate lentiviruses examined here . There are, however, conserved regions on the core polypeptides since human and monkey sera crossreact with the herterologous lentiviruses.

The glycoproteins of LAV/HTLV-III and STLV-IIImac seem to be more distinct antigenically. Most LAV/ HTLV-III specific human sera react weakly or not at all with the heterologous glycoprotein (tab. 3) (18). Vice versa green monkey and macaque sera did not react with the LAV/HTLV-III glycoprotein but green monkey sera precipitated the STLV-IIImac glycoprotein (fig. 2) and the p24 core polypeptide. Interestingly one serum from an AIDS patient with African connection reacted equally well with the glycoprotein of the human and macaque virus. This same pattern of reactivity was displayed by sera from 3 horses infected with EIAV.

Our serological data indicate that lentiviruses closely

related to LAV/HTLV-III are rare among primates. The two isolates found in cercopithecinae are more closely related to each other than to the prototype LAV/HTLV-III. However sera of EIAV-infected horses recognized the glycoproteins of the two primate lentiviruses. Therefore, a non-primate origin of the human AIDS virus should also be considered.

ACKNOWLEDGEMENTS

We thank Drs. J. Schmitt, Tübingen, Y. Cole, Kibale, Uganda, W. Kern, Hamburg, R. Kurth, Frankfurt, D. Neumann-Haefelin, Freiburg and J. Ullstrup, Oslo for gifts of human and non-human primate sera. Sera of EIAV infected horses were kindly provided by Dr. O. Kaaden, Hannover. The technical assistance of L. Ahlborn, H. Buss, E. Hotfilter, R. Jung, A. Jurdzinski, S. Mader, and D. Schreiner is gratefully acknowledged. We thank C. Schalt for typing the manuscript.

REFERENCES

1. Hunsmann G,and Hinuma Y (1984) Human adult T-cell leukemia virus and its association with disesase. In advances in viral oncology (ed.) G. Klein Raven Press. New York 5:147.
2. Weiss R, Teich N, Varmus H, and Coffin J (1984) "Molecular Biology of Tumor Viruses" and supplement (1985), Cold Spring Harbor Laboratory, Cold Soring Harbor, N.Y.
3. Alter HJ, Eichberg JW, Masur H, Saxinger WC, Gallo R, Macher AM, Lane HC, and Fauci AS (1984). Transmission of HTLV-III infection from human plasma to chimpanzees: An animal model for AIDS. Science 226:549-552
4. Jurkiewicz E, Nakamura H, Schneider J, Yamamoto N, Hayami M, and Hunsmann G, (1986) Structural analysis of p19 and p24 core polypeptides of primate lymphotropic retroviruses.
5. Homma T, Kanki PJ, King NWjr, Hunt RD, O Connel MJ, Letvin NL, Daniel MD, Derosiers RC, Yang CS, Essex M. (1984) Lymphoma in macaques: Association with virus of human T lymphotropic Family.
6. Voevodin AF, Lapin BA, Yakovleva LA, Ponomaryeva , Organyan TE, and Razmadze EN (1985) Antibodies reacting with

human T-lymphotropic retrovirus (HTLV-I) or related antigens in lymphomatous and healthy hamadryas baboons. Int J Cancer 36:579.

7. Letvin NL, Daniel MD, Seghal PK, Desrosiers RC, Hunt RD, Waldron LM, MacKay JJ, Schmidt DK, Chalifoux LV, King NW (1985). Induction of AIDS-like disease in macaque monkeys with T-cell retrovirus STLV-III. Science 230:71-73.
8. Kanki PJ, Kurth R, Becker W, Dreesman G, McLane MF, and Essex M (1985) Antibodies to simian T-lymphotropic retrovirus type III in Africa green monkeys and recognition of STLV-III viral proteins by AIDS and related sera. Lancet 1:1330-1332.
9. Norman C, (1985) Africa and the origin of AIDS. Science 230:1141.
10. Schneider J, and Hunsmann G (1978) Surface expression of murine leukemia structural polypeptides on host cell and the virion . Int J Cancer 22:204.
11. Daniel MD, Letvin NL, King NW, Kannagi M, Sehgal PK, Hunt RD, Kanki PJ, Essex M, and Desrosiers RC (1985) Isolation of T-cell tropic HTLV-III-like retrovirus from macaques. Science 228:1201-1204.
12. Schneider J, Yamamoto N, Hinuma Y, and Hunsmann G (1984) Precursor polypeptides of adult T-cell elukaemia virus: Detection with antisera against isolated polypeptides gp68, p24, and p19. J Gen Virol. 65:2249.
13. Hunsmann G, Schneider J, Schmitt J, and Yamamoto N (1983) Detection of serum antibodies to adult T-cell leukemia virus in non-human primates and in peopole from Africa. Int J Cancer 32:329.
14. Bayer H, Bienzle U, Schneider J, and Hunsmann G (1984) HTLV-III antibody frequency and severity of lymphadenopathy. Lancet 2:1347.
15. Schneider J, Yamamoto N, Hinuma Y, and Hunsmann G (1984) Sera from adult T-cell leukemia patients react with envelope and core polypeptides of adult T-cell leukemia virus. Virology 132:1.
16. Jentsch KD, Hunsmann G, and Nickel P (1986) Inhibition of HTLV-III reverse transcriptase by suramin-related compounds. J Gen Virol, submitted.
17. Popovic M, Sarngadharan MG, Read E, and Gallo RC (1985) Detection, isolation, and continous production of cytopathic retroviruses (HTLV-III) from patients with AIDS and pre-AIDS. Science 224:497.
18. Kanki PJ, McLane MF, King jr NW, Letvin NL, Hunt RD, Sehgal P, Daniel MD, Desrosiers RC, and Essex M (1985)

Serological identification and characterization of a macaque T-lymphotropic retrovirus closely related to HTLV-III. Science 228:1199-1201.

Viruses and Human Cancer, pages 333–344

STRUCTURAL AND BIOLOGICAL FEATURES OF STLV-I, AND SERO-EPIDEMIOLOGICAL STUDY OF STLV-I AND -III

Masanori Hayami

Department of Animal Pathology
The Institute of Medical Science
The University of Tokyo
Shirokanedai, Minato-ku, Tokyo 108

ABSTRACT Non-human primates of African and Asian origins have been found to be naturally infected with STLV-I, simian retroviruses closely related to HTLV-I. Ten STLV-I antigenetically cross-reactive with HTLV-I, were isolated from various species of non-human primates. They contain virus sequences homologous to HTLV-I. However, the restriction mapping turned out the difference between STLV-I and HTLV-I, and also among STLV-I. With these findings quite difference of geographycal distribution of STLV-I and HTLV-I in Japan suggests the unlikeliness of recent inter-species transmission. Molecularly cloned STLV-I of African origin had higher homology (95%) than that of Asian macaque (90%) in LTR region with that of HTLV-I. Leukemogenic potential of STLV-I is indicated by findings of spontaneous adult T-cell leukemia (ATL)-like disease.
Recently, STLV-III related to HTLV-III/LAV was isolated from some species of non-human primates. Prevalence of antibody to STLV-III was clustered only in cercopithecus genus of African origin, when about 1,600 serum from 49 species of non-human primates were examined. Two possible ways of virus transmission, horizontal transmission from males to females and vertical transmission from mothers to infants were suggested in STLV-III-infected monkeys by sero-epidemiological study.

ATL-RELATED VIRUS IN NON-HUMAN PRIMATES

It was first reported by Miyoshi et al. in 1982 (1) that Japanese monkeys had antibody crossreactive to HTLV-I, and C-type viruses similar to HTLV-I were detected in their cultured lymphocytes. Thereafter, various monkeys besides Japanese monkeys were shown to be naturally infected with HTLV-I or its related viruses (2-6), suggesting a possibility of virus transmission between humans and non-human primates. Then it is important to compare human viruses and simian viruses by isolating them for clarifying the origin of HTLV-I. In addition, simian virus infection in non-human primates was expected to be useful as natural and experimental model for elucidating the mode of HTLV-I transmission and mechanism of ATL onset.

Sero-epidemiological Survey in Various Non-human Primates of the World

In total, 3,706 sera from 41 species of non-human primates in the world were examined for antibody (4,5). All the antibody-positive ones were detected in catarrhines, the Old World monkeys (779/3,420) or anthropoid apes (12/126) which originated in Africa and Asia. No antibody was detected in any platyrrhines, the New World monkeys (0/120) such as squirrel monkeys and marmosets in south America, or prosimians (0/40) such as tree-shrews in Asia and grand galagos in Africa. In the Old World monkeys, high incidences of sero-positive monkeys were detected in the Macaca genus in Asia, such as Japanese monkeys (670/2,650), rhesus monkeys (20/98), Formosan monkeys (7/26) and cynomolgus monkeys (39/394) etc., the Cercopithecus genus such as African green monkeys (21/44) and Syke's monkeys (5/20) and the Papio genus such as anubis baboon (4/28) and hamadryas baboon (3/18) in Africa.

However, in the Old World monkeys, langurs (0/15) and lutongs (0/10) belonging to Colobus genus which live on the tree in the same forest where seropositive macaques live on the ground were all negative for this virus. In anthropoid apes, chimpanzees in Africa (11/26) and siamangs (1/23) in Asia are also infected with the virus.

Thus, bordered between Cercopithecinae and Colobinae, the higher primates are positive but the lower ones are negative for the virus. This fact is of interest suggesting the association between origin of the virus and evolution of primates.

Sero-epidemiological Survey in Japanese Monkeys

The prevalence of the virus infection was examined by screening 2,670 sera from Japanese monkeys in 42 troops naturally living in different parts of Japan and compared with sero-epidemiological data in humans in this country (5).

High incidence of antibody (25%) was found in widely distributed troops of Japanese monkeys. The frequency of seropositive monkeys increased with age, reaching the maximum in adult animals. However, the geographic distributions of seropositive monkeys and of humans are quite different. High incidence of antibody in monkeys was found throughout Japan, not only in endemic areas of human ATL. In the ATL-endemic southwestern part of Japan, the percentage incidence of seropositive healthy human adults was 6-37%, whereas in half the troops of monkeys in Japan, it was more than 50%. In some troops, the percentage incidence reached more than 80%.

The higher seropositive frequency in monkeys than in humans and the quite different geographical distribution patterns of seropositive humans and monkeys indicate the improbability of direct transmission of the virus between Japanese people and Japanese monkeys.

This improbability was also supported by absence of antibody in persons working in facilities where naturally infected monkeys were kept, and by no detection of antibody in people in South-east Asia where many seropositive monkeys were found.

Establishment of Simian Virus-producing Cell Lines from Various Non-human Primates

Then, isolation of the viruses from various non-human primates for their comparison with HTLV-I was attempted by establishment of the cell lines carrying HTLV-I related viruses.

Eighteen lines of virus-producing lymphoid cells were successfully established from the peripheral blood of 10 species of non-human primates that had antibodies against HTLV-I antigens (7). They are derived from the chimpanzee, siamang, African green monkey, Sykes' monkey, pig-tailed macaque, red-faced macaque, Formosan monkey, Japanese monkey, bonnet monkey and maura monkey. Morphologically, the lymphoid cell lines appeared to be mainly prolymphocytes or lymphoblasts with a round nucleus. These cell lines reacted with human ATL patient sera and monoclonal antibodies to p19 and p24 of HTLV-I antigens. The cellular DNAs were found to contain the provirus sequences homologous to HTLV-I by Southern blot hybridization. Extracellular type-C virus particles and RNA-dependent DNA polymerase were found in cultured medium of all cell lines. Simian virions cross -reactive to HTLV-I confirmed by immunocolloidal gold electron microscopy were morphologically similar to each other and to HTLV-I (8).

All the cell lines gave a positive reaction with a monoclonal antibody to Tac antigen, which was shown to be induced on human lymphocytes by their infection with HTLV-I.

Though, the surface markers of some established cell lines were not determined, almost all cell lines were of T-cell origin, and had helper/inducer T-cell characteristics, whereas all lines derived from African green monkey and Sykes' monkey had suppressor/cytotoxic T-cell characteristics. As almost all human lymphoid cell lines carrying HTLV-I were reported to have helper/inducer T-cell surface antigens, the reaction of simian cells derived from these species (Cercopithecus sub-genus) with Leu2a was unique.

Restriction Analysis of Provirus Genomes of Simian Viruses

For understanding the mode of viral transmission and the origin of HTLV-I, integrated provirus genomes related to HTLV-I in the cell lines were analyzed (9). On the blotting analysis of SstI digests of DNA of 5 cell lines, a strong band of 8.5 kbp common to all monkey cell lines was detected with a representative HTLV-I probe, which was later found to cover the total genome produced by trimming the proviruses at LTRs. Digests with PstI or PstI plus SstI using viral gene-specific probes of HTLV-I, LTR, *gag*, *env* and *pX* demonstrated that all the simian proviruses have homology with all regions of HTLV-I, that is, the *gag*, *env*, *pX* and LTR regions, of which location is just same as in HTLV-I. However, they showed some differences in sites for restriction enzymes. Restriction maps of these simian viruses are different from HTLV-I and each other except those of three species of macaque, Japanese monkey, red-faced monkey and pig-tailed monkey.

The similarity in the restriction maps of these macaque virus showed that these viral sequences have been highly conserved, although the viruses are transmitted exogenously. The high conservation of viral sequences is consistent with the previous findings that two independent isolates from Japanese and Caribbean patients showed identical sites for some restriction enzymes (10). Thus, difference of restriction maps between human viruses and simian viruses indicated the unlikeliness of recent transmission of virus from monkeys to humans.

Sequence Homology of a Simian Virus Genome with HTLV-I

These observations suggested that simian viruses are a member of the HTLV-family, but the extent of relatedness of HTLV-I and these viruses were unclear. Then, the relation was examined by analyzing the nucleotide sequences of the simian provirus genomes (11,12). Of pig-tailed monkey, each nucleotide sequence of env, pX, LTR region has about 90% homology with that of HTLV-I.

Nucleotide sequences of LTR of chimpanzee and African green monkey showed 95% homology with that of HTLV-I and each other, however, each LTR of these two viruses derived from African origin have 90% homology with that of pig-tailed monkey of Asian origin. Closer relatedness of LTR in simian viruses of African origin to HTLV-I than to those of Asian origin is interesting for considering the origin of HTLV-I.

The 90% or 95% homology is significantly lower than those among HTLV-I isolates, where only about 1% nucleotide alterations were observed (13,14). These similarities and differences confirmed that the simian virus is a member of the HTLV-family, but distinct from HTLV-I. Therefore, this simian viruses are now called simian T-cell leukemia/lymphotropic virus type I (STLV-I) by their similarity to HTLV-I. Consequently, these findings also exclude the possibility of recent interspecies transmission of HTLV-I from simians to humans; that is, simians cannot be a reservoir of HTLV-I in humans. Comparative analysis of the nucleotide sequence of STLV-I with previously reported sequences of HTLV-I (13), HTLV-II (15,16) and BLV (17,18) suggest that STLV-I originated from a common ancestor of this group of viruses and that STLV-I diverged from HTLV-I much later than HTLV-II during their evolution. Thereafter, each STLV-I seems to have been maintained and to have diverged separately in different species or genuses of non-human primates.

Leukemogenic Potentials of STLV-I

These structural similarities between STLV-I and HTLV-I suggested that STLV-I has leukemogenic potential in monkeys as HTLV-I has in humans. Though human and simian lymphocytes were immortalized and/or transformed *in vitro* when cocultured with these STLV-I-producing cells (7), no definite case of ATL-like disease in monkey had not been reported till such a case was recently found in an African green monkey (19). This female monkey at estimated age of 8, was imported from Uganda 4 years ago and since then has been seropositive. At necropsy, systemic lymphoadenopathy and remarkable splenomegaly were noted. Just before death, many abnormal lymphocytes constituting 80% of the total leukocytes were seen including cells with a deeply indented or lobulated nucleus that were strikingly similar to those seen in human ATL. The leukemic cells reacted with monoclonal antibodies to OKT 11, Leu 2a and Tac antigen. To examine the possibility that STLV-I was associated with this ATL-like leukemia, provirus integration into the leukemic cells of peripheral blood lymphocytes, lymph nodes and the spleen was analysed. The restriction patterns are consistent with that of one previous isolate, STLV-I from African green monkey and showed the tumor consisting of monoclonally expanded cells in which one copy of the STLV-I proviral genome has been integrated. These findings suggest that STLV-I infection resulted in the development of leukemia as HTLV-I in human ATL (20).

Though this is only a single case, there were several cases considered to be at pre-leukemic or chronic ATL state in this African green monkey colony. These STLV-I-infected monkeys were clinically healthy but significant increase in total numbers of leukocytes and atypical lymphocytes consisting of lobulated lymphocytes was seen. These atypical lymphocytes had Leu2a^{+} surface marker and were monoclonaly expanded cells integrated with one copy of STLV-I just as seen in those of the overt ATL-like case. The higher frequency of ATL-like disease detected here in African green monkey than in humans indicated usefulness of this animal model for clarifying the mechanism of ATL onset.

AIDS-RELATED SIMIAN VIRUS (STLV-III) IN NON-HUMAN PRIMATES

STLV-III or SIV, simian retrovirus related to so-called AIDS virus (HTLV-III/LAV or HIV) naturally infected in and isolated from some monkey species (21-23), were expected to be useful for clarifying the origin of HTLV-III/LAV and for animal model of AIDS just as STLV-I is expected in ATL. We have also isolated STLV-III from 8 African green monkeys and now characterizing them at protein and genomic levels. Here, our recent sero-epidemiological study is described.

Serological Survey in Various Non-human Primates of the World

Antibody presence was examined in 1,613 serums from 49 species of non-human primates of prosimians (4 samples), the New World monkeys (54 samples), the Old World monkeys (1,459 samples) and anthropoid apes (96 samples). By immunofluorescence assay using STLV-III -AGM-infected Hut 78 cells as antigen which was kindly supplied by Dr. M. Essex, seropositive ones were detected in African green monkey (90/343, 26%), Sykes' monkey (4/18, 22%), Mandril (2/20) and cynomolgus monkey (2/427). But all other serums were negative including sooty mangabeis (16 samples), rhesus monkeys (99 samples) and pig-tailed monkeys (14 samples) from which STLV-III isolation was reported (21, 22,24,25,26).

Most of these African green monkey serums were taken at their natural habitats in Ethyopia, Uganda and Kenya, and some of the serums were collected in 4 different facilities in Japan. The higher positive ratio (20-40%) of each serum group obtained from different places in the world indicated STLV-III infection is popular in this species as shown previously by Dr. P.J. Kanki (23). The clustering of seropositivity to STLV-III in African green monkey and Sykes' monkey which belong to *Cercopithecus* subgenus in Africa is in contrast to antibody prevalence of STLV-I which is commonly seen in apes and the New World monkeys in Asia and Africa as described before.

Also all these serums from various monkeys were examined for antibody to HTLV-III. Only one African green monkey serum positive for STLV-III reacted but no other serums of the same species including ones positive to STLV-III (27). No positive reaction was seen in serums of other monkey species. This observations indicate that immuno-cross-reactivity between HTLV-III and STLV-III is not so strong as that between HTLV-I and STLV-I.

The Possible Way of Transmission of STLV-III in African Green Monkey Colony

To clarify the way of transmission of STLV-III in monkeys which can be experimentally handled, is useful for clarifying that of HTLV-III/LAV in humans.

A small scale of sero-epidemiological study was done in an African green monkey colony in Tsukuba Primate Center. In this colony, 45/103 (44%) monkeys were seropositive, and STLV-III was isolated from some of these seropositive monkeys, indicating that seropositive monkeys were also virus carriers as seen in HTLV-III/LAV-infected persons. Relation of antibody presence between mating couples, and between mothers and infants were examined (Table 1-a,b). In mating couples, when males were positive, their females were also all positive, however, when females were positive, their males were not always positive. All seropositive infants were born from positive mothers, however, from sero-negative mothers, no positive infants were born.

TABLE 1
RELATION OF SEROPOSITIVITY IN AN AFRICAN GREEN MONKEY COLONY

a)

		Male +	Male -	
Female	+	8	21	29
	-	0	9	9
		8	30	38

b)

		Mother +	Mother -	
Infants	+	13	0	13
	-	40	14	54
		53	14	67

c)

		STLV-I +	STLV-I -	
STLV-III	+	22	19	41
	-	9	63	72
		31	82	113

a) Mating couples
b) Mothers and their infants
c) Antibodies to STLV-I and STLV-III

From these observations, the horizontal transmission from males to females and the vertical transmission from mothers to infants are suspected. To confirm these possible ways of virus transmission, further epidemiological follow up in this colony, virus isolation from samen and milk from virus-carrying monkeys and experimental infection orally or via vagina should be done.

In this African green monkey colony, there were many monkeys doubly infected with STLV-I and STLV-III (Table 1-c), of which meaning is unclear. High susceptibility of HTLV-I-infected cells to HTLV-III/LAV was reported (28) and also it is possible that STLV-I can easily infect animals immuno-suppressed by STLV-III infection. However, immune-suppression estimated by mitogen response of lymphocytes was not observed in these STLV-III-infected African green monkeys, but was seen in STLV-I-infected ones, hypothesizing that animals immunosuppressed by STLV-I have higher susceptibility to STLV-III infection, which must also wait for further study (Table 2).

TABLE 2
RESPONSE OF PERIPHERAL BLOOD LYMPHOCYTES OF STLV-I AND -III INFECTED AFRICAN GREEN MONKEYS

STLV-III[a]	+	+	-	-
STLV-I[b]	-	+	+	-
ConA	3o.7[c]	5.0	5.0	34.5
	(12)[d]	(13)	(3)	(8)
PHA	7.3	2.1	2.2	9.8
	(7)	(8)	(3)	(8)
PWM	15.9	3.6	1.1	31.0
	(6)	(8)	(3)	(6)

[a]Anti-STLV-III antibody
[b]Anti-STLV-I antibody
[c]Stimulation index
[d]Tested numbers of monkeys

ACKNOWLEDGMENTS

These research works presented here are cooperative studies with Tsukuba Primate Center, NIH of Japan (Director Dr. Shigeo Honjo), Department of Tumor virus, Cancer Institute, Japan (Director, Dr. Mitsuaki Yoshida), Institute of Primate, Kyoto University, Japan (Director, Dr. Ken Nozawa) and author's laboratory (Drs. Hajime Tsujimoto, Kohichi Ishikawa and Yoshihiro Ohta).

REFERENCES

1. Miyoshi I, Yoshimoto S, Fujishita M, Taguchi H, Kubonishi I, Niiya K, Minezawa M (1982). Natural adult T cell leukemia virus infection in Japanese monkeys. Lancet 2:658.
2. Hunsman G, Schneider J, Schmidt L, Yamamoto N (1983). Detection of serum antibodies to adult T-cell leukemia virus in non-human primates and in people from Africa. Int J Cancer 32:329
3. Miyoshi I, Fujishita M, Matsubayashi K, Miwa N, Tanioka Y (1983). Natural infection in non-human primates with adult T-cell leukemia virus or a closely related agent. Int J Cancer 32:333.
4. Hayami M, Ishikawa K, Komuro A, Kawamoto Y, Nozawa K, Yamamoto K, Ishida T, Hinuma Y (1983). ATLV antibody in cynomolgus monkeys in the wild. Lancet ii:620.
5. Hayami M, Komuro A, Nozawa K, Shotake T, Ishikawa K, Yamamoto K, Ishida T, Honjo S, Hinuma Y (1984). Prevalence of anti-adult T cell leukemia virus (ATLV) antibody in Japanese monkeys and other non-human primates. Int J Cancer 33:179.
6. Yamamoto N, Hinuma Y, zur Hausen H, Schneider J, Hunsman G (1983). African green monkeys are infected with adult T cell leukemia virus or a closely related agent. Lancet 1:240.
7. Tsujimoto H, Komuro A, Iijima K, Miyamoto J, Ishikawa K, Hayami M (1985). Isolation of simian retroviruses closely related to human T-cell leukemia virus by establishing of lymphoid cell lines from various non-human primates. Int J Cancer 35:377.
8. Haga S, Tanaka H, Tsujimoto H, Hayami M (1985). Conventional and immunocolloidal gold electron microscopy of eight simian retroviruses closely related to human T-cell leukemia virus type I. Cancer research 46:293.

9. Komuro A, Watanabe T, Miyoshi M, Hayami M, Tsujimoto H, Seiki M, Yoshida M (1984). Detection and characterization of provirus genomes of simian retroviruses homologous to human T-cell leukemia virus. Virol 138:373.
10. Watanabe T, Seiki M, Yoshida M (1984). HTLV type I (US isolate) and ATLV (Japanese isolate) are the same species of human retrovirus. Virol 133:238.
11. Watanabe T, Seiki M, Tsujimoto H, Miyoshi I, Hayami M, Yoshida M (1985). Sequence homology of the simian retrovirus (STLV) genome with human T-cell leukemia virus type I (HTLV-I). Virol 144:59.
12. Watanabe T, Seiki M, Hirayama Y, Yoshida M (1986). Human T-cell leukemia virus type I is a member of the African subtype of simian viruses.(STLV). Virol 148:385.
13. Seiki M, Hattori S, Hirayama Y, (1983). Human T-cell leukemia virus : Complete nucleotide sequence of the provirus genome integrated in leukemia cell DNA. Proc Natl Acad Sci 80:3618.
14. Seiki M, Hattori S, Yoshida M (1982). Human adult T-cell leukemia virus : Molecular cloning of the provirus DNA and the unique terminal structure. Proc Natl Acad Sci 79:6899.
15. Shimotohno K, Wacksman W, Takahashi Y, Golde DW, Miwa M, Sugimura T, Chen ISY (1984). Nucleotide sequence of the 3' region of an infections human T-cell leukemia virus type II genome. Proc Natl Acad Sci 81:6657.
16. Haseltine WA, Sodroski J, Patarca R, Briggs D, perkins D, Wong-Staal F (1984). Structure of 3' terminal region of type II human T lymphotropic virus : Evidence for new coding region. Science 225:419.
17. Rice NR, Stephens RM, Couez D, Deshamps J, Kettmann R, Burny A, Gilden R (1984). The nucleotide sequence of the env gene and post-env region of bovine leukemia virus. Virol 138:82
18. Sagata N, Yasunaga T, Tsuzuku-kawamura J, Ohishi K, Ogawa Y, Ikawa Y (1985). Complete nucleotide sequence of the genome of bovine leukemia virus : Its evolutionary relationship to other retroviruses. Proc Natl Acad Sci 82:677.
19. Tsujimoto H, Seiki M, Nakamura H, Watanabe T, Sakakibara I, Sasagawa A, Honjo S, Hayami M, Yoshida M (1985). Adult T-cell leukemia-like disease in monkey naturally infected with simian retrovirus related to human T-cell leukemia virus type I. Jpn J Cancer Res (Gann) 76:911.

20. Yoshida M, Seiki M, Yamaguchi K, Takatsuki K (1984). Monoclonal integration of HTLV in all primary tumors of adult T-cell leukemia suggests causative role of HTLV in the disease. Proc Natl Acad Sci 81:2534.
21. Kanki PJ, McLane MF, King NW, Letvin NL, Hunt RD, Sehgal P, Daniel MD, Desrosiers RC, Essex M (1985). Serological identification and characterization of a macaque T-lymphotropic retrovirus closely related to HTLV-III. Science 228:1199.
22. Daniel MD, Letvin NL, King NW, Kannagi M, Sehgal PK, Hunt RD, Kanki PJ, Essex M, Desrosiers RC (1985). Isolation of T-cell tropic HTLV-III-like retrovirus from macaques. Science 228:1201.
23. Kanki PJ, Alroy J, Essex M (1985). Isolation of T-lymphotropic retrovirus related to HTLV-III/LAV from wild-caught African green monkeys. Science 230:951.
24. Murphey-Corb M, Martin LN, Montelaro RC, Rangan SRS, Baskin GB, Gormus BJ, Wolf RH, Andes WA (1986). " SAIDS induced by an HTLV-III related simian retrovirus " in Abstract of 15th annual meetings of UCLA Symposia, Park City, p 194.
25. Fultz PN, McClure HM, Anand R, Srinivasan A (1986). " Isolation of a T-lymphotropic retrovirus from healthy mangabey monkeys " ibid. p 188.
26. Benveniste R, Arthur L, Tsai C-C, Sowder R, Henderson L, Orozlan S (1986). " Characterization of a lymphotropic lentivirus isolated from a pig-tailed macaque with lymphoma ". ibid. p 186.
27. Hayami M, Ohta Y, Hattori T, Nakamura H, Takatsuki K, Kashiwa A, Nozawa K, Miyoshi I, Ishida T, Tanioka Y, Fujiwara T, Honjo S (1985). Detection of antibodies to human T-lymphotropic virus type III in various non-human primates. Jpn J Exp Med 55:251.
28. Koyanagi Y, Harada S, Yamamoto N (1985). Correlation between high susceptibility to AIDS virus and surface expression of OKT-4 antigen in HTLV-I-positive cell lines. Jpn J Cancer Res (Gann) 76:799.

Viruses and Human Cancer, pages 345–353

LEUKEMOGENESIS BY BOVINE LEUKEMIA VIRUS (BLV)

A. Burny[·,o], Y. Cleuter[·], C. Dandoy[·,+], R. Kettmann[·,o], M. Mammerickx[*], G. Marbaix[·], D. Portetelle[·,o], A. Tartar[+], A. Van Den Broeke[·] and L. Willems[·,o]

. Department of Molecular Biology, University of Brussels, 1640, Rhode-Saint-Genèse, Belgium
o Faculty of Agronomy, 5800 Gembloux, Belgium
* National Institute for Veterinary Research, 1180 Uccle, Belgium
+ Pasteur Institute, Lille, France.

ABSTRACT Bovine leukemia virus (BLV) is horizontally transmitted among cattle via the cell route. A titration experiment showed that a treshold amount of provirus-carrying lymphocytes has to be administered subcutaneously in order to propagate infection.
BLV can transform lymphocytes in cattle, sheep and goats. Replication and transforming properties of the virus depend upon expression of a nuclear $p34^x$ protein and perhaps a $p14^x$ protein, both products being coded for by two overlapping, open reading frames located between the envelope gene and the 3' long terminal repeat (LTR) of the provirus. A clone of 2353 base pairs long was obtained from a λgt 10 library constructed from mRNA of BLV-infected bat lung cells. mRNA made in SP6 cloning vector, codes for $p34^x$ but not $p14^x$. This material is well suited to better define the role of $p34^x$ and $p14^x$ in the transforming capacity of BLV.
Infectivity and syncytia induction are mediated via the envelope glocyprotein gp51. Three epitopes named F, G and H are involved. Their identification is under way using the synthetic peptide approach.

INTRODUCTION

The infectious character of enzootic bovine leukemia was more and more evident when the viral agent was identified in short-term cultures of peripheral lymphocytes of cattle in PL [1]. It is a non typical type-C retrovirus, found to be unrelated to any known retrovirus family until the discovery of the human T-cell lymphotropic viruses. (HTLV-I in 1980 [2] and HTLV-II in 1982 [3]). The genome of the 3 viruses contains the long terminal repeat (LTR), gag, pol and env sequences characteristic of all retroviruses. In addition, a region designated X by Seiki et al. [4] in HTLV-I, is located between env and the 3'LTR. A similar X region was identified in HTLV-II and BLV, genomes [5-8]. One at least of the protein products of the X region is putatively responsible for the transactivation of BLV-LTR in BLV infected cells [9,10].

The present knowledge concerning leukemogenesis by BLV can be summarized as follows : (i) BLV is a mandatory part of the transforming machinery. Tumors that arise in infected animals always harbor BLV foot prints. (ii) The BLV genome carries no typical cellular derived oncogene. (iii) The provirus has no preferential site for integration making very improbable a cis-acting function to explain cell transformation [11]. (iiii) Expression of the provirus does not appear to be necessary for the maintenance of the transformed state [12]. (iiiii) Oncogenic deleted proviral copies have retained the left hand LTR and the 3' half of the provirus, thus emphasizing the crucial role of these proviral regions.

The longest open reading frame of the X region of BLV is supposed to be transcribed as a double spliced 2.1 kb mRNA [14]. Its putative protein product has been recognized as a 38000 MW protein, precipitated by a rabbit polyclonal serum raised against a synthetic peptide coded for by the LOR sequence. The protein product of the second open reading frame is a 14 kD molecule tentatively called the leuk gene product [15] and identified as the product of a second open reading frame of the BLV X region [17].

In this report we shall also illustrate the data obtained in our search of the minimal infectious dose of circulating lymphocytes from BLV-infected animals. Obviously, the dose varies with the donor (Table 2). Three hundred times more cells were required from animal n° 2 than from animal n° 1.

Investigations aiming at identification of epitopes F, G and H on BLV envelope glycoprotein gp51 extensively used the synthetic peptide approach. The data emphasize that even if the crucial epitopes are conformational structures requiring glycosylation of the protein backbone to adopt their native configuration, synthetic peptides made in well chosen areas of gp51 are recognized by sera of animals immunized by whole BLV particles.

RESULTS

I. Minimal infectious dose of BLV-infected lymphocytes

Two donor animal were compared. Cow n°1 was a PL case with 21978 lymphocytes per mm^3. Cow n°2 was a BLV carrier with no modification of the blood picture (Table 1).

TABLE 1
BLOOD PARAMETERS OF THE ANIMALS USED AS DONORS IN THE INOCULATION EXPERIMENTS

	Number of lymphocytes / mm^3	Percentage of lymphocytes in the white cells	Elisa gp51 antibody titer
Cow n° 1	18681	85	14580
Cow n° 2	2853	42	180

The outcome of the two experiments is illustrated in table 2. 926 Lymphocytes of animal n° 1, in PL, are mandatory to achieve infection. Seroconversion occured after 18 days. Lymphocytes from animal n° 2 were infectious but only at very high dose. 325300 Lymphocytes were the last dilution (in a 1/10 series) that led to seroconversion within a delay of 35 days. 32530 Lymphocytes could not establish infection even after 111 days.

TABLE 2
TITRATION EXPERIMENT IN SHEPP OF INFECTIOUS BOVINE LYMPHOCYTES

Donor animal	Inoculum Number of lymphocytes injected	Days after subcutaneous inoculation							
		0	7	14	18	21	35	140	492
		anti gp51 serum titer							
Cow n°1	926	0	0	0	60	180	14580	43740	131220
Cow n°2	325 300	0	0	0	0	0	1620	14580	

Other experiments have largely confirmed these observations. The longst delay before seroconversion was observed in a goat; it reached 84 days.

II. In vitro synthesis of BLV $p34^x$ and search for antibodies to this protein.

(i) Cloning of cDNA susceptible to code for $p34^x$ mRNA

A BLV-producing cell line (Bat lung cells) was used as a source of mRNA. The cDNA library was constructed in λgt 10 and screened with several probes susceptible to identify recombinant phages with p34 DNA sequences but lacking the upstream non-coding region [13,14]. Among 6 such clones, one was chosen and characterized. Subcloned in SP64 vector, sequencing data revealed a molecule of 2353 bp.

(ii) Translation of $p34^x$ mRNA in reticulocyte lysates

The DNA fragment was transcribed by E. coli DNA polymerase and 1 μg of RNA was used to program a 20 μl rabbit reticulocyte lysate. The blotted $p34^x$ protein was revealed by 50 μl of serum to be tested and 5 x 10^4 cpm of ^{125}I-labeled protein A (Specific activity : 3 x 10^7 cpm/μg) (fig. 1).

It is evident that the anti synthetic peptide antibody reacts quite strongly as well as most sera from tumor cases. Sera from PL cases remained negative. The same type of experiment performed with BLV-infected sheep sera showed a clearly positive outcome with practically all sera tested whether or not in the tumor stage of the disease (not shown). Further experiments indicated that $p34^x$ is found in the nuclei of BLV-infected cells, confirming that the in vitro product has the same size as the molecule made in vivo and thus ruling out major posttranslational modifications of the protein.

III. Investigation of the BLV envelope protein (gp51) antigenic sites by the synthetic peptide approach

A study of the antigenic structure of BLV envelope glycoprotein with virus-neutralizing and non-neutralizing monoclonal antibodies has allowd to associate viral infectivity with 3 out of 8 gp51 epitopes named F, G and H. Moreover, monoclonals to F, G and H epitopes inhibit induction of syncytia by BLV gp51 in sensitive cell cultures. Limited digestion of the protein with urokinase showed that a 15000 MW peptide fragment was recognized by monoclonal antibodies

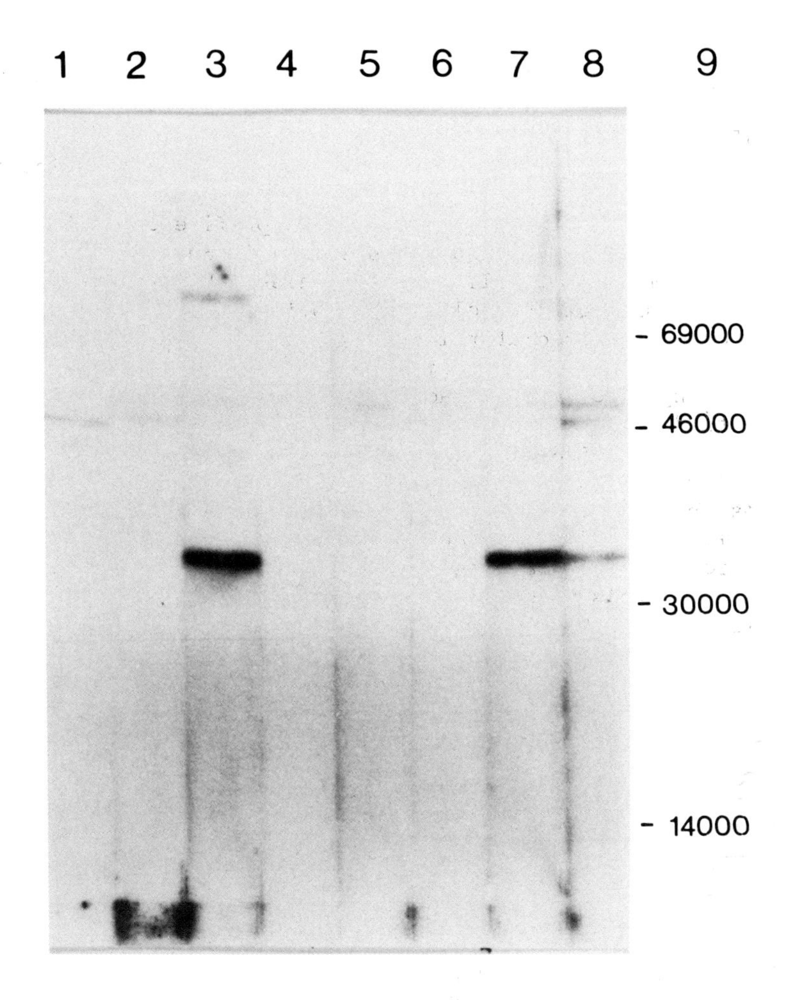

LEGEND TO FIGURE 1. Reactivity of various sera with $p34^{x}$ in Western blots.

Lane 1; normal rabbit serum; lane 2: rabbit anti p24 serum; lane 3: rabbit serum raised against synthetic peptide RFPRDTSEPPLS of the $p34^{x}$ protein [7]; lane 4: normal bovine serum; lanes 5 and 6: sera of cows 285 and 928 in persistent lymphocytosis; lanes 7 and 8: sera of bovine tumor cases 15 and 82; lane 9: molecular weight markers.

with F, G or H specificities. In addition, iodination of gp51 followed by urokinase digestion showed that most of the ^{125}I radioactivity migrated in SDS-PAGE analysis with the 15000 MW fragment. The latter result is a definite proof that the 15000 MW peptide represents the NH_2- moiety of gp51 as 6 out of the 8 cysteine residues lie in the first 130 amino acid residues of the protein.

The Kyte-Doolittle hydropathy profile of gp51 indicated that 5' amino acid sequences 21-28, 56-69, 144-155 and 260-268 formed hydrophilic centers susceptible to be exposed at the outside of the molecule [7] (Oligopeptides were synthesized and coupled by glutaraldehyde to either thyroglobulin or keyhole limpet hemocyanin as a carrier). In the four cases, the injected rabbit reacted very well to the antigenic preparation and synthesized antibody that recognized the oligopeptide in an ELISA test. Conversely an anti-whole BLV rabbit serum was reacted against the four peptides and shown to exhibit a decreased reactivity towards oligopeptides 144-155 (++++), 250-268 (+), and 59-69 (+). Only a very limited reaction (+/-) was observed with fragment 21-28.

At this stage of the investigation, it is concluded :

(1) Regions 59-69, 144-155 and 260-268 elicit antibodies in the rabbit.

(2) Some configurations at least of coupled oligopeptides are recognized by a rabbit immunized with whole BLV particles.

A systematic search for BLV peptides covering the amino half of BLV gp51 should allow the identification of the crucial epitopes involved in biological activities of the virus.

DISCUSSION

The demonstration that a sufficient number of BLV-infected lymphocytes is required to achieve a successful transmission of the virus explains the discrepancies observed in the past concerning spreading or not of the disease within a herd after introduction of a single infected animal. From the data reported here, it is evident that animal n° 2 cannot be considered as infections in natural conditions whilst animal n° 1 would probably behave as a very efficient transmitter. The same probably holds true in the HTLV-I and HTLV-II systems and perhaps even in the case of AIDS viruses. Detecting the presence of a retrovirus in a biological fluid may have little epidemiological significance unless the virus concentration and the amount of fluid transmitted to a recipient are known.

The late seroconversion of animal who received the lowest dose of inoculum might bear significance in the field of AIDS viruses epidemiology. It has been reported by several groups that contacts of AIDS patients may be virus positive but antibody negative. Animal 2 mimicks the situation of AIDS contacts who can be infected by doses of virus that are so low that seroconversion is much delayed.

The common denominator of the biochemistry of BLV, HTLV-I, HTLV-II is the presence of a viral gene coding for a transactivating protein, acting at the level of transcription. Other proteins, also of viral origin , might contribute also. In HTLV-III/LAV (and probably visna, CaEV,...) activation occurs at the post-transcriptional level (Haseltine et al., this symposium). It is believed that the transactivating machinery acts also upon normal cellular genes whose regulatory sequences can be recognized by the activator (the tat gene product). The data that we obtained so far in the BLV system indicate that BLV tat gene product ($p34^x$) is not required to maintain the transformed stage. It follows that $p34^x$ must induce at low frequency (because tumors are rare and incubation time long !) a second event that in turn induces, at low frequency, a third ... One of the event of the cascade, must be self-driven, once it has been initiated (situation of a gene product that stimulates its own production). The initiating cause (the virus) can disappear, the cell is condemned to evolve and to perhaps reach the final stage of transformation. We can also imagine that one of the mandatory step along the pathway of cell transformation is chromosomal rearrangement. Once rearrangement has occured, virus expression can be switched off; the breakpoint of no-return has been reached. Perhaps, as in yeast 21, abnormal chromosome segregation leading to aneuploïdy, non-disjunction, ... can occur due to unbalanced synthesis of histone dimer sets. Other peculiarities such as non-expression of immunoglobulins, switching off of viral structural genes,... are evidently inclusive steps along the pathway of B cell transformation by BLV.

ACKNOWLEDGMENTS

This work was helped financially by the Belgian Ministry of Agriculture and the Fonds Cancérologique de la CGER. CD holds a PREST fellowship from the Ministère de la Politique scientifique. RK is Chercheur Qualifié and LW is Aspirant du Fonds National de la Recherche Scientifique. AV holds a fellowship from the Lady Tata Memorial Trust.

REFERENCES

1. Miller JM, Miller LD, Olson C, Gillette KG (1969). Virus-like particles in phytohemagglutinin-stimulated lymphocyte cultures with reference to bovine lymphosarcoma. J Natl Cancer Inst 43:1297.
2. Poiesz BJ, Ruscetti FW, Gazdar AF, Bunn PA, Minna JD, Gallo RC (1980). Detection and isolation of type C retrovirus particles from fresh and cultured lymphocytes of a patient with cutaneous T-cell lymphoma. Proc Natl Acad Sci USA 77:7415.
3. Kalyanaraman VS, Sarngadharan MG, Robert-Guroff M, Miyoshi I, Blayney D, Golde D, Gallo RC (1982). A new subtype of human T-cell leukemia virus (HTLV-II) associated with a T-cell variant of hairy cell leukemia. Science 218:571.
4. Seiki M, Hattori S, Hirayama Y, Yoshida M (1983). Human adult T-cell leukemia virus : Complete nucleotide sequence of the provirus genome integrated in leukemia cell DNA. Proc Natl Acad Sci USA 80-3618.
5. Haseltine WA, Sodroski JG, Patarca R, Briggs D, Perkins D, Wong-Staal F (1984). Structure of 3' terminal region of type II human T lymphotropic virus : Evidence for new coding region. Science 225:419.
6. Shimotohno K, Wachsman W, Takahashi Y, Golde DW, Miwa M, Sugimura T, Chen ISY (1984). Nucleotide sequence of the 3' region of an infectious human T-cell leukemia virus type II genome. Proc Natl Acad Sci USA 81:6657.
7. Rice NR, Stephens RM, Couez D, Deschamps J, Kettmann R, Burny A, Gilden RV (1984). The nucleotide sequence of the env and post-env region of bovine leukemia virus. Virology 138:82.
8. Sagata N, Yasunaga T, Ohishi K, Tsuzuku-Kawamura J, Onuma M, Ikawa Y (1984). Comparison of the entire genomes of bovine leukemia virus and human T-cell leukemia virus and characterization of their unidentified open reading frames. EMBO J 3:3231.
9. Derse D, Caradonna SJ, Casey JW (1985). Bovine leukemia virus long terminal repeat : A cell type-specific promoter. Science 227:318.
10. Rosen CA, Sodroski JG, Kettmann R, Burny A, Haseltine WA (1985). Trans-activation of the bovine leukemia virus long terminal repeat in BLV-infected cells. Science 227:320.

11. Gregoire D, Couez D, Deschamps J, Heuertz S, Hors-Cayla MC, Szpyrer J, Szpyrer C, Burny A, Huez G, Kettmann R (1984). Different bovine leukemia virus-induced tumors harbor the provirus in different chromosomes. J.Virol 50:275.
12. Kettmann R, Cleuter Y, Grégoire D, Burny A (1985). Role of the 3' long open reading frame region of bovine leukemia virus in the maintenance of cell transformation. J Virol 54:899.
13. Seiki M, Hikikoshi A, Taniguchi T, Yoshida M (1985). Expression of the px gene of HTLV-I. General splicing mechanism in the HTLV family. Science 228:1532.
14. Mamoun R, Astier-Gin T, Kettmann R, Deschamps J, Rebeyrotte N, Guillemain B (1985). The px region of the bovine leukemia virus is transcribed as a 2.1 kilobase mRNA. J Virol 54:625.
15. Ghysdael J, Kettmann R, Burny A (1979). Translation of bovine leukemia virus virion RNAs in heterologous protein-synthesizing system. J Virol 29:1087.
16. Ghysdael J, Kettmann R, Burny A (1978). Translation of bovine leukemia virus genome information in heterologous protein synthesizing systems programmed with virion RNA and in cell-lines persistently infected by BLV. Ann. Rech Vet 9:627.
17. Yoshinaka Y, Oroszlan S (1985). Bovine leukemia virus post-envelope gene coded protein: Evidence for expression in natural infection. Biochem Biophys Res Commun 131:347.
18. Portetelle D, Bruck C, Mammerickx M, Burny A (1983). Use of monoclonal antibody in an ELISA test for the detection of antibodies to bovine leukemia virus. J Virol Meth 6:19.
19. Bruck C, Rensonnet N, Portetelle D, Cleuter Y, Mammerickx M, Burny A, Mamoun R, Guillemain B, Van der Maaten MJ, Ghysdael J (1984). Biologically active epitopes of bovine leukemia virus glycoprotein gp51 : their dependence on protein glycosylation and genetic variability. Virology 136:20.
20. Bruck C, Portetelle D, Mammerickx M, Mathot S, Burny A (1984). Epitopes of bovine leukemia virus glycoprotein gp51 recognized by sera of infected cattle and sheep-Leukemia Res 8:315.
21. Meeks-Wagner D, Hartwell LH (1986). Normal stoichiometry of histone dimer sets is necessary for high fidelity of mitotic chromosome transmission. Cell 44:43.

Viruses and Human Cancer, pages 355-369

MORPHOLOGICAL CORRELATES OF GENITAL HPV INFECTION: VIRAL REPLICATION, TRANSCRIPTION AND GENE EXPRESSION(1)

Christopher P. Crum(2), Daniel Friedman(2), Gerard Nuovo(2) and Saul J. Silverstein(3)

Departments of Pathology, Obstetrics and Gynecology, Microbiology and the Cancer Research Center, Columbia University College of Physicians and Surgeons, New York, N.Y.

ABSTRACT Current studies indicate a strong correlation between specific morphological changes and the presence of certain HPV strains in precancerous squamous epithelium of the cervix, vulva and vagina. HPV type 16 is the most commonly detected HPV type in cervical lesions in our experience, and 85% of these lesions exhibit some morphological features associated with aneuploid epithelium (CIN). However, over 50% of these lesions containing HPV 16 DNA exhibit, in addition, foci of epithelium indistinguishable from condyloma, although in our experience, only one HPV type(16) is detected in the majority of these lesions. DNA-DNA in-situ hybridization analysis of these lesions containing HPV 16 DNA has demonstrated nucleic acids in areas resembling both condyloma and CIN, with the greatest concentration in mature cells containing cytoplasmic maturation. Ten percent of lesions containing HPV 16 produce detectable capsid antigens, and we have confirmed the presence of these antigens in the same areas which hybridize in-situ for HPV DNA. Recent studies using biotin and S-35 labeled RNA probes constructed in GEM-1 vectors

1 Supported by grants from the NIH CA23767 (SJS) and ACS(SJS and CPC). Dr. Crum is a recipient of a Physician Scientist Award from the National Institutes of Health AI 00628

2&3 Present address: Department of Obstetrics and Gynecology, 630 W 168th St., New York, N.Y. 10032

4 Present address: Department of Microbiology, 701 W 168th St., New York, N.Y. 10032

indicate that early HPV genes are expressed primarily in the upper (more mature) regions of the neoplastic epithelium. Thus maturation appears to exert a positive influence on a variety of HPV functions in neoplastic epithelium, including DNA replication, early and late gene expression. It is possible that patterns of gene expression may vary between lesions associated with different HPV types or different morphologies. This possibility is being explored.

INTRODUCTION

The human papillomavirus has emerged as the major suspect in the genesis of intraepithelial and invasive neoplasms of the female genital tract. The possible causal role of HPV in squamous neoplasia is supported by a number of clinical, histological and molecular studies(1-6). From 1965-1978, the reported incidence of genital warts in women increased nearly 700%(7). In 1976, Meisels and Purola and Savia reported that at least 1% of randomly screened women had Papanicolaou smears containing koilocytotic atypia, a presumed cytopathic effect of HPV infection(8-9). Subsequently, molecular hybridization studies suggested that a significantly greater percentage of women, many without abnormal cytology, were harboring HPV in their genital tracts(10,11 and Lorincz A, personal communication).

The close association of HPV with precancerous genital lesions was first appreciated histologically when approximately 20-30% of cervical (CIN) or vulvar (VIN)intraepithelial neoplasms were observed to also contain epithelial changes identical to classical condylomata(2,12). In addition, there exist lesions with features of both neoplasia and condyloma in the same epithelium(2,13,14). These "transition lesions" provided the early basis for assuming that two different pathways of progression occurr in association with HPV infection. In one, the classical condyloma infrequently progresses to invasive cancer(2,15). In the other a CIN or VIN lesion developes which often resembles condyloma but has a significant risk of persisting or progressing to higher grades of neoplasia or invasive cancer. Prospective and retrospective followup studies of similar lesions in the cervix have revealed that condylomata usually regress or persist, in contrast to CIN, which almost always persist and occasionally progresses to invasive cancer if not treated(2,15).

Precancerous lesions of the cervix are a potentially

valuable model for studying the relationship between the expression of HPV genetic information and the development of neoplasms. Since cancers of the cervix are presmumably preceded by a precursor lesion, analysis and comparison of differences in gene expression between benign, precancerous epithelium and invasive cancers might provide clues to which viral functions are critical for progression. Until recently, a critical analysis of such factors was difficult, since precancerous lesions are small, and the yield of RNA is insufficient for detailed analysis. However, the dramatic improvements in sensitivity of the in-situ hybridization technique provided by RNA probes, coupled with the potential to grow precancerous epithelium in athymic nude mice promise both sufficient RNA for molecular analysis as well as the ability to detail specific epithelial sites of RNA transcription and gene expression(16,17).

Our own studies of cervical precancers centered first on defining the criteria for their diagnosis, followed by studies of the distribution of HPV DNA in the epithelium. In this manuscript we will summarize this work and describe our recent experience analyzing precancerous lesions for HPV RNA transcripts.

MORPHOLOGIC CORRELATES OF HPV INFECTION

Definition of Cervical Precancers: The traditional definition of cervical precancers employed the dysplasia-carcinoma-in situ classification, in which abnormal intraepithelial proliferations of the cervix are graded according to the degree to which they diverged from the appearance of normal epithelium. Lesions containing cytological atypia and partial maturation are designated as mild or moderate dysplasias, whereas those lesions containing cytological atypia with less maturation are classified as severe dysplasia-carcinoma in situ. It was generally assumed that the higher grade lesions had the greater potential to behave as true precursors, ie. progress to invasive cancer if untreated. Nevertheless, some lower grade dysplasias also progress through the spectrum to invasive cancer, prompting the introduction of the CIN classification by Richart and colleagues. They demonstrated that a portion of low grade precancers are biologically similar (aneuploid) to the higher grade lesions(18). Fu et al subsequently confirmed that aneuploid precancers have a high risk of either persisting following biopsy or progressing to invasive cancer if not treated(15). Aneuploidy has been correlated with the presence of abnormal mitotic figures and greater degrees of nuclear atypia, particularly in the lower third of the epithelium(13,15,19).

With the discovery of flat condyloma of the cervix, it was possible to distinguish two poles in the precancerous spectrum which represent biologically different lesions. Classical condylomata correlate with very mild dysplasias, and usually have a diploid or polyploid DNA distribution(8,15). In contrast, "true CIN lesions" or aneuploid precursors conform to higher grade dysplasias and carcinomas in situ. This division is not absolute, however, because very well differentiated CIN lesions can bear a close similarity to condylomata and contain koilocytotic atypia(13,15). Based on these observations, we have devised a classification for investigative purposes, where three groups of precancerous lesions are defined. The first, condyloma, is characterized by variable maturation and koilocytotic atypia and minimal cellular atypia in the lower half of the epithelium. The second, koilocytotic CIN, resembles condyloma but exhibits nuclear atypia in the lower half of at least a portion of the lesion, with or without abnormal mitotic figures. The third, CIN without koilocytotic atypia (or simply CIN) does not exhibit features of condyloma, consisting primarily of poorly differentiated cells. The purpose of this approach is to distinguish lesions morphologically into

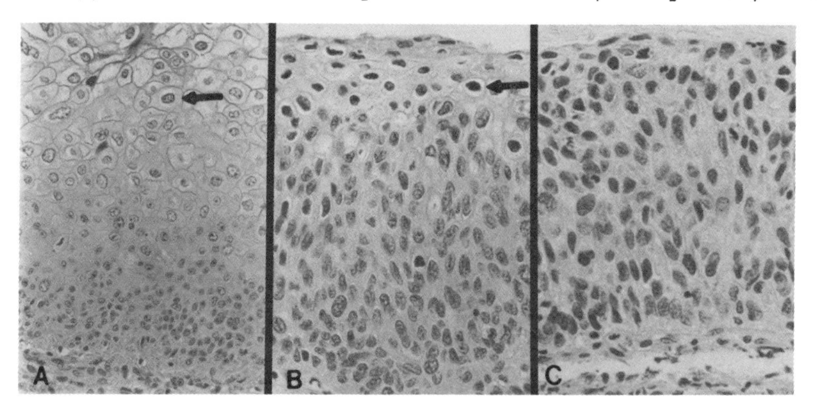

Figure I. Histology of CIN and Condyloma. A. Flat condyloma containing HPV 6 or 11 DNA. B&C. CIN lesions with and without koilocytosis, both of which contained HPV 16 DNA. Koilocytes are noted by the arrows. Both B and C exhibit prominent nuclear atypia in the lower regions of the epithelium.

those most likely to be diploid or polyploid(condyloma) and those most likely to be aneuploid(CIN), reserving a category of koilocytotic CIN for those CIN lesions which might be confused with condyloma(20)(Figure 1).

Correlation of morphology with HPV DNA strain: The discovery, isolation, and subclassification of HPV types resolved many of the questions concerning the relationship of HPV to these different forms of precancers. HPV types 6 and 11, isolated from genital warts, were frequently detected in benign condylomata, particularly in the vulva, and comprised from 60 to 80% of HPV found in these lesions(3,21). In contrast, HPV 16 was found in up to 70% of koilocytotic CIN lesions, 40% of other CIN lesions, and 80% of VIN lesions (Bowenoid papulosis)(4,6,22-24). In two studies from the German group nearly 80% of invasive squamous cell cancers contain either HPV 16 or 18 DNA sequences(4,5). Recently, additional HPV DNA types, such as types 31, 34 and 35, have been isolated and found in either precancerous epithelium and cancers(Type 31) or exclusively in cancer(Type 35)(Lorincz, unpublished data). In addition, types 16 and 18 have been detected in endocervical adenocarcinomas and adenosquamous carcinomas, suggesting that HPV may not be limited strictly to squamous neoplasms(25).

Table 1. Correlation between lesion type and HPV DNA strain(22 and Crum CP, unpublished data)

HPV	No.	NSE	Condy	CINK	CIN
6/11	27	11%	77%	8%	4%
16	40	5%	10%	55%	30%
ND	16	15%	66%	6%	13%

NSE = negative squamous epithelium; Condy = condyloma; CINK = cervical intraepithelial neoplasia with koilocytotic atypia; ND = HPV DNA type other than types 6,11, or 16.

It is important to emphasize that the absolute distinction of lesions containing HPV 16 from those containing other HPV DNA's in the cervix is marred by certain inconsistencies in histological expression of HPV infection(Table 1). From 5-10 per cent of cervical lesions containing HPV type 16

DNA cannot be distinguished from condyloma(Crum, C.P., unpublished data). This might be explained in part by errors in sampling of lesions. We have found that although most cervical lesions containing HPV 16 DNA exhibit features of CIN, over one half will also contain areas resembling classical condyloma(26). This phenomenon occurs despite the fact that, in our experience, 90% of CIN lesions containing HPV 16 DNA appear to contain only this HPV DNA. Thus it would appear that interaction between HPV 16 and the host epithelium may include the development of a condyloma, at least in a portion of the lesional epithelium(26). The possibility that HPV 16 DNA might be found in a classical condyloma is supported by a number of studies reporting HPV 16 DNA in lesions exhibiting features of condyloma only(27 and Winkler B et al and Ferenczy A et al, personal communication). Most of these studies, however, analyzed patients with multiple lesions of the lower genital tract, and often more than one HPV type was isolated from the lesions from individual patients. A final and insurmountable obstacle to morphologically defining all HPV 16 infections is the fact that HPV DNA, either type 16 or others, has been detected in biopsies of normal appearing squamous epithelium or in cervical scrapings from patients with normal Papanicolaou smears(4,5,24).

LOCALIZING HPV IN INTRAEPITHELIAL NEOPLASIA

DNA-DNA in-situ hybridization: In-situ hybridization has made it possible to determine the distribution of HPV nucleic acids in both condylomata and CIN lesions. The techniques used have employed probes labeled with biotin, S-35 and tritium(28-31). In condylomata, the HPV DNA is distributed in a pattern closely resembling that of capsid antigens, in that the greatest concentration of DNA is located in the surface nuclei, including the koilocytotic nuclei(29). However, a greater proportion of nuclei exhibit DNA sequences than capsid antigens. Furthermore, nucleic acids can be identified in cells close to the basal layer. A similar distribution pattern is found with HPV 16 sequences in CIN, where the greatest concentration of DNA is present in the surface nuclei(30). In both condylomata and CIN, the distribution of their respective HPV DNA appears to be a function of maturation and probe sensitivity. Even with biotin probes, where the sensitivity threshold ranges from 100 to 1000 copies per cell, occasional positively hybridizing nuclei can be discerned in cells near the basal layer(30). However, serial sections stained with hematoxylin and eosin stains usually display evidence of maturation in some of these cells. Ostrow et al demonstrated weak hybridization for HPV in basal cells of CIN lesions as well as invasive squamous cells using tritium labled probes and long exposures(31). Thus, DNA-DNA hybridization has confirmed that HPV DNA may be distributed in all layers of the neoplastic epithelium. Furthermore, it has demonstrated that koilocytotic CIN lesions produce the most HPV DNA by virtue of the prominent replication in koilocytotic cells(30). This might explain the high index of cases of koilocytotic CIN lesions in which HPV 16 nucleic acids are detected(22). However this technique has not shed light upon the functional relationship between HPV and the lesions with which they are associated.

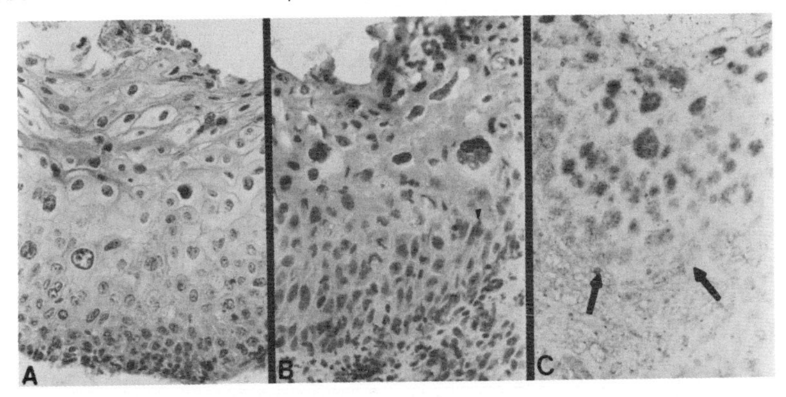

Figure 2. In-situ hybridization of CIN for HPV 16 DNA sequences. A. H and E stained section of a CIN with koilocytotic atypia. B. An adjacent frozen section and C. following hybridization with a biotin labeled HPV 16 DNA probe as previously described(30). Hybridization is most prominent in the upper layers of the epithelium, with some cells staining near the basal layer (arrow).

Late gene expression in condylomata and cin: One of the first techniques used to characterize HPV in epithelium was facilitated by the cross-reactivity between HPV and bovine papillomavirus (BPV) capsid proteins(32). Antisera generated to BPV disrupted with sodium dodecyl sulfate was found to cross react with HPV. Immunohistochemistry with this sera demonstrated that intact HPV virions or late genes could be detected in tissue sections. The sensitivity of the technique varied between studies, but a consistent feature was that intraepithelial neoplasms stained less frequently than condylomata(13,33-35). In our experience, approximately 15% of CIN lesions stained, most of which contain koilocytotic atypia(13). Fu et al have reported similar frequencies of staining for capsid antigens in aneuploid cervical lesions(35). Recently, using Southern blot hybridization to identify a series of lesions containing HPV-16 related sequences, we were able to detect capsid antigens in 3/29(10%)(26). In general, staining for HPV capsid antigens is less widely distributed in CIN lesions containing HPV 16 DNA as compared to condylomata containing other HPV's and is markedly reduced as a function of cytological atypia. In contrast, hybridization for HPV 16 nucleic acids may be prominent not only in areas containing maturation, but also in areas of epithelium with considerable atypia(30)(Figures 3-5).

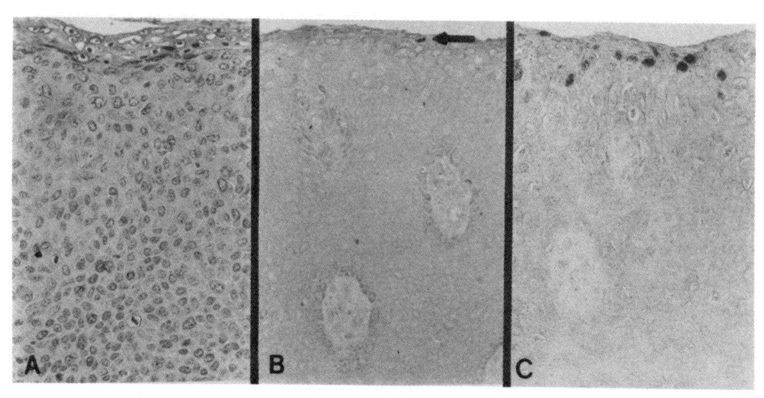

Figure 3. Comparison of DNA replication with capsid antigen production. A. H&E stained section of intraepithelial neoplasm. B. A serial section stained for capsid antigens displays a single positively staining nucleus(arrow). C. Following stringent hybridization with a biotin-labeled HPV-16 probe, several positively hybridizing nuclei are present.

<u>Demonstration of HPV RNA transcription in CIN</u>: Although several investigators have characterized HPV transcripts from invasive cancers, relatively little data is available on HPV transcription in condylomata or CIN lesions. This is primarily because the cervical lesions are often too small to provide sufficient material for RNA analysis. Broker and colleagues have recently analyzed a genital condyloma containing HPV 11, describing a variety of spliced mRNA's from the early region(Broker, T. and Chou, L., personal communication).

One approach which would make it possible to detect HPV transcripts as well as determine their location in CIN is RNA-RNA in-situ hybridization. Vectors containing polylinkers flanked by the Sp6 and T7 promoters have simplified the production of single stranded RNA probes which can be transcribed in the sense or antisense orientation(36,37). A second advantage is the ease with which nucleic acid sequences can be detected in formalin-fixed, paraffin-embedded tissues, using either biotin-labeled or isotope-labeled probes(28-31). Recently, HPV 11 RNA transcripts have been detected in condylomata using S-35 labeled RNA probes. Preliminary studies have detected HPV RNA sequences in condylomata and CIN lesions(Stoller, M. and Broker, T., personal communication). We are currently analyzing a series of lesions containing HPV-16 related DNA sequences for the presence of RNA using both S-35 and biotin-labeled HPV-16 RNA probes from subgenomic fragements of HPV 16 cloned into the GEM-1 vector.

The techniques are a modification of those detailed by Angerer and colleagues(16 and R. Angerer, personal communication). Tissues are sectioned and placed on slides containing poly-D-lysine, deparaffinized, hydrated and digested with proteinase K (1ug/ml) for 30 min. at room temperature. They are then dehydrated through graded alcolhols and dried. RNA probes are constructed using either T7 or Sp6 RNA polymerase. Following hybridization, the slides are incubated sequentially in RNAse (50 ug/ml) at Room temperature then 2X SSC and 0.2X SSC at 55 degrees C. Slides labeled with radioactive probes are dried in graded alcohols, dipped in Kodak NTB-2 emulsion (1:1 in 600 mM ammonium acetate), exposed for 1-5 days at 4 deg. C. and developed. Slides incubated with biotin-labeled probes are processed as previously described(30).

A comparison of the two labels for detecting HPV 16 DNA and RNA is illustrated in a vulvar intraepithelial neoplasm in figure 4. An interesting finding is the intense staining(hybridization) produced with the biotin probe(figure 4C), which compares favorably with a S-35-labeled probe of high specific activity (2 X 10-8 cpm/ug) and a 4 day exposure(figures 4A&B). As depicted in figure 5, a large portion of the hybridization signal is due to the presence of RNA transcribed from the early region of HPV 16.

Preliminary analysis of lesions containing HPV 16 related RNA transcripts indicates that the greatest amount of HPV RNA is produced in cells from the upper regions of the epithelium where the most cytoplasmic maturation takes place(Figures 4 and 5). This observation suggests that both early and late(capsid antigens) HPV gene expression is influenced by maturation. This may result from the abundance of DNA template in the same cells or possibly an independent effect of maturation on RNA transcription.

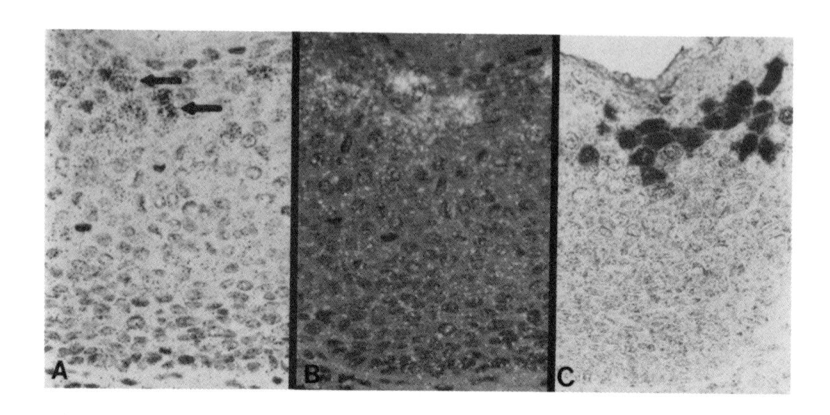

Figure 4. Comparison of nucleic acid hybridization in sections of a vulvar intraepithelial neoplasm hybridized with three HPV 16 RNA probes. Templates consisted of PST I fragments(A,B&D) representing the entire early region of the HPV genome. The sections were denatured to expose both the RNA and DNA sequences in the tissue. A. An H&E stained section following hybridization with the S-35-labeled probes contains prominent grains over a few superficial cells (arrows) B. The same field viewed via epifluorescence. C. A serial section following hyridization with biotin-labeled RNA probes from the same template(s). There is a strong hybridization signal over several cells.

If the above observations prove consistent, they raise questions about the precise relationship between HPV transcription and precancerous changes in squamous epithelium. It is possible that certain transcripts are produced in the immature cells as opposed to the mature epithelium. However, we have hybridized sections separately with each of the probes representing the A(E1-E2), B(Non coding region, E6-E7), and D(E2,E4-5,L1) PST I fragments of HPV 16 and have detected the same epithelial distribution of RNA transcripts with each probe. Another possibility is that specific early transcripts are produced in association with certain morphological changes ie. those associated with aneuploidy. However, it is also possible that the same reading frames are

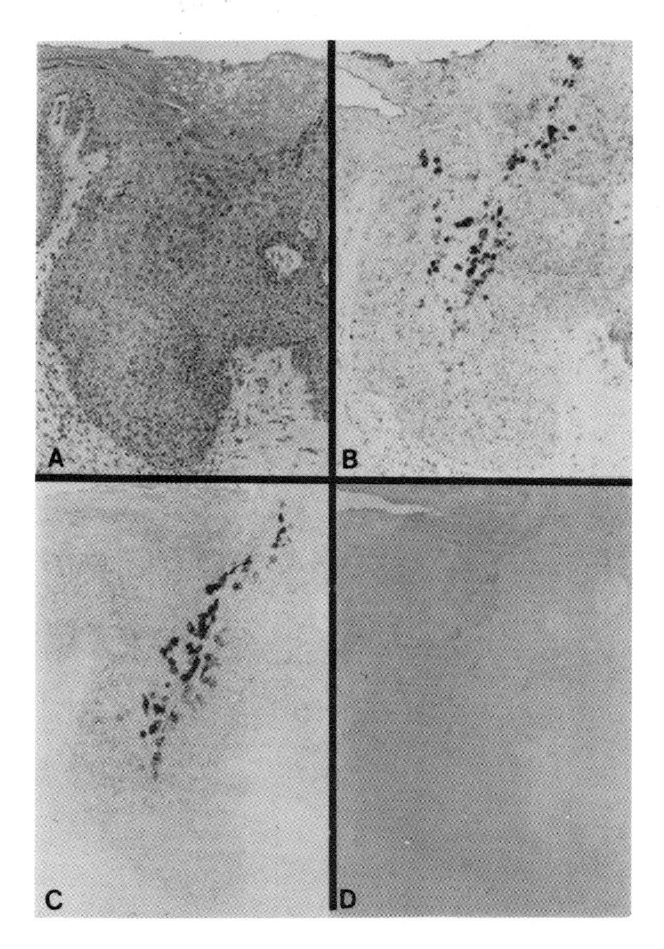

Figure 5. Hybridization profiles in serial sections of the case in Figure 4 analyzed with biotinylated RNA probes. A. H&E stained section of a vulvar precancer (VIN). B. A serial section heated (to denature the tissue DNA) and hybridized with a mixed probe constructed from the PST I(A,B,and D) fragments of HPV 16. C. A serial section which was not heated prior to hybridization demonstrates strong hybridization with slightly less nuclear staining. D. A serial section treated as in C, with the exception that it was incubated with probes constructed in the opposite orientation. There is virtually no hybridization present, which indicates that the signal in panel C is almost exclusively due to RNA-RNA hybridization.

transcribed and translated in CIN and condyloma, but that the translational products have unique functions in the lesions associated with different HPV types. The first possibility can be addressed by comparing HPV transcription in a series of cases exhibiting a wider range of morphological changes as well as different HPV DNA strains. The second is more problematic, and will require a more thorough understanding of the early gene products associated with HPV infections of all types, and their impact on both HPV and host gene functions.

ACKNOWLEDGEMENTS

We are indebted to Drs. Lutz Gissman, Harald zur Hausen and their colleagues for sharing their HPV DNA clones and information concerning the genomic organization of the genital HPV's; Dr. Robert Angerer and colleagues for sharing their protocols on RNA hybridization; Dr. Richard U. Levine for generously providing case material, and Dr. Ralph M. Richart for his continued support.

REFERENCES

1. zur Hausen H (1977). Human papillomaviruses and their possible role in squamous cell carcinomas. Curr Top Micro Immunol 78:1
2. Meisels A, Morin C (1981). Human papillomavirus and cancer of the uterine cervix. Gynecol Oncol 12:s111.
3. Gissman L, Wolnick L, Ikenberg H et al (1983). Human papillomavirus type 6 and 11 DNA sequences in genital and laryngeal papillomas and in some cervical cancers. Proc Nat'l Acad Sci USA 80:560
4. Durst M, Gissman L, Ikenberg H, zur Hausen H (1983). A papillomavirus DNA from a cervical carcinoma and its prevalence in cancer biopsy samples from different geographic regions. Proc Nat'l Acad Sci USA 80:3812
5. Boshart M, Gissman L, Ikenberg H et al (1984). A new type of papillomavirus DNA, its presence in genital cancer biopsies and in call lines derived from cervical cancer. EMBO J 3:1151
6. Crum CP, Ikenberg H, Richart RM, Gissman L (1984). Human papillomavirus type 16 and early cervical neoplasia. N Engl J Med 310:880

7. Center for Disease Control. Non-reported sexually transmitted diseases-United States (1979). MMWR 28:61
8. Meisels A, Fortin R (1976). Condylomatous lesions of the cervix and vagina I. Cytologic patterns. Acta Cytol 20:505
9. Purola E, Savia E (1977). Cytology of gynecologic condyloma accuminatum. Acta Cytol 21:26
10. Wagner D, Ikenberg H, Bohm N, Gissman L (1984). Type specific identification of human papillomavirus in cells obtained from cervical swabs by DNA in-situ hybridization. A cytologic virologic correlation study. Obstet Gynecol 64:767
11. Schneider A, Kraus H, Schumann R, Gissman L (1985). Papillomavirus infection of the lower genital tract. Detection of viral DNA in gynecologic swabs. Int J Cancer 35: 443
12. Crum CP, Egawa K, Barron B, Fenoglio CM, Richart RM (1983). Human papillomavirus infection (condyloma) of the cervix and cervical intraepithelial neoplasia: a histopathologic and statistical analysis. Gynecol Oncol 15:88
13. Winkler BW, Crum CP, Fujii T, Ferenczy A, Boon ME, Braun L, Lancaster WD, Richart RM (1984). Koilocytotic lesions of the cervix: the relationship of mitotic abnormalities to the presence of papillomavirus antigens and nuclear DNA content. Cancer 53:1081
14. Meisels A, Roy M, Fortier R (1981). Human papillomavirus infection of the cervix: the atypical condyloma. Acta Cytol 25:7
15. Fu YS, Reagan JW, Richart RM (1981). Definition of precursors. Gynecol Oncol 12:s220
16. Cox KH, DeLeon DV, Angerer LM, Angerer RC (1984). Detection of mRNA's in sea urchin embryos by quantitative in-situ hybridization using assymetric RNA probes. Dev Biol 101:485
17. Kreider J, Howett MK, Wolfe SA, Bartlett GL, Zaino RJ, Sedlacek TV, Mortel R (1985). Morphological transformation <u>in-vivo</u> of human uterine cervix with papillomavirus from condyloma accuminata. Nature 317:639
18. Richart RM (1973). Cervical intraepithelial neoplasia: a review. In Somers SC (ed): "Pathology Annual," New York: Appleton Century Crofts, p 301
19. Crum CP, Egawa K, Fu YS, Levine RU, Richart RM, Townsend DE, Fenoglio CM (1982). Intraepithelial squamous lesions of the vulva: biological and histologic criteria for the distinction of condyloma from vulvar intraepithelial neoplasia. Am J Obstet Gynecol 144:77

20. Crum CP, Levine RU (1984). Human papillomavirus infection and cervical neoplasia: a new perspective. Int J Gynecol Pathol 3:376
21. Gissman L, deVilliers EM, zur Hausen H (1982). Analysis of human genital warts (condylomata accuminata) and other genital tumors for human papillomavirus type 6 DNA. Int J Cancer 29:143
22. Crum CP, Mitao M, Levine RU, Silverstein S (1985). Cervical papillomaviruses segregate within morphologically distinct precancerous lesions. J Virol 54:675
23. Ikenberg H, Gissman L, Gross G, Grussendorf-Conen EI, zur Hausen H (1983). Human papillomavirus type 16-related DNA in genital Bowen's disease and Bowenoid papulosis. Int J Cancer 32:563
24. Ferenczy A, Mitao M, Nagai N, Silverstein, S, Crum CP (1985). Latent papillomavirus and recurring genital warts. N Engl J Med 313:784
√ 25. Smotkin D, Berek JS, Fu YS, Hacker NF, Major FJ, Lagasse LD, Wettstein FO (1986). Human papillomavirus DNA in adenocarcinoma and adenosquamous carcinoma of the uterine cervix.(submitted)
26. Mitao M, Nagai N, Levine RU, Silverstein S, Crum CP (1986). Human papillomavirus type 16 infection of the cervix. A morphological spectrum with evidence of late gene expression. (submitted)
27. McCance DJ, Clarkson PK, Dyson JL et al (1985). Human papilloamavirus types 6 and 16 in multifocal intraepithelial neoplasias of the female lower genital tract. Br J Obstet Gynecol 92:1093
28. Beckman AM, Myerson D, Daling JR, Kiviat N, Fenoglio CM, McDougall J (1985). Detection of HPV DNA in carcinomas by in-situ hybridization with biotinylated probes. J Med Virol 16:265
29. Gupta J, Gendelman H, Nagashfar Z, Gupta P, Rosenshein N, Sawada E, Woodruff JD, Shah K (1985). Specific identification of human papillomavirus type in cervical smears and paraffin sections by in-situ hybridization with radiolabeled probes: a preliminary communication. Int J Gynecol Pathol 4:211
30. Crum CP, Nagai N, Levine RU, Silverstein SJ (1986). In-situ hybridization analysis of HPV 16 DNA in early cervical neoplasia. Am J Pathol (in press)
31. Ostrow RS, Zachow K, Weber D, Okagaki T, Fukushima M, Clark BA, Twiggs LB, Faras AJ (1985). Presence and possible involvement of HPV DNA in premalignant and

malignant tumors. in: Howley P, Broker T (eds): "Papillomaviruses: Molecular and Clinical Aspects," New York, Alan R Liss, p101

32. Jenson AB, Rosenthal JD, Olson C, Pass F, Lancaster WD, Shah K (1980). Immunologic relatedness of papillomaviruses from different species. JNCI 64:495
33. Kurman R, Shah KH, Lancaster WD, Jenson AB (1981). Immunoperoxidase localization of papillomavirus antigens in cervical dysplasias and vulvar condylomas. Am J Obstet Gynecol 140:931
34. Crum CP, Braun L, Shah KV, Fu YS, Fenoglio CM, Richart RM, Townsend D (1982). Vulvar intraepithelial neoplasia: Correlation of nuclear DNA content and the presence of a human papillomavirus (HPV) structural antigen. Cancer 49:468
35. Fu YS, Braun L, Shah KV, Lawrence WD, Robboy J (1983). Histologic, nuclear DNA and human papillomavirus study of cervical condylomas. Cancer 52:1705
36. Davanloo P, Rosenberg AH, Dunn JJ, Studier FW (1984). Cloning and expression of the gene for bacteriophage T7 RNA polymerase. Proc Nat'l Acad Sci USA 81:2035
37. Melton DA, Krieg PA, Rebagliati MR, Maniatis T, Zinn K, Green MR (1984). Efficient in-vitro synthesis of biologically active RNA and RNA hybridization probes from plasmids containing a bacteriophage SP6 promoter. Nucl Acid Res 12:7035

Viruses and Human Cancer, pages 371–385

HUMAN PAPILLOMAVIRUS-11 INFECTION OF XENOGRAFTED HUMAN TISSUES[1]

John W. Kreider and Mary K. Howett

Departments of Pathology and Microbiology, College of Medicine, The Milton S. Hershey Medical Center, Hershey, Pennsylvania 17033

ABSTRACT Despite exciting advances in the application of the methods of molecular biology to papillomavirus research, a number of obstacles have hindered the study of human papillomaviruses (HPVs). It has not been possible to transform human tissues with HPVs, produce virions in the laboratory, or to isolate sufficient quantities of complete virions to study capsid proteins. We describe here a system for transforming rabbit, cattle, and human epithelial tissues with their respective papillomaviruses. In each case, the morphology of the infected tissue was identical to naturally-occurring lesions. HPV-11 replicated in infected grafts of human cervix and skin. This system offers an opportunity to study mechanisms of morphological transformation of human tissues with HPVs. Complete virions may be produced in the laboratory. The most important aspect of the system is that it is now possible to test the contribution of HPVs and co-factors in the causality of human cancers.

INTRODUCTION

HPVs have recently been implicated in the development of cancers of uterine cervix (1), verrucous carcinomas of larynx (2), and malignant and premalignant lesions of skin (3,4). This reasoning was predicated on the presence of HPV genomes in the pre-malignant and malignant lesions. Direct

[1]This work was supported by USPHS P01-CA 27503 and the Jake Gittlen Memorial Golf Tournament.

proof of causality has been lacking. A major problem in the test of the hypothesis that HPVs are a necessary or sufficient cause of human cancers has been the lack of a suitable experimental system. We have previously described unsuccessful attempts to induce papillomas with human wart virus in human skin grafted to exteriorized hamster cheekpouches (5), or to the dorsum of thymectomized, antithymocyte-sera treated mice (6). A similar negative study using athymic mice hosts was reported (7). We demonstrated that the athymic mouse would provide a satisfactory environment for the induction of both papillomas and carcinomas in rabbit skin infected with Shope papillomavirus (8). This observation encouraged us to again test the idea that the athymic mouse might permit transformation of heterografts of alien species with respective papillomaviruses, particularly bovine skin with bovine papilloma virus (BPV), and, more importantly, human tissues with HPVs. For this purpose, we obtained fetal calf skin and infected it with BPV and grafted it to the dorsum of athymic mice. A variety of human tissues which have been associated with HPV infection were exposed to an extract of human vulvar condylomata containing HPV-11, and grafted beneath the renal capsule of athymic mice (9). We also studied the susceptibility of human skin and cervix xenografts to infection with herpes simplex type 2 virus (HSV-2). Several months later, most of the human grafts were converted to the typical appearance of human condylomata. The results indicate that this system may be useful in a number of lines of investigation with HPVs.

MATERIALS AND METHODS

Condylomata Extract Preparation

A total of 20 g of vulvar condylomata acuminata was obtained from 15 patients (9). The tissue was homogenized with a Virtis machine at 25,000 rpm for 30 min in a total of 50 ml of PBS. The homogenate was centrifuged at 1,000 g, and the supernatant and pellet stored at -70° C.

Identification Of HPVs Present In Original Condylomata Extract And In HPV-Transformed Grafts

HPV-6,11,16, and 18 DNA clones were obtained from Drs. Harald zur Hausen (Freiburg, Federal Republic of Germany) and Peter Howley (Bethesda, MD). They were transfected into *E. coli*, strain HB101 (10), and grown in bulk. DNA was extracted, purified in CsCl gradients and identity verified with restriction endonucleases and electrophoresis. Purified plasmids were nick-translated with ^{32}P with an Amersham kit. DNA extracted from the infecting condyloma extract and the infected grafts were examined by Southern blot analysis.

HSV-2 Preparation

Virus was produced in human diploid cells and supernatants obtained after freezing and thawing. HSV-2 was titrated by plaque assay on primary rabbit kidney cells and diluted to a concentration of $1X10^7$ pfu/ml. HSV-2 was inactivated with ultraviolet light at 46 ergs/mm^2/sec for 10 minutes (11).

Athymic Mice

Athymic mice (nu/nu on a BALB/c background were obtained from Harlan Sprague Dawley, Inc., Madison, WI. Flexible film isolators housed the mice. They were provided with sterile air, water and laratory chow. All surgical manipulations were conducted in similar housings through gloved sleeves.

Graft Preparation

We obtained skin, cervix, and larynx from either surgical excisions or from autopsies. The tissue was immediately placed in Minimum Essential Medium with only gentamycin (800 ug/ml) as supplement. The tissues were stretched, either by pinning over sterile gauze pads, or compression by clamping in a hemostat. The convex surface thus created was sliced horizontally with a scalpel or a double-edged razor blade, to produce a split-thickness graft. The graft sheets were then cut into squares of approximately 1X2 mm and 0.5 mm thickness. Grafts were incubated for one hour in either undiluted condylomata extract, HSV-2, or both viruses, or in PBS.

Grafting Methods

Under Nembutal anesthesia, the kidneys were

individually delivered through dorsal paravertebral incisions. The kidneys were prevented from retracting back into the peritoneal cavity by gentle compression of the incision. The renal capsule was nicked with a scalpel and a toothless forceps such as those commonly employed in electron microscopy was used to blunt dissect and then to insert a graft beneath the capsule. A sterile, cotton-tipped applicator stick was used to prevent the graft from slipping out of the incision. Animals received drinking water supplemented with trimethoprim (0.01 mg/ml) and sulfamethoxazole (0.05 mg/ml) for the duration of the experiment.

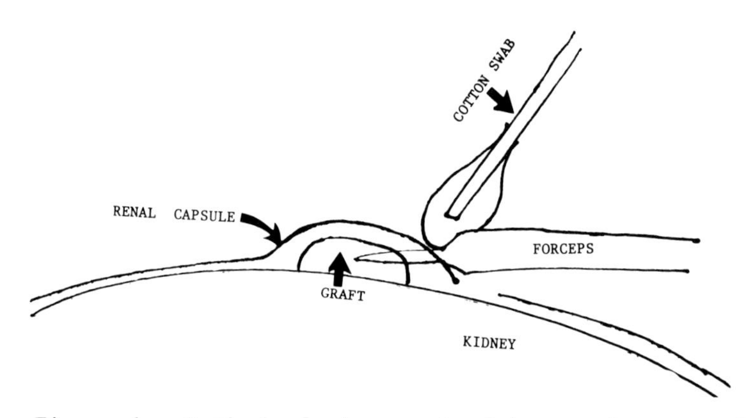

Figure 1. Method of placement of human tissue grafts beneath the renal capsule of athymic mice.

RESULTS

Infection Of Fetal Bovine Skin With BPV

Split-thickness skin grafts were cut from the dorsal skin of a Charolais cattle fetus of approximately 40 cm crown-rump length. The grafts were incubated in a suspension of BPV kindly provided by Dr. Peter Howley, NCI. After one hour incubation at 37° C, the tissues were grafted to the dorsum of athymic mice. Two such experiments were performed, including a total of 6 BPV-infected grafts and 4 PBS-treated, control grafts. The macroscopic appearance of representative grafts at 49 days of growth is presented in Fig. 2. The control grafts produced coarse, pigmented hairs of bovine origin (Fig. 2A). The BPV-infected grafts were indurated, elevated above the surrounding mouse skin, and often bore cattle hair (Fig. 2B). H&E sections of some of these infected grafts demonstrated the usual fibropapillomas associated with BPV infection. These results indicated that BPV could transform normal fetal bovine skin grafts in athyic mice to the morphology typical of the infection in cattle.

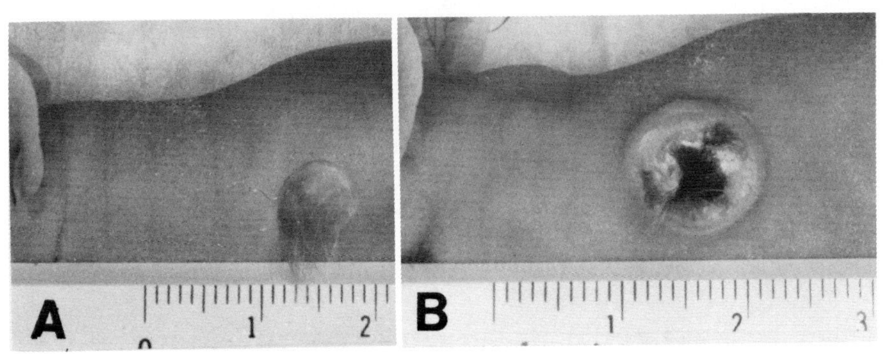

Figure 2. Fetal bovine skin grafts after 49 days of growth on athymic mice. A, Normal, PBS-treated skin grafts; B, BPV-infected skin graft.

Infection Of Human Skin With HSV-2

The purpose of this experiment was to determine if HSV-2 would infect human skin grafts and induce the histologic and cytologic lesions typical of human HSV-2 infections. Split-thickness human skin grafts were cut from mastectomy specimens and established orthotopically on the dorsolateral thorax of athymic mice. They were injected intradermally with $1X10^6$ pfu in 0.1 ml. Controls were injected with a similar volume of PBS. Six days later, erythema and vesicle formation were obvious at the injection site. The mice were killed and sections obtained. Microscopically, all of the sections from the infected grafts bore vesicles of 1-2 mm diameter, filled with exfoliated epithelial cells containing intranuclear, eosinophilic inclusion bodies (Fig. 3A). Some of the sections were reacted with anti-HSV antibody and immunoperoxidase reagents. These sections demonstrated positive staining in the nucleus and cytoplasm of cells with inclusions

(Fig. 3B). None of these changes were seen in the control, uninfected grafts. These results clearly demonstrate that human skin grafts on the athymic mouse can support productive HSV-2 infections, identical to spontaneous lesions of patients.

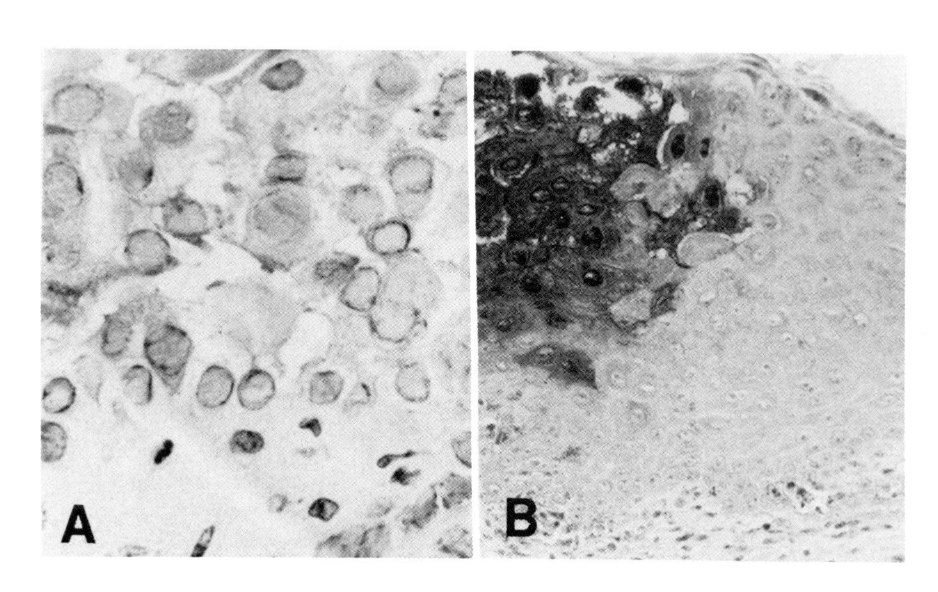

Figure 3. Human skin grafts infected with HSV-2. A, Intranuclear, eosinophilic inclusion bodies in vesicle, H&E; B, Immunoperoxidase stain for herpes antigen demonstrates intranuclear and cytoplasmic localization. Normal human epidermis is to the right of the section.

Identification of HPV Genomes Present In The Condylomata Extract

Dot blot analysis of DNA extracted from our original condylomata extract revealed strong hybridization to HPV-11 and lesser reactivity to HPV-6 (9). We conducted Southern blot analysis on DNA solubilized from the pellet of the condylomata extract. The results (12) revealed that EcoRl did not cleave the DNA but it was cleaved by ThaI just as was the HPV-11 positive control. These data suggested that HPV-11 but not HPV-6 DNA was detected in the condylomata pool.

Morphological Transformation of Human Cervix by HPV-11

The objective of these experiments was to determine if infection of normal human cervix with HPV-11 extracted from vulvar condylomata acuminata could transform the cervix to the typical appearance of condylomata or "flat warts" of the uterine cervix. Observations on the first series of 4 patients have been previously published (9). In that series, cervical grafts were treated with either condylomata extract containing HPV-11, UV-irradiated HSV-2, or both viruses concurrently. The results demonstrated that 9 grafts treated with UV HSV-2 alone did not differ in appearance or behaviour from 13 which were treated with PBS. No replication of HSV-2 was detected. In contrast, 7 of 22 grafts infected with condylomata extract showed conspicuous koilocytosis, nuclear hyperchromasia, wrinkling, and binucleation, all hallmarks of condylomatous change (13). Strong and specific reactions of the papillomavirus group-specific antigen (GSA) were found in the nuclei of the koilocytotic cells (14). The addition of UV HSV-2 to the condylomata extract treatment resulted in condylomatous change in 8 of 14 grafts. No changes could be attributed to the herpes exposure. Since the initial series of 4 patients, we have studied an additional 6 patients. Infected grafts were allowed to grow for 6-8 months. The incidence of condylomatous transformation was essentially the same as the first series. No malignancies were observed. In an earlier series of untreated cervical grafts we have observed normal cervical grafts surviving for 1-2 years. We conclude that this system does allow the routine, condylomatous transformation of human cervix with HPV-11. The resultant lesions are identical to those which appear spontaneously in patients.

Morphological Transformation By HPV-11 Of Human Skin From Various Anatomical Sites

The purpose of these studies was to determine if human skin could be transformed with HPV-11 and if the anatomic site from which the skin was obtained were an important determinant of success. We secured skin from foreskins, vulva, abdomen, and lower leg. They were infected with HPV-11 and grafted beneath the renal capsule. After 3-5 months of growth, they were removed, measured, sectioned, and stained with either H&E or for the papillomavirus GSA.

Where the samples were large enough, DNA was extracted and dot blots for HPV genome detection were conducted. The results (Table 1) demonstrated that there was a high frequency of transformation with foreskin. The infected grafts were 2-3 times the diameter of the uninfected controls (Fig. 4). Most of the tested grafts showed koilocytotic changes (Fig. 5A) and were positive for the GSA (Fig. 5B) and in dot blots they reacted very strongly with the HPV-11 probe and much less with the HPV-6 probe. Southern blots were conducted on the infected grafts (data not shown). This revealed that the transformed graft DNA was not cleaved by *EcoR1* but was cleaved by *Thal* in a pattern similar to that of the HPV-11 positive control. In the first foreskin studied, the treatment with UV-irradiated HSV-2 alone had no effect and did not alter the appearance of HPV-11 infected grafts.

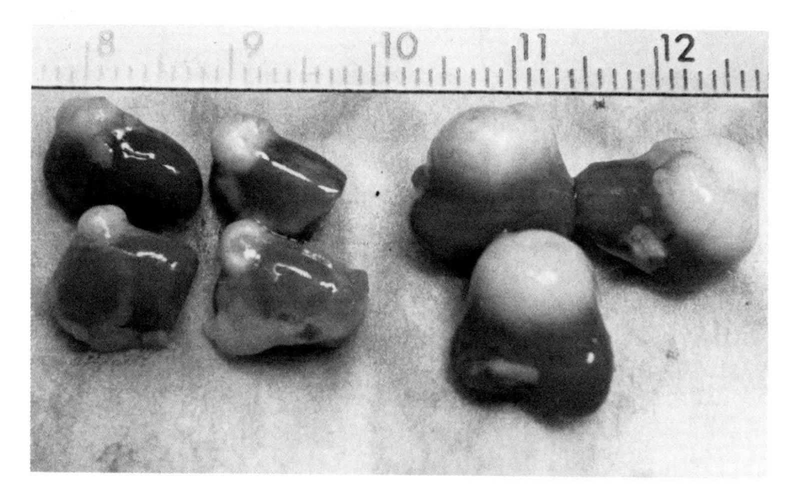

Figure 4. Human foreskin grafts 5 months after being placed beneath the renal capsule of athymic mice. The four grafts on the left were PBS-treated. The three on the right were infected with HPV-11.

Vulvar skin cysts did not increase in size after infection, but 4 of 13 were both morphologically transformed and GSA-positive. Abdominal skin infected with HPV-11 also did not increase in size, and no morphological transformation was found. However, 2 of the 9 infected grafts were positive for GSA, and 3 of 3 grafts tested were positive for HPV-11 DNA. Lower leg skin also showed no proliferative increase after infection, but 2 of 20 grafts were morphologically transformed and 3 of 5 tested were positive by dot blot for HPV-11 DNA. These results indicate that human skin is susceptible to transformation with HPV-11 but that skin from various sites varies dramatically in its response to infection.

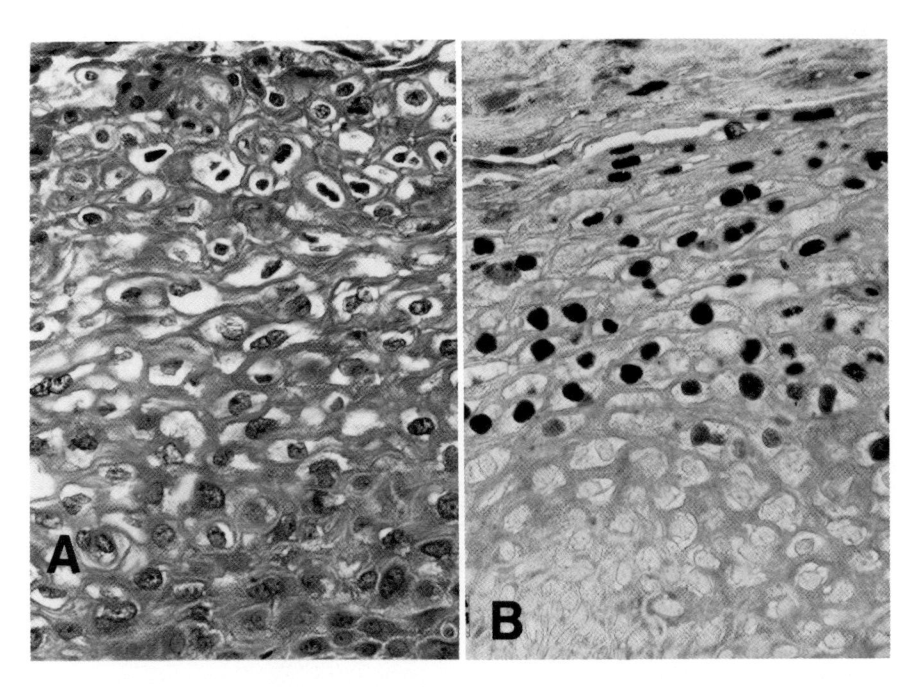

Figure 5. Human foreskin grafts 5 months after being placed beneath the renal capsule of athymic mice. A, Koilocytotic change of HPV-11 foreskin graft. Nuclei are wrinkled, sometimes doubled, and surrounded by clear space, H&E; B, Immunoperoxidase stain for the GSA of papillomaviruses.

TABLE 1
MORPHOLOGICAL TRANSFORMATION OF HUMAN SKIN GRAFTS FROM VARIOUS ANATOMICAL SITES INFECTED WITH HPV-11 AND TRANSPLANTED BENEATH THE RENAL CAPSULE OF ATHYMIC MICE

		Graft Treatments[a]			
Age/Sex	Site	PBS	HSV	C.A.	HSV + C.A.[b]
29 yo/M	foreskin	0/2/6	0/5/6	3/5/5	5/9/12
8 mo/M	foreskin	0/6/6	-	4/6/6	-
23 mo/M	foreskin	0/11/12	-	9/11/11	-
81 yo/F	vulva	0/10/10	-	4/12/13	-
80 yo/F	abdomen	0/9/10	-	0/9/12	-
80 yo/F	lower leg	0/13/22	-	2/20/24	-

[a]No. grafts transformed/No. grafts which survived/No. grafts attempted.
[b]HSV, UV-irradiated Herpes simplex Type 2; C.A., condylomata acuminata extract.

Morphological Transformation By HPV-11 Of Human Vocal Cord

The purpose of these experiments was to determine if human vocal cord grafts, placed beneath the renal capsule, could be transformed with HPV-11. True vocal cords were obtained at autopsy from three patients who were deceased <10 hours. These included a 53 year old male, a 4 month old female, and a 5 year old female. Split-thickness grafts were cut from the appositional surfaces of both cords. They were then treated with either PBS or HPV-11 and grafted beneath the renal capsule of the athymic mice. They were removed after 3-4 months of growth. The cysts which had formed were usually filled with clear mucous. An accumulation of keratin was present in 3 of the 21 HPV-11 infected grafts. Only one of the infected grafts was substantially larger than the controls. Microscopic sections of the 18 PBS-treated grafts revealed that they were lined by pseudostratified, ciliated, columnar epithelium. The cells often contained mucous-filled vacuoles (Fig. 6A). In contrast, 9 of the 21 HPV-11

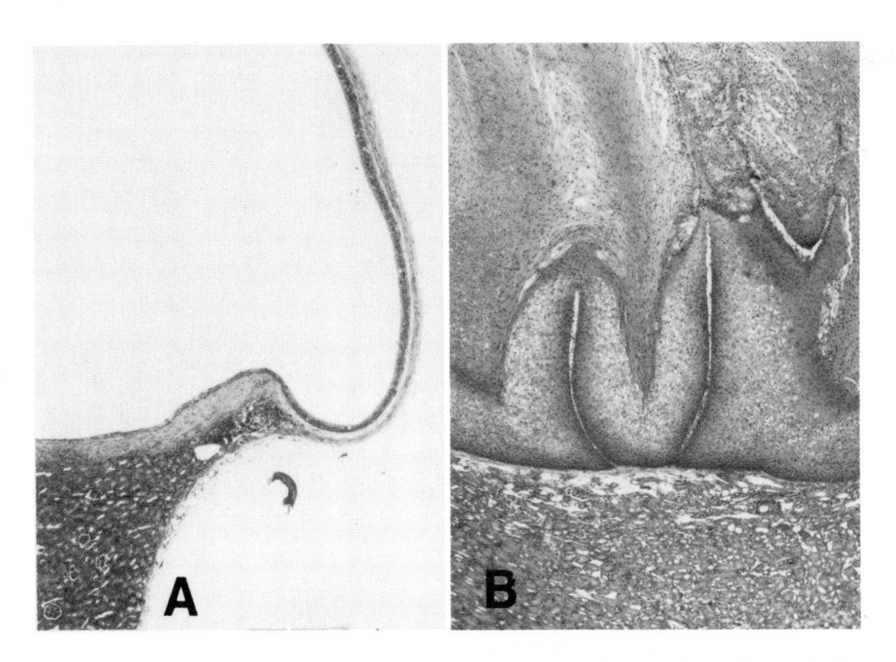

Figure 6. Human vocal cord grafts 3 months after being placed beneath the renal capsule. A, PBS-treated control graft. Cyst is lined by pseudostratified ciliated, columnar epithelium, which is adherent to renal capsule, H&E; B, HPV-11 infected vocal cord graft. Typical squamous papilloma with koilocytosis, papillae formation and accumulation of exfoliated keratinized cells above the squamous layers.

infected grafts showed focal, nodular proliferations of squamous epithelium, sometimes at multiple loci within a single cyst. The squamous cells in these foci were sometimes koilocytotic with clear spaces surrounding the nuclei. Multiple nuclei were often found within single cells, but the nuclei were not hyperchromatic or wrinkled as they were in comparable infected grafts of cervix or skin. We have defined this appearance as a "squamous metaplasia", and we believe that it is a transitional morphology, preliminary to the formation of the squamous laryngeal papillomas. In 3 of the 21 HPV-11 infected grafts, true squamous papillomas were found (Fig. 6B). Sections of these lesions were stained for the group specific antigen, but only equivocal reactions were observed. There was insufficient material for dot blot analysis of the HPV genomes present. The observations were essentially similar for all patients studied, with no correlation with age. These data suggest that HPV-11 can morphologically transform human vocal cord epithelium. This tissue is more resistant to infection than cervix or skin. The rate of growth of the transformed cells is slower. Observations on material from additional patients is in progress.

DISCUSSION

The application of molecular biology to the study of papillomaviruses has led to a number of important new insights into the potential of HPVs in inducing human disease. Despite those advances, investigations with HPVs have met with a number of persistent obstacles. There is no transforming system, since HPVs do not infect experimental animals and ethical considerations greatly limit study of infections in human subjects. _In vitro_ systems have been described (15,16) but these have used murine fibroblasts and the relationship of the induction of foci to human lesions remains to be established. There is no laboratory method for producing useful quantities of HPVs in the laboratory. Since HPVs other than HPV-1 replicate poorly or not at all in spontaneous human lesions, it has not been possible to examine virion capsid antigens. Further, the lack of a preclinical system for the assay of the effectiveness of agents which might prevent or inhibit the development of HPV lesions has retarded drug development.

Our investigations demonstrate that the athymic mouse will support the morphological transformation by papillomaviruses of xenografts of epithelial tissues from rabbits, cattle, and humans. In every respect, the resulting infections are morphologically and functionally identical to the naturally occurring lesions of those species. In particular, human skin, cervix, and larynx infected with HPV-11 produced lesions were the same morphology as the spontaneous diseases. Further, we found that HSV-2 will induced typical cytopathic effects in human skin.

These observations indicate that the system which we describe here can solve many of the technical problems which have heretofore hampered HPV investigations. The ethical restrictions of studies employing human subjects are bypassed, but the system employs human cells and human papillomaviruses. The interactions of HPVs with other viruses, chemical or physical initiators and promotors may be followed. In view of zur Hausen's suggestion (17) that HPV and HSV-2 may be important cofactors in the induction of cervical cancer, it is significant that the nude mouse allows both viruses to infect human skin and cervix. We have not yet found any effects which could be attributed to HSV-2, but this negative result does not deny the hypothesis.

The endpoints of HPV interaction in this system included infected cyst growth rates, histological appearance, expression of viral capsid antigen, and presence of HPV genome. The susceptibility of different human tissues varied widely, even in the case of human skin from disparate sites. Foreskin was the most exuberant responding tissue, and has a number of attributes which commend it for HPV investigations. It is abundant, and experimental variables such as age, sex, and site are constant. There are no preexistent diseases. Microbial flora and viral infections are negligible. Most importantly, foreskin vigorously responded to HPV infection by markedly accelerating rates of cell proliferation and production of keratinized squames. This suggests that it might be a good source for HPV replication. HPV-11 undoubtedly replicates in infected grafts since there was a remarkable number of GSA-positive nuclei in the differentiating epithelial layers. The numbers were far greater than those which are found in spontaneous human condylomata. Further, we found that dot blots of infected graft DNA reacted very intensely with HPV-11 probes. Thus, the system offers an opportunity to produce infectious laboratory stocks of HPV virions with which to conduct additional experiments and for the study of capsid antigens. This system also provides a convenient method for preclinical bioassay of agents which might prevent or treat HPV infections. This could forestall clinical trials of ineffective drugs.

Finally, the nude mouse with HPV-infected xenografts provides a unique circumstance in which a causal role of particular HPVs and co-factors in human cancer could be directly tested. Present data strongly support an association between certain HPVs, especially types 16 and 18 in human cervical cancer. This system provides an arena in which we can define the respective roles of suspect HPVs and cofactors in the initiation, promotion, and neoplastic progression of human cervical cancer.

ACKNOWLEDGMENTS

Technical support was provided by Sue Anne Wolfe and Janet Weber. We are indebted to Ms. W.C. Calabash for expert typing of the manuscript.

REFERENCES

1. Durst M, Gissman L, Ikenberg H, zur Hausen H (1983). A papillomavirus DNA from a cervical carcinoma and its prevalence in cancer biopsies from different geographic regions. Proc Natl Acad Sci 80:3812.
2. Abramson AL, Brandsma J, Steinberg B, Winkler B (1985). Verrucous carcinoma of the larynx. Possible human papillomavirus etiology. Arch Otolaryngol 111:709.
3. Ikenberg H, Gissman L, Gross G, Grussendorf-Conen E-I, zur Hausen, H (1983). Human papillomavirus type 16-related DNA in genital Bowen´s disease and in Bowenoid papulosis. Int J Cancer 32:563.
4. Orth G, Jablonska S, Jarzabek-Chorzelska M, Obalek S, Rzrsa G, Favre M, Croissant O (1979). Characteristics of the lesions and risk of malignant conversion associated with the type of human papillomavirus involved in epidermodysplasia verruciformis. Cancer Res 39:1074.
5. Kreider JW, Haft HM, Roode PR (1971). Growth of human skin on the hamster. J Inv Derm 57:66.
6. Pass F, Niimura M, Kreider JW (1973). Prolonged survival of human skin xenografts on antithymocye serum-treated mice: failure to produce verrucae by inoculation with extracts of human warts. J Inv Derm 61:371.
7. Cubie HA (1976). Failure to produce warts on human skin grafts on ´nude´ mice. Br J Derm 94:659.
8. Kreider JW, Bartlett GL, Sharkey FE (1979). Primary neoplastic transformation _in vivo_ of xenogeneic skin grafts on nude mice. Cancer Res 39:272.
9. Kreider JW, M.K. Howett MK, Wolfe SA, Bartlett GL, Zaino RJ, Sedlacek TV, Mortel R (1985). Morphological transformation of human uterine cervix with papillomavirus from condylomata acuminata. Nature 317:639.
10. Mandel M, Higa A (1970). Calcium dependent bacteriophage DNA infection. J Mol Biol 53:159.
11. Duff R, Rapp F (1971). Oncogenic transformation of hamster cells after exposure to herpes simplex virus type 2.

Nature 233:48.

12. Kreider JW, Howett MK, Lill NL, Bartlett GL, Zaino RZ, Sedlacek TV, Mortel R (1986) Induction in vivo of condylomatous transformation of human skin with HPV-11 from condylomata acuminata. Submitted for publication.
13. Reid R, Crum CP, Herschman BR, Fu YS, Braun L, Shah KV, Agronow SJ, Stanhope CR (1984). Genital warts and cervical cancer. III. Subclinical papillomavirus infection and cervical neoplasia are linked by a spectrum of continuous morphologic and biologic change. Cancer 53:943.
14. Jenson AB, Rosenthal JR, Olson C, Pass F, Lancaster WD, Shah KV (1980). Immunologic relatedness of papillomaviruses from different species. J Natl Canc Inst 64:495.
15. Watts SL, Phelps WC, Ostrow RS Zachow KR, Faras AJ (1984). Cellular transformation by human papillomavirus DNA in vitro. Science 225:634.
16. Yasumoto S, Burkhardt AL, Doniger J, Dipaolo JA (1985). Human papillomavirus type 16 DNA-induced malignant transformation of NIH 3T3 cells. J Virol 57:572.
17. zur Hausen H (1982). Human genital cancer: Synergism between two virus infections or synergism between a virus infection and initiating events? The Lancet 2:1370.

Viruses and Human Cancer, pages 387–396

THE PREDICTIVE VALUE OF PAP SMEAR, COLPOSCOPY, AND CERVICAL BIOPSY IN DETERMINING HUMAN PAPILLOMAVIRUS (HPV) TYPE-16 POSITIVITY

Joel Palefsky,[1,2] Barbara Winkler,[3] Carolina Braga,[3] Victor Nizet,[2] and Gary Schoolnik[2]

Department of Medical Microbiology, Stanford University School of Medicine, Stanford, CA, and Departments of Pathology, and Obstetrics, Gynecology, and Reproductive Sciences, University of California, San Francisco, CA

ABSTRACT. The presence of human papillomavirus (HPV) type-16 DNA in intraepithelial lesions of the cervix is currently thought to be associated with an increased risk of malignant transformation. However, convenient DNA typing procedures are not yet available to the clinician. In an effort to determine how the results of Pap smear and colposcopy correlate with the presence of HPV-16 DNA, the colposcopic findings, pathology, and HPV-DNA typing of cervical specimens from 41 women attending the Colposcopy Clinic at the Medical Center of the University of California, San Francisco were analyzed. In all patients, colposcopic and pathological findings were limited to the cervix. Nine of 41 (22%) patients were positive for HPV-16 DNA on cervical swabs analyzed by Southern blotting. Of these, 9/9 patients (100%) had abnormal colposcopy, whereas only 11/32 HPV-16 DNA-negative patients (34%) had abnormal colposcopy ($p < .001$). Eight of the 9 HPV-16 DNA-positive patients (89%) had abnormal Pap

[1]Fellow of the Medical Research Council of Canada

[2]Present address: Department of Medical Microbiology, Sherman Fairchild Science Building, D-035, Stanford University School of Medicine, Stanford, CA 94305

[3]Present address: Departments of Pathology (HSW-501), and Obstetrics, Gynecology and Reproductive Sciences, School of Medicine, University of California, San Francisco, CA 94143

smears ranging from cervical intraepithelial neoplasia (CIN), grade 1, to CIN, grade 3, while only 6/32 HPV-16 DNA-negative patients (19%) had abnormal Pap smears all with CIN 1 ($p < .001$). The positive predictive values for the presence of HPV-16 DNA with abnormal colposcopy and Pap smear were 45% and 57%, respectively, while the negative predictive values with normal colposcopy and Pap smear for the absence of HPV-16 DNA were 100% and 96%, respectively.

Directed biopsies of colposcopic abnormalities were performed in 20 patients. Histologic diagnoses were physiologically normal squamous metaplasia in 10, CIN 1 in 5, CIN 2 in 4, and CIN 3 in 1. Five of the 10 (50%) patients with CIN were HPV-16 DNA-positive. One patient with a normal biopsy was also positive. Of the 14 biopsied patients who were HPV-16 DNA-negative, 9/14 (64%) were histologically normal and only 5/14 (36%) had CIN (p = NS). The positive predictive value of a histologic diagnosis of CIN for the presence of HPV-16 DNA was 50%, while the negative predictive value of a normal histologic diagnosis for the absence of HPV-16 DNA was 90%. These data suggest that HPV-16 DNA typing would be of only limited value in routine screening of low risk populations for whom the negative predictive value of a normal Pap smear, colposcopic examination, and histologic diagnosis would be expected to be substantially higher than in this study of high risk patients. In contrast, screening for HPV-16 DNA may be useful in the clinical management of high risk women and women with CIN.

INTRODUCTION

Genital human papillomaviruses (HPV) have been strongly implicated in the pathogenesis of cervical intraepithelial neoplasia (CIN) and cervical carcinoma. Furthermore, the evidence suggests that HPV may have differing oncogenic potentials and that specific HPV types can be identified with cervical cancer precursors and a high risk of malignant transformation (1). HPV type-16, in particular, has been associated with cervical carcinogenesis because of its high prevalence in histologically defined CIN and in invasive cervical carcinomas (1,2).

Clinical problems in differential diagnosis and prognosis have recently been investigated utilizing cloned

HPV-DNA of different types as molecular probes. To study the potential of HPV-16 DNA as a marker for cervical neoplasia, Crum et al (2) correlated lesional cytohistology with HPV-DNA typing. It was found that the presence of HPV-16 DNA in CIN lesions could be used to discriminate benign cervical condyloma from histologically similar, but prognostically more serious, CIN. To evaluate the significance of HPV latency, Ferenczy et al (3) correlated the post-therapy recurrence of ano-genital condylomata with the isolation of HPV-DNA from clinically normal skin adjacent to the treated lesions. It was found that the absence of HPV-DNA correlated well with therapeutic cure, whereas the presence of latent HPV beyond the treatment area could be used to predict recurrence. These studies suggest that HPV-DNA typing may be a useful, objective tool in patient management and may provide important information about the natural history of HPV-associated genital lesions.

The present study addresses the applications of HPV-DNA typing to the evaluation and management of women with a history of CIN. The specific aim of this study was to determine how the standard cervical screening techniques of Pap smear and colposcopy correlate with concurrent typing for HPV-16.

MATERIALS AND METHODS

Clinico-Pathologic Evaluation

The patient population consisted of women attending the colposcopy clinic of the Ambulatory Care Center of the University of California, San Francisco. All of the patients had a biopsy-documented history of previous or intercurrent CIN. Patients with cervical abnormalities less severe than CIN 1, and patients with invasive cervical carcinoma were excluded from this analysis.

At the time of examination, a Pap smear was obtained for routine cytologic evaluation and a second cervical swab of the exo- and endocervix was taken concurrently for the purpose of HPV-DNA analysis. Colposcopic examination was performed in all patients and directed cervical biopsies obtained in women with colposcopically visible cervical abnormalities. All Pap smears and biopsies were reviewed by one investigator (BW). Cytologic, colposcopic, and histologic evaluations were classified according to standard criteria (4).

Molecular Analysis

Cervical swabs obtained from women giving informed consent were frozen at -70°C until ready for analysis. DNA was extracted using a modification of the method of Chirgwin (5) and Scott (Bethesda Research Laboratories) to simultaneously extract both DNA and RNA. DNA extracted from cervical swabs was ethanol precipitated, centrifuged, and digested with restriction endonuclease PST1. The DNA was loaded onto a 1% agarose gel for electrophoresis. Contents of the gel were transferred to nitrocellulose according to the method of Southern (6). The filters were sequentially hybridized at T_m-25°C and T_m-10°C to nick-translated whole genomic probes of HPV types 16, 18, and a 6/11 mixture (kindly provided by A. Lorincz,, Bethesda Research Laboratories).

The results of the clinicopathologic analyses and molecular analyses were tabulated and correlated. Predictive values were calculated using Fisher's exact two-tailed test.

RESULTS

Correlation with Pap Smear Results

Forty-one women were entered into the study. Pap smears were interpreted as normal in 27/41 (66%) and abnormal in 14 (34%). Patients with normal cervical examinations were either post-therapy or had a history of lesion regression following biopsy. The abnormal smears were graded as follows: 8 - CIN 1, 5 - CIN 2, and 1 - CIN 3. The Pap smear correlation with HPV-DNA typing is summarized below in Table 1.

TABLE 1
CORRELATION BETWEEN PAP SMEARS AND HPV-DNA TYPING

Pap Smear Result	No. Cases	HPV6/11	HPV-16	HPV-18
Normal	27	0	1 (4%)	0
CIN 1	8	0	5 (63%)	1 (12.5%)
CIN 2	5	0	2 (40%)	0
CIN 3	1	0	1 (100%)	0
	41	0	9 (22%)	1 (2%)

Of the 27 women with a single normal Pap smear, only 1 (4%) had HPV-16 DNA isolated. This patient had a biopsy-documented CIN 1 persisting during follow-up for 7 mos. On the day of entry into the study, this patient had a colposcopically visible lesion and was scheduled for cryosurgical treatment. The normal Pap smear in this case, therefore, represents a false-negative cytology. Because the standard for accurate Pap smear screening is 3 consecutive smears, we recorded the number of smears in the group of 27 patients with normal cytologic results. At the time of entry into our study, the majority of women met these standards. Sixteen women had had 3 consecutive normal smears, 4 had 2 consecutive normal smears, and 7 had only 1 normal smear. Of these patients, the one who was HPV-16 DNA-positive was in the group who had had two consecutive normal smears. In correlating Pap smear results with HPV-16 typing, the positive predictive value of a single abnormal smear for the presence of HPV-16 DNA was calculated to be 57%. In contrast, the negative predictive value of a single normal smear for the absence of HPV-16 DNA was 96%.

Correlation with Colposcopic Findings

Twenty-three of the total 41 patients (56%) were found to have colposcopic abnormalities of varying severity. The cervical swabs of 9 of these 23 women (39%) were positive for HPV-16 DNA. No colposcopic abnormalities were seen in 18 of the 41 women (44%). All of the colposcopically normal women were HPV-16 DNA negative. The positive predictive value of an abnormal colposcopy for the presence of HPV-16 DNA was 45%. On the other hand, the negative predictive value of normal colposcopic findings for the absence of HPV-16 DNA was 100%.

Correlation with Histologic Diagnosis

Directed biopsies were obtained concurrently with HPV-DNA typing in 20 of the 41 patients. Diagnoses in these 20 cases were interpreted as physiologically normal squamous metaplasia in 10, CIN 1 in 5, CIN 2 in 4, and CIN 3 in 1. The correlation between pathology and HPV-DNA typing is summarized below in Table 2.

TABLE 2
CORRELATION BETWEEN HISTOLOGIC DIAGNOSIS AND HPV-DNA TYPING

Diagnosis	No. Cases	HPV 6/11	HPV-16	HPV-18
Squamous metaplasia	10	0	1 (10%)	0
CIN 1	5	0	3 (60%)	1 (20%)
CIN 2	4	0	2 (50%)	0
CIN 3	1	0	0 (0)	0
	20	0	6 (30%)	1 (5%)

Typing for HPV 6/11 and 18

Thirty-one of the total 41 patients were analyzed for HPV 6/11 and all 31 samples were negative. Results of HPV 6/11 hybridization are pending in the remaining 10 patients. HPV-18 DNA typing was performed for all 41 cervical swab samples and was positive for HPV-18 DNA in only 1 woman (2%). The cervical swab of this woman was simultaneously positive for HPV-16 DNA. The cytologic and biopsy diagnoses in this case were CIN 1.

Three of the 5 patients (60%) with CIN 1 were HPV-16 DNA-positive as were 2/4 (50%) of cases with CIN 2, and 1/10 (10%) of women with normal biopsies. Two of 4 CIN 2 patients, the 1 CIN 3 patient, and 9/10 histologically normal biopsies were negative for HPV-16 DNA. The positive predictive value of a histologic diagnosis of CIN for the presence of HPV-16 DNA was 50%. In contrast, the negative predictive value for the absence of HPV-16 DNA in a normal biopsy was 90%.

The results of the clinicopathologic and HPV-DNA analyses are summarized below in Table 3.

TABLE 3
SUMMARY OF RESULTS

	No. Cases	Pap Smear		Colpo[a]		No. cases	Bx[b]	
		ABNL[c]	NL[d]	ABNL	NL	with Bx	CIN	NL
HPV-16+	9 (22%)	8 (57%)	1 (4%)	9 (45%)	0	6 (30%)	5 (50%)	1(10%)
HPV-16-	32 (78%)	6 (43%)	26 (96%)	11 (55%)	21 (100%)	14 (70%)	5 (50%)	9(90%)
	41	14	27	20	21	20	10	10

[a]Colpo - Colposcopic evaluation

[b]Bx - Biopsy histologic evaluation

[c]ABNL - abnormal

[d]NL - Normal

DISCUSSION

There has been increasing concern about infection of the cervix by HPV-type 16 because of the consistent association of HPV-16 and high-grade CIN and invasive cervical carcinoma (1,2). It has also been postulated that the presence of HPV-16 DNA in cervical samples may serve as a marker for neoplastic transformation and can be used to identify women who are at high risk for the development of cervical cancer (2). Because of the more widespread availability of convenient HPV-DNA typing procedures, the efficacy of routine screening of cervical samples for HPV-DNA of specific types has become a clinical issue. Typing for HPV-16 DNA is of particular interest because of the high prevalence of this HPV-type in women with cervical cancer precursors. To address the issue of the efficacy of routine typing for HPV-16 DNA, the following questions have to be answered: 1) What is the frequency of HPV-16 infection of the cervix in epidemiologically defined groups of women at low risk and high risk for the development of CIN? 2) What is the significance of the detection of HPV-16 DNA to the natural history of genital HPV infection (especially in patients with a normal cervical examination)? 3) What are the prognostic and therapeutic implications of the finding of HPV-16 DNA in a given patient? 4) Is testing for HPV-16 DNA a cost-effective screening tool when compared to the usual technique of Pap smear screening?

In an effort to begin to answer these questions, we undertook a study of routine HPV-DNA typing in a colposcopy clinic population. This current report details the results of the concurrent sampling of the cervix for clinicopathologic and molecular analyses of a preliminary group of 41 women enrolled in the University of California, San Francisco, Colposcopy Clinic. All 41 women had a history of past or intercurrent CIN documented by cervical biopsy. The prevalence of HPV-16 DNA in the total study group was 22%. In the women with an abnormal Pap smear, the prevalence of HPV-16 DNA was 57%; with colposcopic abnormalities, the prevalence was 45%; and with CIN on biopsy, 50%. The high prevalence of HPV-16 DNA in these women is not surprising because they have been defined, a priori, as high risk because of their history of CIN.

Knowledge of the presence of HPV-16 DNA in the cervices of high risk women with a defined intraepithelial lesion may be useful in planning management of these lesions. A more aggressive therapeutic approach to the eradication of an HPV-16 DNA-positive lesion may be justified as it can be supposed that these lesions are more likely to persist or to progress to invasive cancer. Similarly, the detection of latent HPV-16 DNA may aid in judging the effectiveness of therapy and in determining the risk of persistent or recurrent disease. The fact that the predictive values of abnormal Pap smear, colposcopy, and biopsy are as low as 45-57% is not surprising given the multiplicity of HPV types and pathogenetic stimuli which may affect the genitalia. Under these circumstances, a screening test for HPV-16 DNA may be particularly important.

Contrasting data, however, emerge with respect to the relevance of typing for HPV-16 DNA in the women in this study with normal cervical examinations, even though they come from a high risk group. In the patients with a normal Pap smear, the prevalence of HPV-16 DNA was only 4%. The corresponding prevalences of HPV-16 in the women with normal colposcopy and biopsy were 0% and 10%, respectively. The negative predictive values of a normal Pap smear, normal colposcopic examination, and normal histologic evaluation for the absence of HPV-16 DNA were thus found to be 96%, 100%, and 90%, respectively. These definitive data from a high risk population suggest that HPV-16 DNA typing would be of only limited value in routine screening of low risk populations in whom the prevalenve of HPV-16 DNA would be expected to be substantially lower. With particular respect to Pap smear screening, the negative predictive value of a single normal Pap smear was 96%. It is obvious that the cost-effectiveness of Pap smear screening is at present much better than for HPV-DNA typing and that the routine screening of the general population, even for HPV-16 DNA, may not be justified. The relevance and efficacy of screening for HPV-16 DNA in the clinical management of high risk women and women with CIN requires further study.

ACKNOWLEDGMENTS

The authors would like to acknowledge the technical support of Ms. Jennifer Kidd, as well as the editorial assistance of Mr. David Geller. The technical advice of Drs. Attila Lorincz and Cynthia Scott, Bethesda Research Laboratories, is gratefully acknowledged.

REFERENCES

1. Gissman L (1984). Papillomaviruses and their association with cancer in animals and in man. Cancer Surv 3:161.
2. Crum CP, Mitao M, Levine RU, Silverstein S (1985). Cervical papillomaviruses segregate within morphologically distinct precancerous lesions. J Virol 54:675.
3. Ferenczy A, Mitao M, Nagai N, Silverstein SJ, Crum CP (1985). Latent papillomavirus and recurring genital warts. N Engl J Med 313:784.
4. Ferenczy A (1982). Cervical intraepithelial neoplasia. In Blaustein A (ed): "Pathology of the Female Genital Tract," 2nd ed. New York: Springer Verlag, p 156.
5. Chirgwin JM, Pryzbyla AE, MacDonald RJ, Rutter WJ (1979). Isolation of biologically active ribonucleic acid from sources enriched in ribonuclease. Biochemistry 18:5294.
6. Southern E (1975). Detection of specific sequences among the DNA fragments separated by gel electrophoresis. J Mol Biol 98:503.

Viruses and Human Cancer, pages 397–414

TRANS-ACTING FUNCTION THAT MAINTAINS EPSTEIN-BARR VIRUS EPISOMES ACTS INTERSPECIES ON A *HERPESVIRUS PAPIO* PUTATIVE *CIS*-ACTING ORIGIN OF REPLICATION

Rick L. Pesano[1] and Joseph S. Pagano[2]

Department of Medicine[2]
and
Lineberger Cancer Research Center[1]
University of North Carolina at Chapel Hill
Chapel Hill, N.C. 27514

ABSTRACT *Herpesvirus papio* (HVP) and Epstein-Barr virus (EBV) are closely related biologically and biochemically; both can produce latent infections. HVP is a lymphotropic virus and is able to immortalize both human and baboon B lymphocytes *in vitro* (1). Southern blot hybridization data show a 40% homology between HVP and B95-8 EBV; however, the relatedness is not restricted to certain regions of the viral DNA but is dispersed (2). HVP-permissive lymphoblastoid cells contain both unit length linear and episomal viral DNA. The circular molecules are approximately the same size as episomes found in EBV growth-transformed Raji cells, 170×10^3 nucleotide base pairs (3). The putative origin of replication in EBV (*oriP*) has been assigned to a 1790 base-pair fragment (*cis*) in the short unique region of the genome which requires a function supplied in *trans* from elsewhere in the genome (4). We report here the identification of the

[1]This work was supported by Public Health Service grant numbers IF32 CA07991-01 and 5 PO1 CA19014-10, awarded by the National Cancer Institute, DHHS, and the Infectious Disease Pathogenesis Training Grant, NIAID, AI07151.

putative origin of replication (*cis*) in HVP, and have assigned it to the HVP *Eco*RI K fragment. Our results indicate that the HVP origin of replication requires both a *cis* and a *trans*-acting function, analogous to that found in EBV. Furthermore, additional data indicate that there is cooperativity of function between the HVP *cis* and the EBV *trans* fragments. Plasmid constructs containing the HVP *cis* fragment can replicate autonomously in cell lines containing an endogenous EBV genome. The results indicate that similar mechanisms and probable conserved sequences operate to maintain latent episomes in EBV and HVP infection.

INTRODUCTION

Epstein-Barr virus (EBV) is an important human pathogen, recognized as the causative agent of infectious mononucleosis, and has been associated with Burkitt's lymphoma and nasopharyngeal carcinoma (5). EBV is the only human herpesvirus system in which a model for viral latency is captured *in vitro*, namely in non-productive, continuously EBV-infected cells (i.e. Raji). *In vitro* infection of human B lymphocytes with EBV frequently results in the establishment of a latent infection and cellular immortalization (6, 7). B-lymphoid cell lines generally express Epstein-Barr virus nuclear antigen (EBNA I) (8), and contain approximately 50 copies of the viral genome. The linear form of the genome, which is the encapsidated form, has been estimated to be 172 x 10^3 nucleotide base pairs (9, 10, 11). The episomal DNA, which appears to be exclusively an intracellular form of the EBV genome, is located in the nucleus and although not covalently linked to cell DNA, may be allied with cellular chromatin (12). Covalently closed circular episomal DNA is identical in length to linear EBV DNA (13) and is formed by the covalent linkage of the termini (14). The EBV episome in latently infected cells is believed to be replicated by a cellular DNA polymerase (15), while a viral DNA polymerase is responsible for replication of the linear form of the genome (16). The presence of the EBV episomes in latently infected cells from seropositive patients and in Burkitt-tumor and nasopharyngeal carcinoma tissue (17, 18, 19) add significantly to the biological importance of the episomal form.

The putative origin of replication in EBV (*oriP*) has been assigned to a 1790 base pair subfragment of *Bam*HI C (*cis*) in the short unique region of the genome which requires a function supplied in *trans* from elsewhere in the genome (20). More precisely, the *cis* fragment is comprised of a 30 bp A + T-rich tandem repeat, approximately 600 bp in length, and a 114 bp region displaying a 65 bp dyad symmetry, both of which were shown to be required for autonomous replication (21). Deletion of the unique internal 960 bp separating these two regions did not appear to affect the *cis* function. The *trans*-acting function is encoded within a 2.8 kbp *Bam*HI-*Hin*dIII subfragment of *Sal*I F, a fragment known to encode Epstein-Barr nuclear antigen I (EBNA I). Deletion of a 700 bp repetitive triplet which encodes part of EBNA I did not affect the *trans* function (22). Additional sequences as far as 2000 bp upstream (possible transcriptional enhancer from the EBNA I open-reading frame) seemed to be required for enhanced expression of the function. Further studies have shown that the EBNA I protein has direct interaction with the essential regions of *oriP* (23). Filter binding assays and DNAase I footprinting have revealed that a fusion protein representing the carboxy-terminal domain of EBNA I protects binding sites within both the 30 bp tandem repeats and the 65 bp dyad symmetry of *oriP*. In addition, a large in-frame deletion and a linker insertion frameshift mutation within the coding region of EBNA I which altered its carboxy-terminus destroyed its *trans*-acting ability (24).

Herpesvirus papio (HVP) and EBV are closely related, both biologically and biochemically. HVP is a lymphotropic virus of baboons and is able to immortalize both human and baboon B lymphocytes *in vitro* (1). The growth-transformed cells contain early antigen, membrane antigen, and an intranuclear antigen designated HV Papio nuclear antigen (HVPNA). Both HVP early antigen and membrane antigen were clearly shown to be cross-reactive with EBV antigens (1, 25). Cross-reactivity of human anti-EBV serum to HVPNA was subsequently demonstrated by the extraction of soluble antigens from lymphoblastoid cells transformed with HVP (26). All EBNA-positive human sera react with both EBNA and HVPNA (27).

DNA-DNA hybridization studies indicate that HVP and B95-8 EBV DNA share a 40% homology that is not restricted to certain regions of the viral DNA, but is, instead, dispersed (2). In addition, the restriction endonuclease digest patterns of B95-8 EBV DNA and HVP DNA are distinctly different. Cell lines derived from lymphocytes of infected baboons contain from 3 to 63 genome copies per cell (28). HVP-permissive lymphoblastoid cells contain both unit length linear and episomal viral DNA. The circular molecules are approximately the same size as episomes found in EBV growth-transformed Raji cells, 170 x 10^3 nucleotide base pairs (3). Physical maps of HVP have shown that the genome contains both terminal and internal repeat sequences (29, 30) similar to those existing in the EBV genome (31). The internal direct repeats separate the genome into short unique (U_S) and long unique (U_L) regions (32, 29).

As a result of experiments in our laboratory, we have identified a region in *Herpesvirus papio* (HVP) with sequence and functional homology to the putative EBV origin of replication. HVP provides us with a unique opportunity to define better the essential region containing the EBV origin of replication as well as the recently identified *trans*-acting function that seems to be necessary for episomal maintenance.

MATERIALS AND METHODS

Cell Culture

D98/HR1 is a somatic hybrid cell line between the human epithelial cell line, D98, and the nonvirus-producing lymphoblastoid cell line P3JHR1 (33), containing multiple copies of the EBV genome. The 594S/F9 cell line, originally isolated from baboon lymphoma cells, is productive for *Herpesvirus papio*. Both cell lines were grown in RPMI 1640 supplemented with 10% heat-inactivated fetal bovine serum, plus 100 U penicillin and 100 µg streptomycin per ml. HeLa cells were grown in Dulbecco modified Eagle medium supplemented with 10% heat inactivated fetal bovine serum plus antibiotics.

The pKan2:EcoK construct is shown in Fig. 1. pKan2 (gift of Dr. B. Sugden) contains the gene encoding aminoglycoside phosphotransferase II from the bacterial transposon Tn5 for selection in mammalian cells. Transcriptional control is afforded by the HSV-1 TK transcription initiation and termination signals. pKan2 was linearized by partial digestion with *Eco*RI. Extracellular

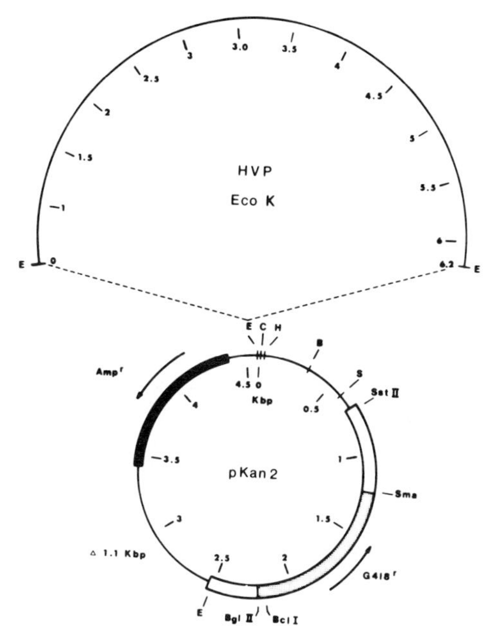

FIGURE 1. The plasmid vector pKan2 (4) containing HVP Eco K. The gene encoding aminoglycoside phosphotransferase II of the bacterial transposon Tn5 (shaded area) provides resistance to G418, and is under the transcriptional regulation of the HSV-1 *TK* gene (open boxes). Restriction endonuclease sites are shown for *Bam*HI, B; *Cla*I, C; *Eco*RI, E; *Hin*dIII, H; *Sal*I, S; *Sma*I, Sma; *Sst*II; *Bcl*I; and *Bgl*II.

HVP DNA was isolated from the lymphoblastoid cell line 594S/F9 and digested with *Eco*RI, ligated into the *Eco*RI site of pKan2 and subsequently transformed into HB101. Individual isolates were screened for the EcoK fragment by *in situ* hybridization using the EBV fragment containing *oriP* as probe.

DNA Transfection

Plasmid DNA was introduced into both D98/HR1 and HeLa cells by electroporation (34). Cells were trypsinized and washed twice with calcium- and magnesium-free phosphate-buffered saline (PBS), and finally resuspended in 1.0 ml of the PBS at a concentration of 2×10^7 cells per ml. The cell suspension was transferred to a plastic cuvette containing 1 cm wide electrodes separated by 4 mm. 10 μg of plasmid DNA was added to the cells and put on ice for ten minutes. 2000 V were transiently passed through the cuvette by a single discharge from a capacitor charged to 2000 volts. The cells were kept on ice for ten minutes, and then diluted with RPMI 1640 medium plus 10% fetal bovine serum and incubated at 37° C for 72 hours. The medium was then supplemented with G-418 at a final concentration of 600 μg per ml (46% active: Gibco Laboratories). Medium supplemented with 600 μg per ml G-418 was replaced every four to five days. G-418-resistant colonies appeared in seven to ten days, and were carried in medium containing 600 μg per ml of G-418.

Analysis of G-418-Resistant Cells

G-418-resistant colonies were pooled and expanded, and low molecular weight DNA was isolated by the method of Hirt (35). The resulting DNA was digested with restriction enzymes and electrophoresed in 0.7% agarose gels. DNA was transferred to a nitrocellulose membrane by the method of Southern (36). The nitrocellulose filter was prehybridized for three hours at 42° C and hybridized with ^{32}P nick-translated plasmid DNA for 14 hours in 50% formamide and 10% dextran sulfate at 42° C.

RESULTS

Identification of *Herpesvirus papio* (HVP) Fragment Homologous to the Fragment of Epstein-Barr Virus Containing *oriP*

Since HVP and EBV share approximately 40% base pair homology, in our first attempts to find a plasmid origin of replication functionally analogous to the EBV *oriP* in the HVP genome, we used the EBV *oriP* fragment as probe to define a homologous HVP fragment. Extracellular viral DNA from HVP was digested with *Bam*HI, *Hin*dIII, and *Eco*RI, electrophoresed on agarose gels and analyzed by the method of Southern. There was hybridization (Tm-30) between EBV *oriP* and two HVP fragments, HVP *Eco*RI K and the terminal fragment HVP *Eco*RI J (Fig. 2 & 3).

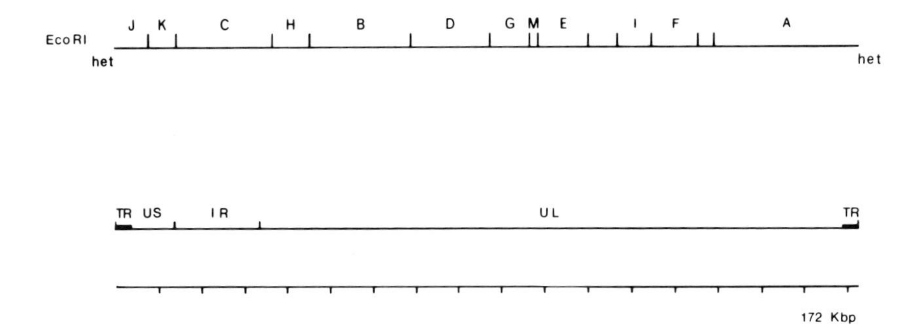

FIGURE 2. *Eco*RI restriction digest map of HVP DNA (30).

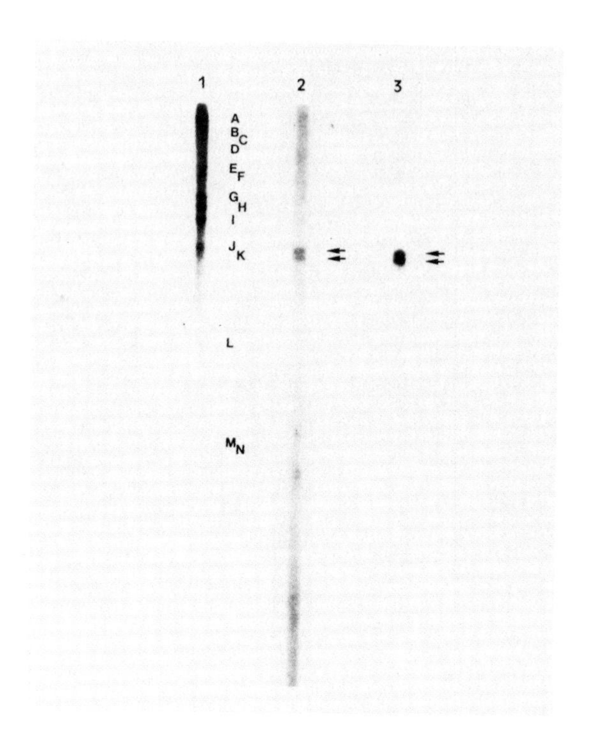

FIGURE 3. Southern blot hybridizations of *Eco*RI fragments of HVP with EBV DNA and HVP DNA as probe. 1) HVP extracellular virus DNA digested with *Eco*RI probed with ^{32}P-HVP total extracellular viral DNA. 2) HVP extracellular virus DNA digested with *Eco*RI probed with ^{32}P-labeled *Eco*RI/*Sst* II subfragment of EBV-*Bam*HI C (EBV *oriP*-containing fragment). 3) HVP extracellular virus DNA digested with *Eco*RI probed with ^{32}P - HVP Eco K. HVP fragments *Eco*RI J and K are indicated by arrows in panels 2 and 3.

HVP *Eco*RI K was subsequently cloned into the G-418-selectable plasmid vector pKan2 (Fig. 1). Southern blots of *Eco*RI-digested HVP were probed with pKan2:Eco K. The resulting signals showed that Eco K hybridized with both Eco K and Eco J, indicating that homologous sequences present in Eco K are also represented in Eco J.

High Frequency of Stable Transfection of HVP Eco K

In order to determine whether the HVP *oriP* is functionally similar to the EBV *oriP*, we transfected the pKan2:Eco K construct into D98/HR1 and selected for G-418-resistant colonies. Table 1 shows that the plasmid construct pKan2:Eco K gave rise to greater numbers of G-418-resistant colonies when compared with other plasmids when electroporated into the EBV - genome-positive cell line D98/HR1. As shown in Table 1, approximately 5000 of 1 x 10^7 D98/HR1 cells electroporated with pKan2:Eco K (0.05%) gave rise to G-418-resistant colonies. No G-418-resistant cell lines were established in the HeLa cell line, which lacks an endogenous EBV genome.

TABLE 1
SELECTION OF G-418-RESISTANT COLONIES IN DIFFERENT CELL LINES AFTER ELECTROPORATION WITH VARIOUS PLASMID CONSTRUCTS

Cell line	Plasmid	Number of G-418-resistant colonies	Plasmid copies per cell
D98/HR1	pKan2	0	---
D98/HR1	pKan2:HVP Eco K	5000	5
HeLa	pKan2	0	---
HeLa	pKan:HVP Eco K	0	---
HeLa	pKan2:HVP Eco K + pKan2:EBV Sal F	5	nd[a]

[a]not determined

HVP *oriP* Can Replicate Autonomously as a Plasmid in EBV-Positive Cells

The above assay, that of selection of G-418-resistant colonies, presumes that the pKan2 plasmid construct remains stable in the cell line. This phenotype can be elicited by either stable integration or the autonomous replication of the plasmid construct. We then asked whether the latter explanation was correct. DNA isolated from G-418-resistant D98/HR1 cells electroporated with pKan2:Eco K was analyzed for autonomously replicating plasmid molecules. Low molecular weight DNA was isolated from pools of G-418-resistant colonies by the method of Hirt (Fig. 4). The resulting low molecular weight DNA was digested with *Eco*RI, electrophoresed and analyzed by the method of Southern using pKan2:Eco K as probe. The major hybridization pattern observed was identical to the mobility of an *Eco*RI digest of the original construct pKan:Eco K, indicating that the plasmids did not undergo rearrangement. Undigested DNA was also analyzed as above, and the resulting hybridization pattern was identical to the mobility of supercoiled plasmid DNA (data not shown). In two independent transfections, one by the calcium phosphate coprecipitation method (37), and the other, electroporation, plasmid molecules were easily detected in Hirt supernatant fluids. No G-418-resistant colonies arose when the pKan2:Eco K plasmid was introduced by electroporation into cells which lacked an endogenous EBV genome.

In order to determine if the putative HVP origin of replication might utilize the EBNA-I function of EBV as a trans-activating factor for replication, HeLa cells were co-transfected with pKan:Eco K and a second plasmid pKan2 *Sal*I F; the *Sal*I F component from EBV DNA encodes EBNA-I. G-418-resistant cells were selected and DNA was extracted from mass cultures and digested with *Hin*dIII to linearize the pKan2:Eco K plasmid. After electrophoresis and Southern transfer, the HVP Eco K fragment was used as probe. The hybridization signal showed a single band migrating at a molecular weight equal to that of linearized pKan2:Eco K (Fig. 4). The single hybridization band is indicative of a non-integrated and presumably autonomously replicating molecule in the cells, suggesting that the EBV EBNA-I encoding region is the *trans* factor allowing maintenance of the plasmid.

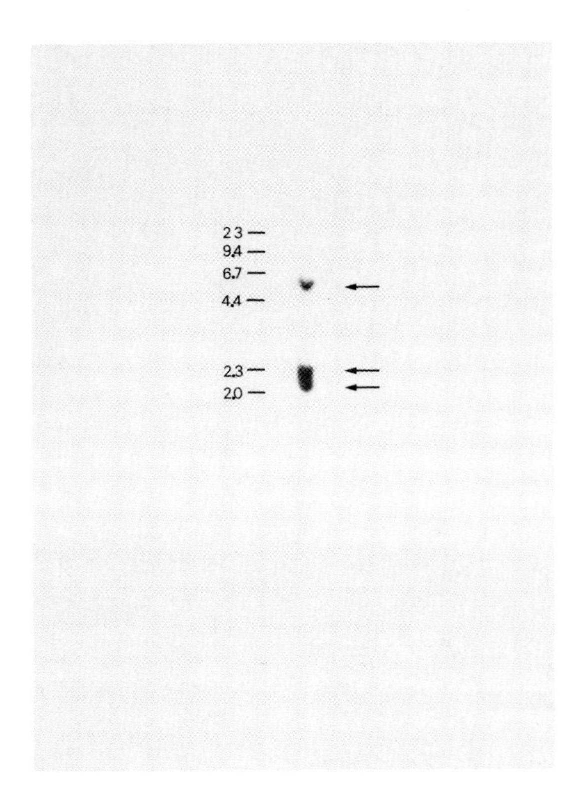

FIGURE 4. pKan2 construct containing HVP Eco K replicates as a plasmid in D98/HR1 cells. D98/HR1 cells were electroporated with pKan2:Eco K and selected with G-418. Colonies (5 x 10^3) were pooled and passaged approximately 20 times in medium containing 600 µg of G-418 per ml and extracted by the method of Hirt (35). The DNA in the supernatant was digested with *Eco*RI and Southern blots were probed with ^{32}P-pKan2:Eco K. Arrows indicate the 6.2 Kb HVP Eco K fragment and the two *Eco*RI fragments of pKan2, 2.5 Kb and 2.1 Kb respectively.

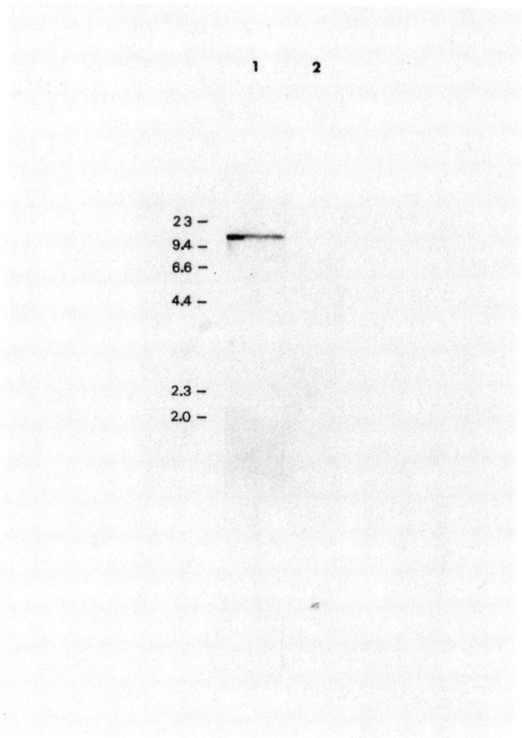

1 2

23 –
9.4 –
6.6 –
4.4 –
2.3 –
2.0 –

FIGURE 5. pKan2 plasmid containing HVP Eco K replicates autonomously when EBNA I is supplied in *trans*. HeLa cells were co-transfected with pKan2:HVP Eco K and pKan2:EBV Sal F (region which encodes EBNA I) and selected with G-418. Colonies were pooled and passaged approximately 20 times. Total cellular DNA was isolated and digested with *Hin*dIII. Southern blots were probed with ^{32}P-HVP Eco K insert DNA.

Plasmids Containing Herpesvirus Papio *oriP* in G-418-Resistant D98/HR1 Colonies

In order to determine the HVP *oriP* plasmid copy number, total cellular DNA was isolated from G-418-resistant D98/HR1 colonies, previously transfected with pKan2:Eco K. The DNA was electrophoresed on agarose gels and analyzed by the method of Southern, using pKan2 as probe. Plasmid copy reconstructions were made at 1, 10 and 100 copies per cell; the resulting hybridization signals indicated a plasmid copy number of approximately 5 per cell (Table 1). Low plasmid copy numbers could represent an inefficient plasmid maintenance due to a divergence in sequence homology from the EBV *oriP*.

DISCUSSION

We report here the identification of a *Herpesvirus papio* (HVP) *cis*-acting element homologous to the Epstein-Barr virus (EBV) *oriP*, which can function as a plasmid origin of replication in cell lines containing an endogenous EBV genome. The HVP element has been localized to a 6.2 Kb fragment within the short unique region of the HVP genome (Fig. 2).

Analogous to the *oriP* in EBV, this element requires a function supplied in *trans* from a latent viral genome for maintenance of the episome. The genomes of some eukaryotic viruses are known to initiate DNA replication at specific sites designated as origins of replication. The question remains whether initiation of replication is confined to specific DNA sequences. Although there is indirect evidence of this in yeast (38), direct proof of sequence-specific initiation has been restricted to small extrachromosomal molecules where replication of the entire molecule originates at a single site.

Our data indicate that a latent EBV function can act interspecies to provide a factor in *trans* to support replication of a *cis*-acting element from HVP. Recent evidence indicates that Epstein-Barr nuclear antigen (EBNA I) is required in *trans* for maintenance of EBV *oriP*-containing plasmid constructs (22). Furthermore, DNA binding studies indicate that EBNA I may have sequence-specific DNA-binding properties, showing preferential binding to the essential regions of EBV *oriP*, those being the 30 bp tandem repeats and the 65 bp dyad symmetry (23). It remains to be determined whether the EBNA I protein from EBV is the essential *trans* element allowing episomal maintenance of the HVP plasmid construct. One can, however, predict that if the EBV *oriP* represents the origin of replication used to maintain the episome in a latent state, and if EBNA I plays a role in maintaining the episomal form, a direct analogy could be made in the *Herpesvirus papio* system. There exists in the HVP transformed cell line a protein designated *Herpesvirus papio* nuclear antigen (HVPNA) which is antigenically cross-reactive with serum from EBV-infected individuals. If HVP Eco K represents the analogous fragment to EBV *oriP*, the HVPNA and EBNA I could be structurally similar enough for EBNA I to function as the *trans* element in a cooperative fashion with the HVP plasmid origin of replication. Co-transfections of the EBV EBNA I encoding fragment (EBV *Sal*I F) and the HVP Eco K fragment resulted in the establishment of a G-418-resistant cell line which contained the papio construct replicating autonomously. Therefore, EBNA I appears to be the responsible *trans*-acting element. We were unable to detect EBNA I in the cell line using standard anti-complement immunofluorescence techniques and human serum containing EBNA antibodies, although this may be due to a lack of sensitivity in the assay since the EBV *Sal*I F fragment may represent a single copy integration event.

Herpesvirus papio provides us with a unique opportunity to better define the essential region containing the EBV origin of replication. Despite a loss of approximately 60% sequence homology between EBV and HVP, a homologous *cis* function appears to be maintained in both EBV and HVP, and the sequence is located colinearly on HVP and EBV genomes. The HVP *trans*-containing segment remains to be defined; however, our data indicate that it probably exists. Eventually, its identification will help to define the general essential genome structure of EBV. By comparing sequences from functionally identical fragments in two related viruses, we should be able to pinpoint the common sequences required for these functions in the two genomes. It remains to be determined whether other distantly related herpesviruses, especially those such as *Herpesvirus saimiri* and *Herpesvirus ateles* that are known to form episomes, retain such functions, and what sequence divergence is tolerated.

ACKNOWLEDGMENTS

We thank Nancy Raab-Traub for many fruitful discussions during the course of this work. We also thank the secretarial staff for typing the manuscript.

REFERENCES

1. Rabin H, Neubauer R, Hopkins R, Dzhibidze E, Sheutsova Z, Lapin B (1977). Transforming activity and antigenicity of an Epstein-Barr like virus from lymphoblastoid cell lines of baboons with lymphoid disease. Intervirology 8:240.
2. Lee Y, Tanaka A, Lau RY, Nonoyama M, Rabin H (1980). Comparative studies of Herpesvirus Papio (Baboon Herpesvirus) DNA and Epstein-Barr virus DNA. J Virol 51:245.
3. Falk L (1979). A review of Herpesvirus papio, a B-lymphotropic virus of baboons related to EBV. Comp Immunol Microbiol Infect Dis 2:229.
4. Yates J, Warren N, Reisman D, Sugden B (1984). A cis-acting element from the Epstein-Barr virus genome that permits stable replication of recombinant plasmids in latently infected cells. Proc Natl Acad Sci USA 81:3806.
5. Epstein M, Achong BG (1979). Morphology of the virus and of virus induced cytopathologic changes. In Epstein M, Achong B (eds.): "The Epstein-Barr Virus," Berlin:Springer-Verlag, p. 24.
6. Henle W, Diehl V, Kohn G, zur Hausen H, Henle G (1967). Herpes-type virus and chromosome marker in normal leukocytes after growth with irradiated Burkitt cells. Science 157:1065.
7. Gerber P, Whang-Peng J, Monroe JH (1969). Transformation and chromosome changes induced by Epstein-Barr virus in normal human leukocyte cultures. Proc Natl Acad Sci USA 63:740.

8. Reedman BM, Klein G (1973). Cellular localization of an Epstein-Barr virus (EBV) - associated complement-fixing antigen in producer and non-producer lymphoblastoid cell lines. Int J Cancer 11:499.
9. Pritchett RF, Hayward SD, Kieff E (1975). DNA of Epstein-Barr virus. I. Comparison of DNA of virus purified from HR-1 and B95-8 cells. J Virol 15:556.
10. Hayward SD, Kieff E (1977). The DNA of Epstein-Barr virus. II. Comparison of the molecular weights of restriction endonuclease fragments of the DNA of strains of EBV and identification of end fragments of the B95-8 strain. J Virol 23:421.
11. Baer R, Bankier AT, Biggin MD, Deininger PL, Farrell PJ, Gibson TJ, Hatfull G, Hudson GS, Satchwell SC, Sequin C, Tuffnell PS, Barrell BG (1984). DNA sequence and expression of the B95-8 Epstein-Barr virus genome. Nature 310:207.
12. Nonoyama M, Pagano JS (1971). Detection of Epstein-Barr viral genome in non-productive cells. Nature New Biol 233:103.
13. Lindahl T, Adams A, Bjursell G, Bornkamm GW, Kascha-Sierich C, Jehn U (1976). Covalently closed circular duplex DNA of Epstein-Barr virus in a human lymphoid cell line. J Mol Biol 102:511.
14. Heller M, Dambaugh T, Kieff E (1981). Epstein-Barr virus DNA. IX. Variation among viral DNAs. J Virol 38:632.
15. Pagano J (1979). In Cummings DJ (ed.): "Extrachromosomal DNA," New York:Academic Press, p. 235.
16. Datta AK, Feighny RJ, Pagano JS (1980). Induction of Epstein-Barr virus-associated DNA polymerase by 12-0-tetradecanoylphorbol-13-acetate:Purification and characterization. J Biol Chem 255:5120.
17. Adams A, Bjursell G, Kaschka-Dierich C, Lindahl T (1977). Circular Epstein-Barr virus genomes of reduced size in a human lymphoid cell line of infectious mononucleosis origin. J Virol 22:373.
18. Kaschka-Dierich C, Adams A, Lindahl T, Bornbamm G, Bjursell G, Klein G (1976). Intracellular forms of Epstein-Barr virus DNA

in human tumor cells in vivo. Nature 260:302.

19. Kaschka-Dierich C, Falk L, Bjursell G, Adams A, Lindahl T (1977). Human lymphoblastoid cell lines derived from individuals without lymphoproliferative disease contain the same latent forms of Epstein-Barr virus DNA as those found in tumor cells. Int J Cancer 20:173.
20. Yates J, Warren N, Reisman D, Sugden B (1984). A cis-acting element from the Epstein-Barr virus genome that permits stable replication of recombinant plasmids in latently infected cells. Proc Natl Acad Sci USA 81:3806.
21. Reisman D, Yates J, Sugden B (1985). A putative origin of replication of plasmids derived from Epstein-Barr virus is composed of two cis-acting components. Mol Cell Biol 5:1822.
22. Yates J, Warren N, Sugden B (1985). Stable replication of plasmids derived from Epstein-Barr virus in various mammalian cells. Nature 313:812.
23. Rawlins D, Milman G, Hayward SD, Hayward GS (1985). Sequence-specific DNA binding of the Epstein-Barr virus nuclear antigen (EBNA-I) to clustered sites in the plasmid maintenance region. Cell 42:859.
24. Lupton S, Levine A (1985). Mapping genetic elements of Epstein-Barr virus that facilitate extrachromosomal persistence of Epstein-Barr virus-derived plasmids in human cells. Mol Cell Biol 5:2533.
25. Rabin H, Neubauer R, Hopkins R (1978). Studies on Epstein-Barr (EBV)-like viruses of old world nonhuman primates. In Bentbelzen (ed.):"Advances in Comparative Leukemia Research," Amsterdam:Elsevein/North Holland Biomedical Press, p. 205.
26. Ohno S, Luka J, Falk L, Klein G (1977). Detection of a nuclear, EBNA-type antigen in apparently EBNA-negative Herpesvirus papio (HVP)-transformed lymphoid lines by the acid-fixed nuclear binding technique. Int J Cancer 21:941.
27. Ohno S, Luka J, Falk L, Klein G (1978). Serological reactivities of human and baboon

sera against EBNA and Herpesvirus papio - determined nuclear antigen (HVPNA). Eur J Cancer 14:955.

28. Falk LA, Henle G, Henle W, Deinhardt F, Schudel A (1977). Transformation of lymphocytes by Herpesvirus papio. Int J Cancer 20:219.

29. Heller M, Gerber P, Kieff E (1981). Herpesvirus papio DNA is similar in organization to Epstein-Barr virus DNA. J Virol 37:698.

30. Lee Y, Nonoyama M, Rabin H (1981). Colinear relationships of Herpesvirus papio DNA to Epstein-Barr virus DNA. Virology 110:248.

31. Kieff E, Dambaugh T, Hummel M, Heller M (1983). Epstein-Barr virus transformation and replication. In Klein G (ed.):"Advances in viral oncology," New York: Raven Press, p. 133.

32. Dambaugh T, Beisel C, Hummel M, King W, Fennewald S, Cheung A, Heller M, Raab-Traub N, Kieff E (1980). Epstein-Barr virus (B95-8) DNA VII:Molecular cloning and detailed mapping of EBV (B95-8) DNA. Proc Natl Acad Sci USA 77:305.

33. Glaser R, Rapp F (1972). Rescue of Epstein-Barr virus from somatic cell hybrids of Burkitt lymphoblastoid cells. J Virol 10:298.

34. Neuman E, Schaefer-Ridder M, Wang Y, Hofschneider PH (1982). Gene transfer into mouse lyoma cells by electroporation in high electric fields. EMBO J 1:841.

35. Hirt B (1967). Selective extraction of polyoma DNA from infected mouse cell cultures. J Mol Biol 26:365.

36. Southern EM (1975). Detection of specific sequences among DNA fragments separated by gel electrophoresis. J Mol Biol 98:503.

37. Graham F, Van der Eb A (1973). A new technique for the assay of infectivity of human adenovirus 5 DNA. Virology 52:456.

38. Kearsey S (1984). Structural requirements for the function of a yeast chromosomal replication. Cell 37:299.

Viruses and Human Cancer, pages 415–425

CONSTRUCTION OF PLASMIDS CONTAINING SYNTHETIC 29 BP BINDING SITES FOR EPSTEIN-BARR VIRUS NUCLEAR ANTIGEN

Gregory Milman and Mark Chernaik

Department of Biochemistry
The Johns Hopkins University
School of Hygiene and Public Health
Baltimore, MD 21205

ABSTRACT The carboxyl-third of EBNA-1 encoded in the Epstein-Barr virus (EBV) BamH1 K-restriction fragment was synthesized in bacteria [Milman et al. (1985). Proc Natl Acad Sci, USA 82:6300]. The bacterially synthesized peptide specifically binds to clustered EBV DNA sequence repeats in the Ori-P region required for plasmid maintenance [Rawlins et al. (1985). Cell 42:859]. To elucidate the binding properties, a synthetic DNA binding site

```
5' GATCTAGGATAGCAT|ATGCTACCCCGGGG     3'
3'     ATCCTATCGTA|TACGATGGGGCCCCCTAG 5'
```

was cloned as a monomer, dimer, or trimer into the BamH1 site of plasmid pUC8. Sequence specific binding of EBNA to the three plasmids was detected by mobility retardation on agarose gels and by protein mediated binding of DNA to nitrocellulose filters. EBNA binding is strongly cooperative even to the monomer sequence. Cleavage of the binding sequence with Ndel produces half a binding site to which EBNA does not appear to bind.

Aided by grant MV-287 from the American Cancer Society and grants ES03131 and GM32950 from the National Institute of Environmental Health Sciences and the National Institute of General Medical Sciences, National Institutes of Health.

INTRODUCTION

Epstein-Barr virus (EBV) replicates as a plasmid in latently infected B-lymphocytes. An 1800 bp region of EBV DNA (ori-P) possesses the cis-acting signals which enable circular DNA containing ori-P to replicate as a plasmid in a number of cell lines (1-3). At each end of ori-P are loci essential to its function (4). The left loci consists of 20 copies of 30 bp tandem repeats which have a 12 bp palindromic consensus sequence TAGCATATGCTA. The right loci contains four copies of repeats nearly identical to the consensus sequence. The DNA in this region can be drawn as a 116 bp stem-loop dyad symmetry structure reminescent of other origins of replication.

Ori-P dependent plasmid replication and maintenance requires that host cells express a single EBV protein (2), Epstein-Barr virus nuclear antigen (EBNA-1) coded in the EBV BamHI K-restriction fragment. A 28 kDalton carboxyl-terminal fragment of EBNA-1 (28K-EBNA) was synthesized in bacteria (5). Filter binding assays and DNase I footprinting (6) demonstrated that the 28K-EBNA polypeptide binds tightly to the repeat sequences in ori-P. Quantitative filter binding assays of EBNA interactions with DNA are complicated if the DNA contains multiple binding sites. Multiple sites also interfer with the analysis of binding structures responsible for mobility changes during gel electrophoresis. This paper describes the construction of a synthetic consensus EBNA binding site and the cloning of monomer, dimer and trimer forms of the synthetic site into the E. coli plasmid pUC8. These constructs provide the tools to better understand the interaction of 28K-EBNA with its DNA binding sequences.

METHODS

Baterially synthesized 28K-EBNA.

The carboxyl-terminal one-third of the Epstein-Barr virus nuclear antigen (EBNA-1) encoded by the BamHI restriction fragment K was synthesized in E. coli and purified as previously described (6). This 28 kilodalton peptide is referred to as 28K-EBNA throughout this paper.

Construction of synthetic DNA binding sequence.

DNA oligonucleotides 5' GATCTAGGATAGCATATGCTACCCCGGGG 3' and 5' GATCCCCCGGGGTAGCATATGCTATCCTA 3' were synthesized on an Applied Biosystems DNA synthesizer. The oligomers were separated from low molecular weight contaminants by chromatography on Sephadex G-50 and further purified by acrylamide gel electrophoresis. The 29-mers were identified by absorption of UV light using a fluorescent TLC plate as a background, and recovered from the acrylamide gel by soaking in gel elution buffer (0.5 M ammonium acetate, 10 mM $MgCl_2$, 1mM EDTA, and 0.1% SDS). The oligomers were kinased with both ^{32}P-ATP and unlabeled ATP in separate reactions. To anneal the oligonucleotides, approximately 1 ug of each oligomer (10^5 CPM) was placed in 20 ul of TE buffer (10 mM Tris-HCl, pH 7.6 and 1 mM EDTA), heated to 90 ^{o}C, and allowed to slowly cool to room temperature. The annealed oligomers were ligated with T4 ligase and then cleaved with BamHI and BglII. The resulting tandem multiple repeats were deproteinated by phenol-chloroform extraction and ethanol precipitation.

Insertion of synthetic sequences in pUC8.

A 2 ug sample of pUC8 DNA was cleaved with BamHI, and deproteined by phenol-chloroform extraction and ethanol precipitation. A 0.2 ug sample of the cleaved pUC8 DNA was mixed with 24 ng of the tandem multiple repeats and ligated overnight at 15 ^{o}C using T4 ligase in a 15 ul volume. The ligated DNA was used to transfect JM101 cells. Transfected cells were spread on L-broth agar plates containing ampicillin, IPTG, and X-gal indicator. Colonies containing pUC8 without inserts were blue and those containing pUC8 with inserted synthetic DNA were white. Approximately 30% of the colonies contained inserts. Rapid lysis DNA plasmid preparations were obtained from plasmids with inserts and the number of inserts was determined by cleavage with PvuII and electrophoresis of the cleaved DNA on 1.5% agarose. Plasmids pR1, pR2, and pR3 were identified containing 1, 2, and 3 copies of the synthetic binding site.

Mobility retardation in agarose electrophoresis.

PvuII digested plasmid DNAs (0.2 ug each plasmid) and 28K-EBNA (.36 ug) were mixed in 4 ul of solution containing 40mM Tris-acetate, pH 8.3 and 1 mM EDTA. The samples were incubated for 1 hr at 24 ^{o}C and then mixed with 1 ul of the same buffer containing 50% glycerol and bromphenol blue indicator. Samples were loaded into wells of a 1.5% low melting point agarose gel in a Hoeffer mini-gel apparatus, and subjected to electrophoresis at 4 watts (approximately 100 volts) until the tracking dye approached the end of the gel (approximately 2 hours). HindIII or PvuII cleaved lambda DNA were used to provide DNA size markers. The gels were soaked for 10 min in 0.5 ug/ml ethidium bromide and then photographed.

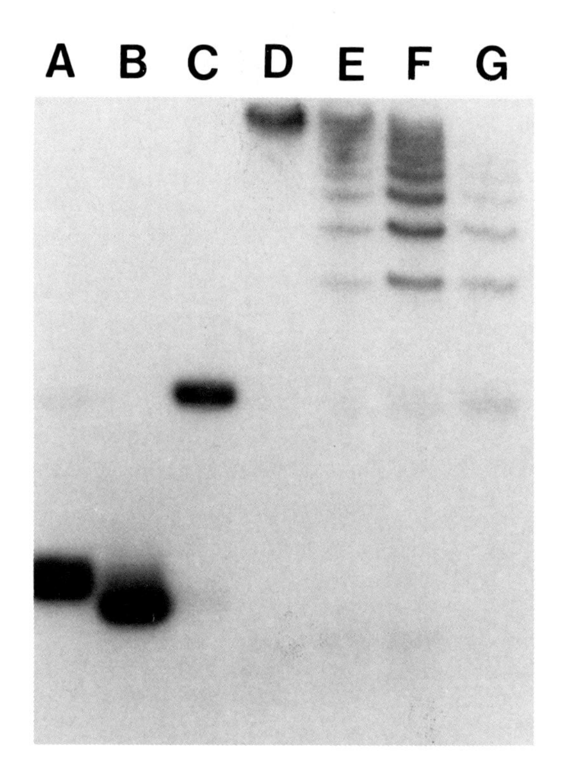

FIGURE 1. Construction of synthetic DNAs. Lanes A and B contain samples of the two initial oligomers; Lane C, the annealed oligomers; Lane D, the ligation reactions; and Lanes E, F and G, the products after cleavage with BglII, BamHI and both restriction enzymes, respectively.

RESULTS

The components used in the construction of the synthetic DNA EBNA binding sites were analyzed by polyacrylamide gel electrophoresis and an autoradiograph of the gel is presented in Figure 1. A restriction map of pUC8 and pR1, containing a single insert, is illustrated in Figure 2.

FIGURE 2. Restriction map of pUC8 and pR1.

```
                              PLASMID pUC8


       PvuII
PvuII     NdeI
|      | |
----------------------------------------------------------------------
1         430                                                    2722
       308

                 PLASMID pUC8 (308 bp PvuII fragment)

                                        EcoRI  HindIII
PvuII                                     BamHI                     PvuII
|                                       |  |   |                    |
------------------------------------------------------------------------
1                                         187                       308
                                        177   207

                 PLASMID pR1 (337 bp PvuII fragment)

                                       186  216
                                       )--I--)
                                         NdeI  HindIII
PvuII                                 EcoRI  BamHI                  PvuII
|                                     |   |  |  |                   |
------------------------------------------------------------------------
1                                     177    216                    337
                                         201    236

                 PLASMID pR1 (258 bp NdeI fragment)

186     216
)---I---)
   NdeIBamHIHindIII                   PvuII                      NdeI
   |   |   |                          |                          |
   ------------------------------------------------------------------
   201 216  236                       337                        459
```

The number of synthetic DNA fragments inserted in pUC8 was determined by comparing the sizes of the small DNA fragment obtained upon digestion with PvuII. The gel pattern of PvuII digests of pUC8 and pR1, pR2, and pR3 containing 1, 2, and 3 tandem copies of repeat sequences is shown in Figure 3.

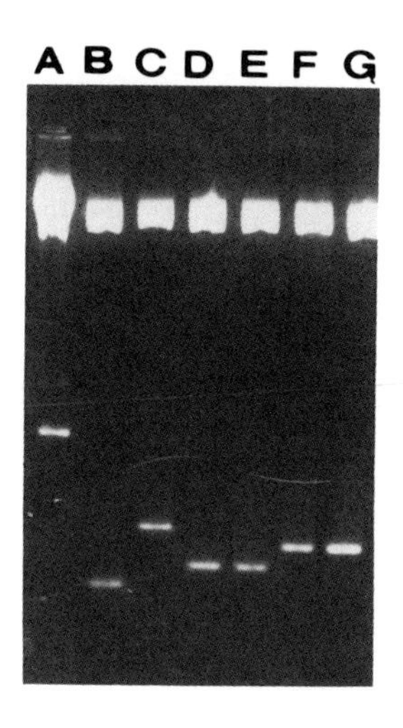

FIGURE 3. Agarose gel electrophoresis of PvuII cleaved plasmids. Lane A, HindIII cleaved lambda with 564 bp band in the middle of the gel; Lane B, pUC8; Lane C, pR3; Lanes D and E, pR1; Lanes F and G; pR2.

The 28K-EBNA binds to the synthetic repeat.

Figure 4 illustrates the binding of 28K-EBNA to a single synthetic repeat. Each binding mixture contained PvuII cleaved pR1 and PvuII cleaved pUC8 DNA as a negative control. As the amount of 28K-EBNA increases from Lanes F to B, there is no apparent binding until 90 ng is added in Lane C. The 28K-EBNA binds only to the pR1 337 bp fragment in Lanes B and C shifting its mobility to the positions indicated by the two left arrows. The relative intensity of the two new mobility bands is the same in Lanes B and C and appears to be independent of 28K-EBNA concentration. The intensity of the upper new mobility band is proportional to the concentration of DNA with EBNA binding sites (data not shown).

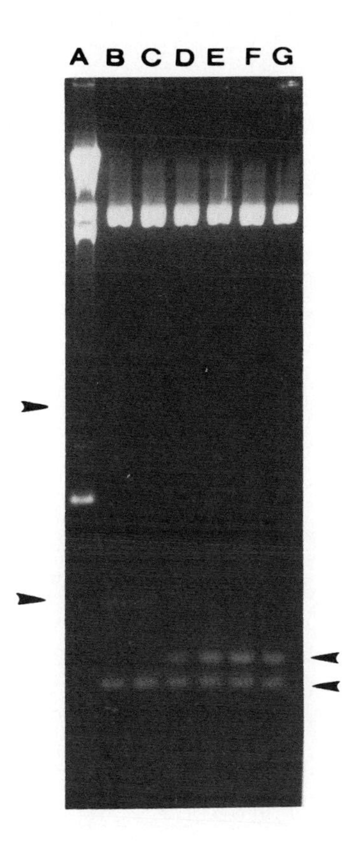

FIGURE 4. Binding of 28K-EBNA to single repeat in pRl. Lane A contains HindIII cleaved lambda DNA, and Lanes B through G contain the plasmid DNAs with Lane G containing no 28K-EBNA. The highest amount of 28K-EBNA (360 ng) is in Lane B, and Lanes C through F contain two-fold decreasing amounts of 28K-EBNA. The bottom right arrow indicates the pUC8 308 bp PvuII fragment and the top right arrow indicates the pRl homologous 337 bp fragment containing the 29 bp EBNA binding site.

The 28K-EBNA does not bind to half of repeat sequence.

The dyad symmetry of the repeat sequence contains an NdeI site at its center (Figure 2). NdeI cleavage of pRl yields a 258 bp fragment which contains half an EBNA binding site. Figure 5 illustrates that 28K-EBNA does not bind to half a site (Lane F) under conditions where it binds to the whole site (Lane B).

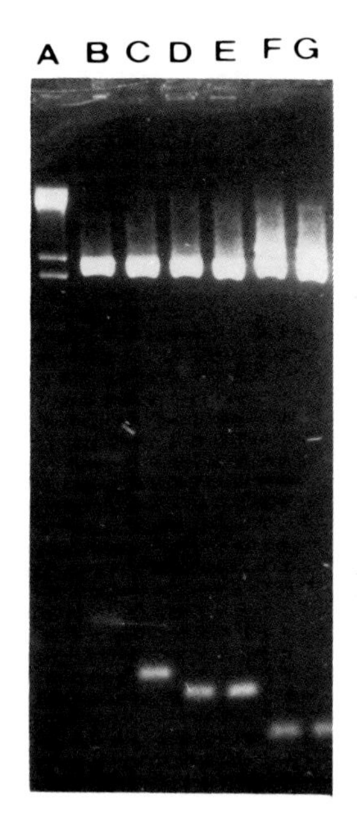

FIGURE 5. NdeI cleaved pRl does not bind 28K-EBNA. Lane A contains HindIII cleaved lambda DNA and Lanes B through C contain binding reactions: Lane B, PvuII cleaved pRl and 28K-EBNA, Lane C, PvuII cleaved pRl; Lane D, PvuII cleaved pUC8; Lane E, PvuII cleaved pUC8 and 28K-EBNA; Lane E, NdeI cleaved pRl and 28K-EBNa; Lane F, NdeI cleaved pRl.

DISCUSSION

The 30 bp tandem EBNA binding sites in ori-P contain different left (AGGA) and right (CCCR) consensus sequences surrounding the 12 bp palindromic repeat. When the synthetic DNA strands (Figure 1, Lanes A and B) are annealed (Lane C) the resulting double-stranded DNA contains similar asymetry. The synthetic DNA repeats associate randomly upon ligation forming the high molecular weight DNA in Lane D. To insure tandem duplication upon ligation, the annealed synthetic DNA was constructed with half a BglII site on the left end and half a BamHI-site on the right end. A non-tandem ligation event forms a complete sites for one of these enzymes. The ligated high molecular weight DNA was cleaved with BglII (Lane E), BamHI (Lane F), and both enzymes (Lane G). All the oligomer sequences in Lane F should be tandem repeats.

The mobility retardation experiment illustrated in Figure 4 demonstrates that 28K-EBNA binds specifically to the 337 bp fragment of pR1 and not to the homologous 308 bp fragment of pUC8. The only difference between these fragments is the inserted 29 bp synthetic binding site. The binding of 28K-EBNA to the 337 bp fragment causes a decrease in the intensity of this band and the appearance of two new bands of slower mobility. The transition between unbound DNA and bound DNA occurs over a two-fold increase in the concentration of 28K-EBNA. The relative intensities of the two new bands do not change with 28K-EBNA concentration indicating that they do not represent different ratios of protein to DNA.

Previous studies (6 and unpublished observations) of 28K-EBNA binding to authentic ori-P sites in the plasmid pHEBO (3) indicated a "step-function" dependence on 28K-EBNA concentration for binding. A sharp concentration dependence might have resulted from binding cooperativity among the multiple binding sites. Figure 4 shows that the same sharp binding dependence on 28K-EBNA concentration was found for a plasmid containing a single binding site. Within a 2-fold increase in 28K-EBNA concentration, unbound DNA became totally bound. The concentration of 28K-EBNA required for binding in these experiments was almost identical to that observed for binding to pHEBO in earlier studies. If the step-function dependence results from cooperativity, the interactions must require only a single binding site.

The dyad symmetry of the single binding site in pR1 implies that a minimum of two molecules of 28K-EBNA are bound per site. Therefore, a half-site might be capable of binding a single molecule. As shown in Figure 5, no binding occured to a half-site under conditons where binding occured to the whole site suggesting that the interaction between symmetrically bound 28K-EBNA molecules may be necessary to stabilize binding.

In summary, 28K-EBNA binds tightly to a 29 bp synthetic binding site inserted in plasmid pUC8. Plasmids containing different numbers of EBNA binding sites sites provide useful substrates to study the kinetics of the binding reaction and may provide a better understanding of the role of EBNA in regulating EBV replication and copy number in eukaryotic cells.

ACKNOWLEDGMENTS

We thank Nina S. Levy for aid in synthesizing the DNA oligonucleotides, Kathryn Stephens for prelimary binding studies with pHEBO, and Steven Hartman for preparing plasmid DNA.

REFERENCES

1. Yates J, Warren N, Reisman D, Sugden B (1984). A cis-acting element from the Epstein-Barr viral genome that permits stable replication of recombinant plasmids in latently infected cells. Proc Natl Acad Sci USA 81:3806.

2. Yates JL, Warren N, Sugden B (1985). Stable replication of plasmids derived from Epstein-Barr virus in a variety of mammalian cells. Nature (London) 313:812.

3. Sugden B, Marsh K, Yates J (1985). A vector that replicates as a plasmid and can be efficiently selected in B-lymphoblasts transformed by Epstein-Barr virus. Mol Cell Biol 5:410.

4. Reisman D, Yates J, Sugden B (1985). A putative origin of replication of plasmids derived from Epstein-Barr virus is composed of two cis-acting components. Mol Cell Biol 5:1822.

5. Milman G, Scott AL, Cho M-S, Hartman SC, Ades DK, Hayward GS, Ki P-F, August JT, Hayward SD (1985). Carboxyl-terminal domain of the Epstein-Barr virus nuclear antigen is highly immunogenic in man. Proc Natl Acad Sci USA 82:6300.

6. Rawlins DR, Milman G, Hayward SD, Hayward GS (1985). Sequence-specific DNA binding of the Epstein-Barr virus nuclear antigen (EBNA-1) to clustered sites in the plasmid maintenace region. Cell 42:859.

Viruses and Human Cancer, pages 427–441

NOVEL HERPES SIMPLEX VIRUS GENOMES: CONSTRUCTION AND APPLICATION

Minas Arsenakis, Kimber Lee Poffenberger and Bernard Roizman

The Marjorie B. Kovler Viral Oncology Laboratories
The University of Chicago
910 East 58th street,. Chicago, IL 60637

ABSTRACT The genome size, structure, and the host range of herpes simplex viruses 1 and 2 (HSV-1 and HSV-2) make these viruses ideal vectors of genetic material, both their own and those of foreign genomes. Techniques have been developed to construct HSV vectors for several purposes. These include the functional analyses of HSV genes, the study of regulation and expression of the promoter-regulatory regions of HSV genes, and the expression of genes from other organisms that are difficult to grow or present extreme handling problems. The foreign genes expressed thus far include the chicken ovalbumin gene, the hepatitis B virus S gene, the Epstein-Barr virus EBNA1 and LYDMA genes, and the HTLVIII/LAV envelope gene. The gene products of these genes were indistiguishable from their authentic counterparts. One requirement for the expression of non herpesvirus genes in the HSV genome is that they be placed under the control of a HSV promoter. Currently efforts are focused on development of host range HSV mutants and on tailoring the expression of inserted genes to specific requirements. It is envisaged that this technology will aid in the development of diagnostic tools and multivalent vaccines that will permit the use of a single preparation for the vaccination against a variety of pathogenic agents.

INTRODUCTION

The HSV genome is a linear double stranded DNA molecule

approximately 150 kilobase pairs (Kbp) in size (1). It consists of two covalently linked components, L and S. Each of the components consists of unique sequences flanked by inverted repeats (2,3). The inverted repeats flanking the L component are each 9 kbp in size and have been designated ab and a'b'. The inverted repeats of the S component are 6.5 kbp in size and have been designated a'c' and ca. (3). Each of the inverted repeats contains at least one gene in its entirety and these are therefore diploid. As a consequence of the inverted terminal repeats during the viral replication cycle the L and the S components invert relative to each other. Thus, viral DNA extracted from purified virions or infected cells consists of 4 equimolar isomeric populations (4,5).

The HSV genome codes for at least 50, but probably in excess of 70 genes whose expression is coordinately regulated and sequentially ordered in a cascade fashion (6). The α genes are the first set of genes to be expressed and their expression is not dependent upon the prior synthesis of other viral gene products. α gene expression is an absolute requirement for the expression of all other HSV genes. The second group of genes to be expressed are the β genes. The products of these genes are mainly nonstructural proteins involved in the replication of the viral DNA. The last group to be expressed are the γ genes. This group comprises two subgroups, γ_1 and γ_2 , they differ in that the γ_2 genes are more stringently dependent on prior DNA replication for their expression than the γ_1 genes. γ genes are mainly structural components of the virion.

The size, structure and wide host range of the HSV-1 genome make it particularly suitable for the maintainance and expression of additional genetic material. Early studies indicated that it was possible to insert up to 7.4 kbp of additional genetic material into the unmodified genome (7). Later studies increased this upper limit to 9.7 kbp (8). Poffenberger et al (9) in this laboratory constructed a mutant from which 15 kbp of the inverted repeats at the L-S junction region has been deleted. This and subsequent large deletion mutants contain space for insertion of approximatelly 25 kbp of additional genetic material.

PRINCIPLES OF CONSTRUCTION OF DELETIONS AND INSERTIONS.

The use of the HSV-1 genome as a vector has been made possible in large by recent development in this laboratory of techniques allowing the insertion and deletion of genetic material in large genomes (10). Essential to the success of this technology are three requirements. First, there must be either non-coding sequences whose integrity is not essential essential or "dispensible" genes that are not required for virus replication. The HSV genome codes for many replication functions that have cellular counterparts that are normally expressed in actively growing cells. As a result the viral counterparts can be deleted or interrupted without seriously affecting the ability of the virus to grow in cell culture. Most of the constructions to date have involved the viral thymidine kinase (TK) gene and the inverted repeats region in the L-S junction. The TK gene function can be deleted without compromising the ability of the virus to replicate in cells maintaining adequate pools of deoxynucleoside triphosphates. The function of the genes located in the inverted repeats flanking the junction between the L and S components is not clearly understood at the present time, but the 3 genes that might be essential, i.e, α0, α4, and $\gamma_1$34.5, are repeated in in an inverted orientation at the termini of the DNA. For example, in the case of the I358 recombinant (9) which contains a 15 kbp deletion in the junction region this results in the loss of one copy of the major regulatory gene α4. The multiplication of recombinant I358 is not affected by the deletion because a second copy of the α4 gene is located in the inverted repeats located at the terminus of the S component.

The second requirement is that suitable "parental" constructs are provided to facilitate the recombination process. In principle deletions or insertions due to recombination occur spontaneously. However the frequency of specific insertions or deletions not based on homologous recombination is extremely low. To increase the frequency of specific recombinational events we confront the viral DNA with a cloned structure consisting of the desired modification flanked by sequences homologous to the target site where the modification is to be introduced. Specifically we transfect susceptible cells with intact viral DNA and a molar excess of the cloned gene structure. The progeny of the transfection are then screened or selected for the desired recombinant.

The third requirement is to facilitate the selection of the desired recombinant viruses from the progeny of the transfection, since even homologous recombination is a relatively rare event. For this purpose we have used the

HSV-1 thymidine kinase (TK) gene as a selectable marker. The choice of the TK gene is based on three considerations. First the enzyme has a broad substrate specificity and it phosphorylates in addition to thymidine a wide varity of thymidine and other nucleoside analogues. Some of these analogues (e.g. Ara T, Acyclovir) are preferentially phosphorylated by the viral TK and are therefore non-toxic to uninfected cells. Other analogues (e.g. BUdR) are phosphorylated in uninfected cells uniquely by the host TK and are therefore non-toxic to cells lacking that enzyme (TK^- cells). In both cases phosphorylation of the analogues by the viral enzyme results in destruction of the infected cell thereby aborting the replication of the virus. The second important property of the enzyme is that, as was mentioned earlier, it is not essential for virus replication in cells maintaining adequate pools of deoxynucleosides. In TK^- cells the only source of TdRMP is via conversion of UdRMP by Thymidilate Synthetase. This pathway can be blocked by drugs (e.g. methotrexate). Thus in the presence of methotrexate only cells infected with viruses carrying the TK gene (TK^+ viruses) can multiply and produce infectious virus. Thus recombinants lacking a functional TK gene can be selected in TK^- cells in the presence of BUdR, and recombinants containing a TK gene inserted so as to interrupt a target gene can be selected in TK^- cells in the presence of methotrexate.

A diagram outlining the insertion and deletion pathways is presented in Figure 1.

APPLICATIONS OF THE INSERTION-DELETION TECHNIQUES IN HSV

Identification of the cis-acting site mediating inversion.

As noted in the introduction HSV DNA extracted from infected cells and from virions consists of four equimolar populations differing solely in the relative orientation of the L and S components (4,5). To identify the cis-acting site mediating the inversion, Mocarski et.al (11,12,13) inserted DNA fragments spanning the L-S junction into the transcribed non-translated domain of the TK gene thus separating the promoter from the structural gene. The recombinant TK^- viruses that contained the additional junction fragments inserted in the TK gene, formed 12 instead of the usual 4 isomers. Thus the insertion of the additional junction sequences caused additional inversions to occur.

This approach permitted the mapping of the cis-acting site mediating the inversion to the 500 bp a sequence located at the termini of the DNA and at the junbction between L and S components. The a sequence has a complex structure consisting of a 20 bp direct repeat (DR1) a 64 bp unique sequence (U_b), a 12 bp direct repeat No. 2 (DR2) which is repeated 19 to 22 times, a 37 bp direct repeat No. 4 (DR4) which is repeated 2 to 3 times, a 59 bp unique sequence (U_c) followed by a copy of the 20 bp DR1 sequence (12). Analyses of deletion mutants of the a sequece indicated that DR2 and DR4 are the inversion specific cis-acting sites (14). The inversion is mediated by viral trans-acting factors currently unknown (11). The mapping of the cis-acting inversion sites lead ultimately to the isolation of the I358 recombinant, deleted in the L-S junction region, which although it does not invert it replicates well in cell culture (9). As discussed later in the text, the I358 recombinant has been particularly usefull as a starting point for a variety of other constructions. The physiological significance of the inversion process remains obscure but it is clear that the function is not required for virus growth in cell culture and experimental animals.

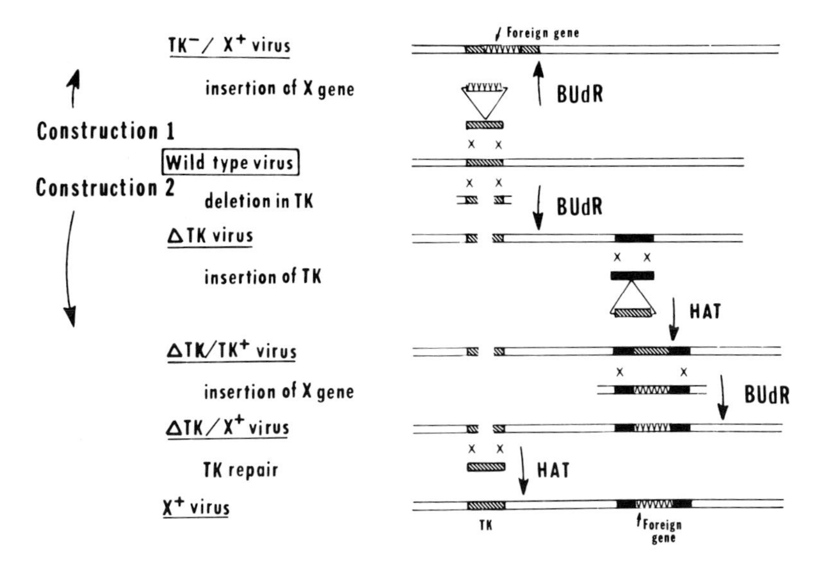

FIGURE 1. Flow diagram of construction of recombinant HSV genomes. The HSV DNA is shown as a double line. The phenotype of the recombinant virus is underlined. The symbol

Regulation of HSV gene expression.

The kinetic class to which a gene belongs can usually be determined from analyses of the patterns of viral protein synthesis under defined experimental conditions. However at times it is usefull to determine the requirements for gene expression and to identify the regulatory domain of the gene by constructing chimeric genes consisting of promoter-regulatory domains of test genes to the coding sequences of an indicator gene. In HSV this procedure is particularly useful in cases where the abundance of the gene product is low (15) or where it is desirable to identify precisely the domains of the regulatory seqences. Use of the DNA sequence insertion deletion techniques described above allows to test the functionality of HSV promoters by fusing them to the structural domain of the TK gene and subsequently inserting these chimeric genes into the translated domain of the TK gene in the viral genome. Thus Post et.al. (16) to identify the promoter-regulatory regions of the HSV α genes constructed chimeras consisting of the 5'sequences located upstream of the translated domains of α genes fused to the translated domains of the TK gene which belongs to the β class of genes. These chimeras were then cotransfected with intact viral DNA from a mutant (HSV-1(F) Δ305) that contained a 700 bp deletion in the coding sequences of the TK gene. Recombinants carrying the chimeric genes were selected in the presence of methotrexate. These studies identified the promoter-regulatory domains of the α genes 4, 0, and 27 (16,17). Similar experimental designs led to the identification of the promoter-regulatory domains of γ_1 and γ_2 genes (18,19,20). It is interest to note that although there exist a variety of systems for gene expression that would permit the identification of promoter domains, the regulation of viral genes contained in cellular chromosomes may be different from those of genes residing in viral chromosomes. Specifically in the case of the chimeras consisting of the promoter domains of γ_2 genes fused to the coding sequences of the TK gene behave as γ_2 genes in the enviroment of the viral genome and as β genes in the enviroment of the cellular genome (18). Thus meaningfull conclusions regarding the regulation of HSV genes can only

---

delta (Δ) indicates deletion. BudR is 5-bromo-2-deoxyuridine. HAT is hypoxanthine, aminopterin (or methotrexate), and thymidine.

be made in the background of the viral genome.

Expression of foreign genes.

The most recent application of the insertion technology has been in the expression of foreign genes. As was mentioned in the introduction, the unmodified HSV genome can accept up to 9.7 kbp of additional genomic material (8). This upper limit has been expanded to 25 Kbp by the availability of mutants with deletions in non-essential genes. An example of such a virus is the non-inverting mutant I358 in which 15 kbp of sequences were deleted from the L-S junction (9). The only requirement for optimal expression of a non herpesvirus gene in the HSV genome is that it is placed under the control of a HSV promoter. Use of the HSV genome for the purpose of obtaining expression of a particular gene has in some cases specific advantages. For example it can be used to express the genes of a pathogenic agent that is difficult to grow and more importantly hazardous to handle. Such examples may be the members of the Human T Cell Leukemia Virus group, and the hepatitis B virus. In other cases it may be desirable to study the function of specific gene products in cells in which the gene is not naturally expressed. The extensive host range of the HSV genome is particularly useful in this respect. Finally it can be envisioned that in the future when appropriate HSV strains are developed for vaccination purposes, this technology should enable the construction of multivalent vaccines that will afford protection against a variety of pathogens. Our understanding of the structure of the HSV genome allows a great degree of flexibility in designing the most suitable expression system to suit a particular need. This flexibility stems from the fact that there is a large selection of HSV promoter regulatory elements from each of the three kinetic classes that were described in the introduction. This allows the manipulation of the system such that expression is either optimized or subject to specific controls, or both depending on the requirements. One effective way to maximize the yield of a particular foreign gene is to place it under the control of the α 4 gene promoter and to subsequently insert the chimeric gene into the genome of a virus that carries a temperature sensitive (*ts*) lesion in the α 4 gene. In the absence of a functional α 4 gene product, the infected cells express α genes only and the transition from α to β and γ gene expression does not ensue (7,21). Consequently such *ts* recombinants should express the foreign gene under α control for as long as the host cell survives. An added advantage of this approach is that in addition to maximizing product yield it also minimizes the yield of infectious HSV virus which simplifies the purification of the product.

All of the constructions to date have involved the TK gene as the target for the insertion and as the selectable marker. The construction of these viruses follows the same procedure as detailed earlier. The repertoire of foreign genes expressed to date includes the chicken ovalbumin gene, the hepatitis B virus S gene (22) the Epstein-Barr virus EBNA1 and LYDMA genes (23), and the human T cell leukemia virus III (HTLVIII/LAV) envelope gene.

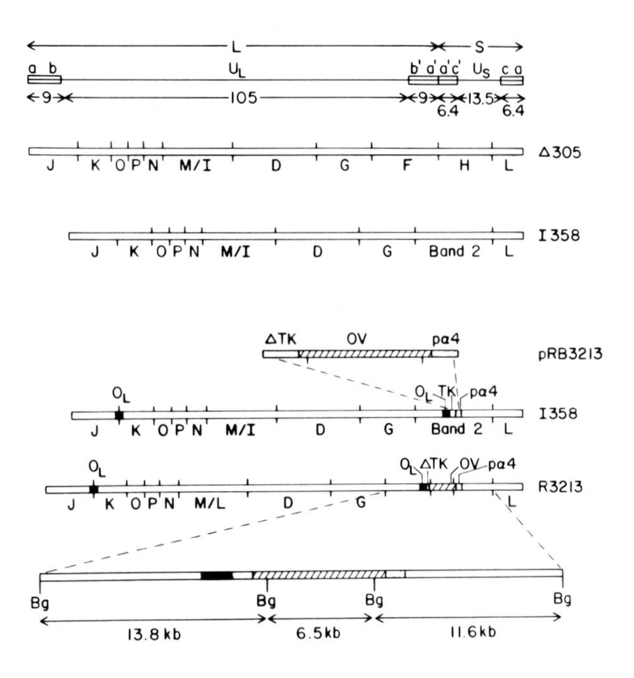

Figure 2. Schematic diagram of the construction and structure of the R3213 recombinant virus. The top line shows the arrangement of the HSV-1 genome. Δ 305 is a TK^- virus that contains a 700bp deletion in the coding sequences of the TK gene and it was the starting point for the construction of the I358 recombinant (ref. 9). The Bgl II restriction enzyme sites are indicated below each construction. Δ TK

The ovalbumin gene was chosen as the model gene in studies pertaining to relative efficiencies of promoters and the fidelity of the regulation conferred upon the foreign gene by the particular HSV promoter. This choice was based on the following considerations: The genomic ovalbumin sequences are contained within a 10 kbp DNA fragment and its domain is much larger than that of most viral genes mapped to date. Unlike that of most HSV genes, the ovalbumin mRNA is spliced. A quantitative immunoassay for its detection entails the use of commercial ovalbumin and anti ovalbumin antibody. Our studies indicate that ovalbumin produced by the recombinant viruses is secreted into the extracellular fluid and is relatively stable. One example of a recombinant expressing the ovalbumin gene is R3213 constructed from the noninverting I358 mutant. The ovalbumin gene in R3213 is under the control of the α4 promoter and is inserted in the L-S junction region. The structure of the R3213 recombinant is illustrated in figure 2. The chimeric α4-ovalbumin gene was regulated as an authentic α gene in as much as its expression did not require the prior expression of other HSV genes. As can be seen in figure 3 treatment of the cells with cycloheximide for 6 hours to block protein synthesis, and then reversing the block in the presence of actinomycin D, to prevent the further synthesis of RNA, resulted in the expression of the ovalbumin gene. Since the I358 recombinant, which is the parental virus to the R3213 recombinant, is ts in the α4 gene, it was possible to ascertain the maximal yield of ovalbumin from the α promoter at the nonpermissive temperature. The results are shown in figure 4. By comparison with the control lanes that contained known amounts of authentic ovalbumin it can be estimated that approximatelly 15 μg of ovalbumin is produced per 4 X 10^6 cells under these conditions. We are currently comparing the relative efficiency of HSV promoters belonging to the various kinetic groups. For these studies we have constructed chimeras consisting of the promoter and transcription initiation sites of α, β γ_1 ,and γ_2 genes to the structural sequences of the ovalbumin gene. These chimeric genes were then inserted in the domain of the TK gene in the HSV genome. Although these studies are not yet complete several features are already apparent. Comparison of the R3213 virus with the

---

represents the deleted TK gene from the Δ 305 virus. OV is the chicken ovalbumin gene. pα4 is the promoter of the α4 gene. Bg denotes a BglII site.

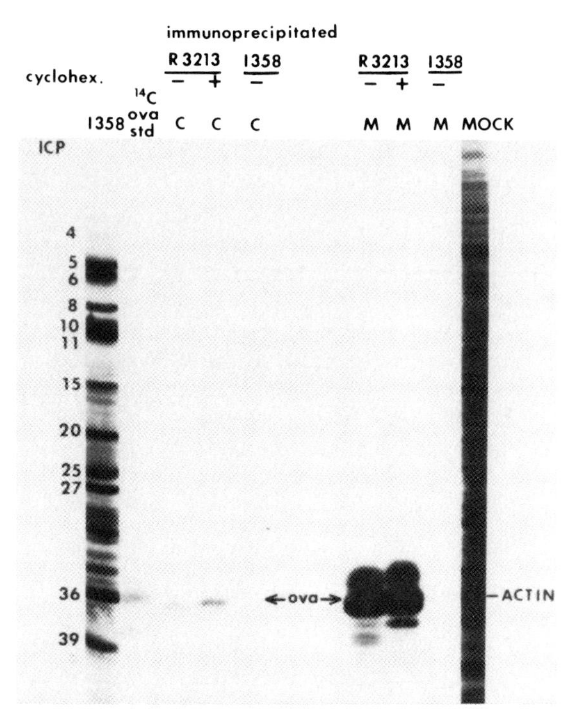

FIGURE 3. Expression of the ovalbumin gene in the R3213 virus. Vero cells were infected with R3213 or I358 virus at a multiplicity of 10 plaque forming units per cell. Infected cell proteins were labeled with ^{35}S methionine immediatelly following virus adsorption untill 18 hours post infection. In the cases were cycloheximide was used the cells were pretreated with the drug for one hour prior to exposure to the virus. The drug was present during the virus adsorption and was maintained untill 6 hours post infection when it was removed and actinomycin D was added to prevent further mRNA synthesis. Infected cell proteins from cycloheximide treated cells were labeled immediatelly after cycloheximide removal for 2 hours. The extracellular medium was collected separatelly from the cells. Ovalbumin expression was assayed by immunoprecipitation with rabbit antiserum to ovalbumin. ICP refers to HSV-1 infected cell proteins and were used as molecular weight markers. C designates cell fraction and M the extracellular medium fraction. ^{14}C labeled ovalbumin was obtained commercially and used as a standard.

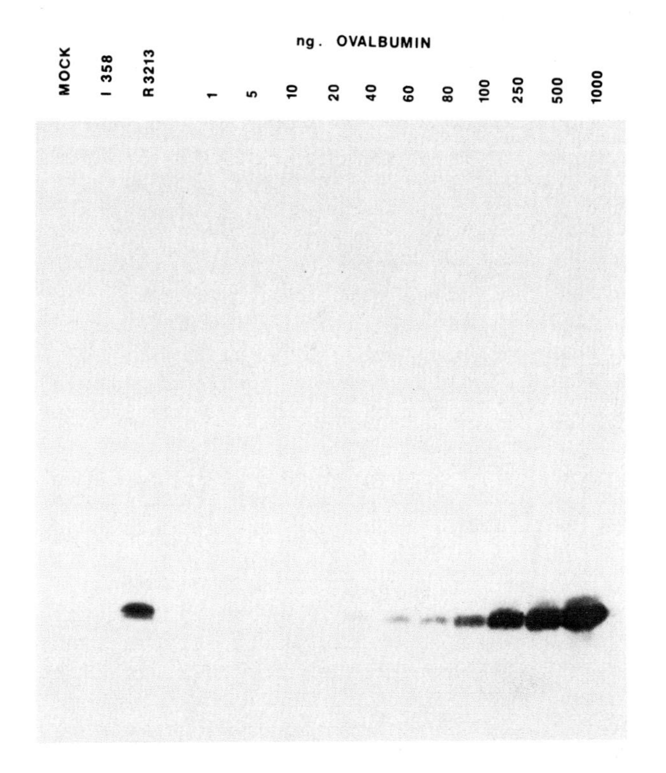

Figure 4. Quantitation of ovalbumin produced by the R3213 virus by immunoblotting. $4X10^6$ Vero cells were infected at 39.5°C with 10 plaque forming units per cell of R3213 virus. The virus infection was allowed to proceed at 39.5°C for 24 hours when the extracellular media were harvested. A portion representing 1/100 of the media were run along with known amounts of authentic commercially obtained ovalbumin on a 9.3% polyacrylamide gel. The separated proteins were transfered electrically to nitrocellulose and blotted with rabbit anti-ovalbumin sera. The reaction was detected with anti-rabbit coupled peroxidase and the amounts shown represent absolute amounts of protein.

virus that has the α 4-ovalbumin inserted in the TK domain shows no apparent difference in the ovalbumin yields. This indicates that the relative expression of a foreign gene is not influenced by the genomic location of the insertion. It is also apparent at this time that the α promoters are stronger than the β promoters. One particularly interesting observation is that a recombinant virus in which the ovalbumin gene is placed under the control of the HSV glycoprotein D (gD) γ_1 gene promoter did not express the ovalbumin gene. It has been previously suggested that the gD gene is regulated both at the transcriptional and translational levels (24). The gD-ovalbumin chimera comprises both the promoter and the transcription initiation site of gD but with very little of the 5' transcribed non translated sequences. The same promoter was used in the studies of Everett (19,20) were it was fused to the rabbit β globin gene. In those studies (19,20) the activity of the gD promoter was assesed only at the transcriptional level. It was found that the sequences tested possesed promoter activity since β globin transcripts were detected and shown to be initiated at the predicted locations in the gD sequences. Preliminary results of our studies indicate that the gD regulated ovalbumin is transcribed but not translated. Studies on the expresion of gD-ovalbumin gene may identify the specific DNA sequence responsible for the post-trascriptional regulation of the authentic HSV gene.

In the case of the hepatitis B virus S gene, expression was obtained by fusion of the S gene structural domain to α and β promoters (22). In each instance the chimeric gene exhibited regulated expression characteristic of the promoter used. The gene product was aggregated into the typical coreless Dane particles commonly seen in the sera of infected individuals. The particles were excreted and accumulated in the extracellular fluid and accumulated.

The Epstein-Barr virus (EBV) EBNA1 and LYDMA genes were expressed from the α 4 promoter and the level of production was fifty to one hundred fold higher than is ordinarily expressed in human lymphocytes carrying the EBV genome (23). Expression of the EBV genes had no effect on the replication of the HSV vector genome and helped establish the functional domain of the EBNA1 gene.

Expression of the HTLVIII/LAV envelope gene (<u>env</u>) was a colaborative effort involving the laboratories of Drs. William Haseltine and Robert Gallo. In this instance the the structural domain of the <u>env</u> gene was fused to the α 4 promoter. The product reacts with monoclonal antibody to the <u>env</u> gene and it appears that it is correctly processed in that both the 120,000 and the 41,000 molecular weight processed forms are detected (unpublished data). The HSV vector system may provide a good source of the HTLVIII/LAV glycoprotein in a native conformation.

CONCLUSIONS

The procedures initially designed to probe the function of specific sequences within the HSV genome have been used to insert and express foreign genes within the genome. The potential use of the HSV as a vector of foreign genes rests on the accumulated experience that (a) the timing and abundance of expression of the foreign gene can be regulated, (b) genetic engineering of the HSV genome should yield genomes suitable for optimal production of the non HSV gene product in cell culture or for human immunization against both HSV and the infectious agents whose genes have been inserted into the HSV vector.

ACKNOWLEDGEMENTS

These studies were supported by grants from the United Public Health Service (CA 08494 and CA 19264) from the National Cancer Institute and by grant MV-2T from the American Cancer Society. K.P was a predoctoral trainee (PHS-5-T32-GM07183-08) and M.A was a postdoctoral fellow of the Damon Runyon - Walter Winchell Cancer Fund.

REFERENCES

1. Kieff ED, Bachenheimer SL, Roizman B (1971). Size, composition and structure of the deoxyribonucleic acid of herpes simplex virus subtypes 1 and 2. J Virol 8: 125.
2. Sheldrick P, Berthelot N (1975). Inverted repetitions in the chromosome of herpes simplex virus. Cold Spring Harbor Symp Quant Biol 39: 667.
3. Wadsworth R, Jacob RJ, Roizman B (1975). Anatomy of herpes simplex virus DNA. II. Size, composition, and arrangement of inverted terminal repetitions. J Virol 15: 1487.
4. Hayward GS, Jacob RJ, Wadsworth SC, Roizman B (1975). Anatomy of the herpes simplex virus DNA: evidence for four populations of molecules that differ in the relative orientations of their long and short segments. Proc Natl Acad Sci U S A 72: 4243.
5. Delius H, Clements JB (1976). A partial denaturation map of herpes simplex virus type 1 DNA: evidence for inversions of the unique DNA regions. J Gen Virol 33: 125.

6. Honess RW, Roizman B (1974). Regulation of herpesvirus macromolecular synthesis. I. Cascade regulation of the synthesis of three groups of viral proteins. J Virol 14: 8.
7. Knipe DM, Ruyechan WT, Roizman B, Halliburton IW (1978). Molecular genetics of herpes simplex virus: demonstration of regions of obligatory and non obligatory identity in diploid regions of the genome by sequence replacement and insertion. Proc Natl Acad Sci U S A 75: 3896.
8. Jenkins FJ, Casadaban MJ, Roizman B (1985). Application of the mini-Mu-phage for target-sequence-specific insertional mutagenesis of the hrpes simplex virus genome. Proc Natl Acad Sci U S A 82: 4773.
9. Poffenberger KL, Tabares E, Roizman B (1983). Characterization of a viable noninverting herpes simplex virus 1 genome derived by insertion and deletion of sequences at the junction of components L and S. Proc Natl Acad Sci U S A 80: 2690.
10. Post LE, Roizman B (1981). A generalized technique for deletion of specific genes in large genomes: α 22 gene of herpes simplex virus 1 is not essential for growth. Cell 25: 227.
11. Mocarski ES, Post LE, Roizman B (1980). Molecular engineering of the herpes simplex virus genome: insertion of a second L-S junction into the genome causes additional genome inversions. Cell 22: 243.
12. Mocarski ES, Roizman B (1981). Site specific inversion sequence of the herpes simplex virus genome: domain and structural features. Proc Natl Acad Sci U S A 78: 7047.
13. Mocarski ES, Roizman B (1982). Structure and role of herpes simplex virus DNA termini in inversion, circularization and generation of virion DNA. Cell 31: 89.
14. Chou J, Roizman B (1985). Isomerization of the herpes simplex virus 1 genome: Identification of the cis-acting and recombination sites within the domain of the a sequence. Cell 41: 803.
15. Ackermann M, Braun DK, Pereira L, Roizman B (1984). Characterization of herpes simplex virus 1 α proteins 0, 4, and 27 with monoclonal antibodies. J Virol 52: 108.
16. Post LE, Mackem S, Roizman B (1981). Regulation of α genes of herpes simplex virus: expression of chimeric genes produced by fusion of thymidine kinase with α gene promoters. Cell 24: 555.
17. Mackem S, Roizman B (1982). Structural features of the herpes simplex virus α gene 4, 0, and 27 promoter-regulatory sequences which confer α regulation on chimeric thymidine kinase genes. J Virol 44: 939.

18. Silver SS, Roizman B (1985). γ_2 thymidine kinase chimeras are identically transcribed but regulated as γ_2 genes in herpes simplex virus genomes and as β genes in cell genomes. Mol Cell Biol 5: 518. 44: 939.
19. Everett RD (1984). A detailed analysis of an HSV-1 early promoter: sequences involved in trans-activation by viral immediate-early gene products are not early-gene specific. Nucl Acids Res 12: 3037.
20. Everett RD (1983). DNA sequence elements required for regulated expression of the HSV-1 glycoprotein D gene lie within 83 bp of the RNA capsites. Nucl Acids Res 11: 6647.
21. Dixon RAF, Schaffer PA (1980). Fine-structure mapping and functional analysis of temperature-sensitive mutants in the gene encoding the herpes simplex virus type 1 immediate early protein VP175. J Virol 36: 189.
22. Shih M-F, Arsenakis M, Tiollais P, Roizman B (1984). Expression of hepatitis B virus S gene by herpes simplex virus type 1 vectors carrying α and β regulated gene chimeras. Proc Natl Acad Sci U S A 81: 5867.
23. Hummel M, Arsenakis M, Marchini A, Lee L, Roizman B, Kieff E (1986). Herpes simplex virus expressing Epstein-Barr virus nuclear antigen 1. Virol 148: 337.
24. Johnson DC, Spear PG (1984). Evidence for translational regulation of herpes simplex virus type 1 gD expression. J Virol 51: 389.

Viruses and Human Cancer, pages 443–450

THE THYMIDYLATE SYNTHASE GENE OF HERPESVIRUS SAIMIRI

Hans Helmut Niller, Nikolaus Nitsche, Rüdiger Rüger, Iris Puchtler, Walter Bodemer, and Bernhard Fleckenstein

Institut für Klinische Virologie,
Universität Erlangen-Nürnberg, Loschgestrasse 7,
D-8520 Erlangen

ABSTRACT *Herpesvirus* (*H.*) *saimiri* is an oncogenic agent of New World primates. The virus codes, unlike most other herpesviruses, for a thymidylate synthase (TS). The enzyme protein of 33.5 K shares extensive amino acid sequence homology with TS of human cells and various bacteria. The viral TS is 70% identical in amino acids with the human enzyme. The TS gene of *H. saimiri* is unusual in structure and in regulation of transcription and translation. It is transcribed into large amounts of a late non-spliced mRNA of 2190 nucleotides (without poly(a)-tail). The untranslated sequences are not homologous with any other known TS genes; the 5'-untranslated mRNA is manyfold interrupted by ATG and stop codons in sequence. The TS-gene of *H. saimiri* appears to be acquired from the host cell genome by an ancestral herpesvirus.

INTRODUCTION

Herpesvirus (*H.*) *saimiri* is a common virus of squirrel monkeys (*Saimiri sciureus*) that seems not to be pathogenic in the natural host. Experimental infection of numerous other New World primate species and rabbits results in rapidly progressing malignant T-cell lymphomas and acute lymphocytic leukemias (1, 2). The virus is also capable of transforming peripheral monkey lymphocytes *in vitro* to continuously growing T-cell lines (3). Lymphoid cell lines derived from tumors and *in vitro*-transformed cells contain multiple copies of non-integrated covalently closed circular viral DNA molecules (4, 5). The virion genome, in contrast, is linear double-stranded DNA of about 160 kb; it consists of an internal 112 kb L-DNA segment (36% GC content) that is flanked at both ends by multiple tandem repeats (H-DNA; 1.44 kb, 70.8% GC content) (6, 7, 8). The circular viral genomes of T-lymphoma cells always contain many copies of H-DNA repeats; however large parts of L-DNA can be missing (3, 4, 9). Only the right and left terminal segments of L-DNA appear always to be preserved. A single *H. saimiri*-specific mRNA of 2.5 kb could be identified in cultured lymphoma cells (10). We found the same mRNA transcribed in lytically infected owl monkey kidney cell cultures. While it had appeared scarcely formed at the initial (immediate early = IE) and the early period of viral replication, this mRNA is synthesized in excessive abundancy during late phase. Here we describe that it encodes a virus-specific thymidylate synthase (TS). The TS gene is not only remarkable in its regulation; it is also unusual in the structure of the transcription unit; and the evolution of the gene within the herpesvirus group deserves particular interest.

RESULTS

1. Primary Structure of the Gene.

The entire TS gene of *H. saimiri* is contained within four adjacent *Hind*III fragments (about 3.3kb) (Fig. 1). The nucleotide sequence was determined by the chemical degradation procedure (11), complemented by the dideoxynucleotide chain termination method (12). Reading frame analysis indicated a single open reading frame of 882 nucleotides, that can form a polypeptide of 33.5 K. No other reading frame for a polypeptide ≧ 45 amino acids was found in the direction of transcription. This was consistent with *in vitro*-translation experiments using hybrid-selected *in vitro*-translation. If mRNA from infected culture cells was hybridized with cloned DNA of the four *Hind*III fragments, eluted from solid phase, and translated in a rabbit reticulocyte lysate, a single virus-specific polypeptide of about 33K appeared in polyacrylamide gels.

A computer search for sequences homologous to this translational reading frame in the actual EMBL and NIH-GenBank data bases and in the NBRF Georgetown University amino acid sequence data bank indicated striking homologies with the thymidylate synthases of *Escherichia coli* and *Lactobacillus casei*. The homology was even more significant, if the

H. saimiri gene was compared with TS DNA of human cells (13). The predicated polypeptide encoded by the H. saimiri gene was found about 70% homologous to human thymidylate synthase, without a single amino acid inserted or deleted within the entire length of colinear sequences. This led us to the hypothesis that the H. saimiri gene expressed in transformed cells and excessively transcribed late in replication codes for a functional viral thymidylate synthase of 33.5 K.

2. Transcription.

Transcription initiation and polyadenylation sites of the putative TS-gene of H. saimiri were localized by nuclease protection analyses. Poly(A)$^+$ RNA was extracted from owl monkey kidney cells 40 hours after infection and was hybridized with appropriated end-labelled restriction fragments. The DNA/RNA hybrids were treated with nuclease S1 or exonuclease VII prior to sizing in agarose gels or denaturing polyacrylamide gel systems. The precise 5'-terminus of viral mRNA was determined by primer extension experiments using a synthetic oligonucleotide of 30 residues. The experiments altogether showed that the TS mRNA of H. saimiri is a non-spliced transcript of 2187 nucleotides, not taking in account the 3'-poly(A)-tail.

The most striking structural feature of H. saimiri TS mRNA is an extremely long 5'-untranslated leader sequence of 1207 nucleotides (Fig. 1). The sequence is very high in average AT composition (76%), significantly higher than the protein coding sequence (62%) and the 3' non-translated sequence (69%). The 5'-ut sequence of the H. saimiri TS mRNA contains 22 ATG triplets that are shortly followed by stop codons. The sequence conservation between the TS gene of H. saimiri and its prokaryotic and eukaryotic counterparts was found confined to the respective open translational reading frames for the enzyme. DNA homologies were not recognized in the untranslated regions of the transcription units or in the transcriptional regulatory sequences of the various thymidylate synthase genes.

Initial studies had indicated that the TS gene is weakly transcribed, if first viral protein synthesis is blocked by cycloheximide, suggesting that the TS gene could by a type of immediate early transcription unit (14). In order to do a first functional analysis of the TS promoter, a fusion gene was constructed consisting of the TS promoter with upstream sequences (Hind III fragment of 786 nt in Fig. 1) and a bacterial chloramphenicol acetyl transferase (CAT) gene. Transient expression experiments with owl monkey kidney cell cultures did not result in appreciable CAT activity, indicating that the TS promoter is not constitutively active. This could indicate that the observed IE expression was due to infection conditions as the TS promoter activity depends on a virion component.

3. Evidence for a Functional Thymidylate Synthase.

Studies of the structure and function of the thymidylate synthase of *Lactobacillus casei* have revealed a number of properties of the reaction intermediates that appear common to all such enzymes and that provide a sensitive and specific method to identify thymidylate synthases. 5'-fluoro-2-deoxyuridine monophosphate (FdUMP) is an inhibitor of TS. It forms a binary complex with the protein, binding to the active site of the enzyme. In the presence of 5,10-methylenetetrahydrofolate (5,10-MeTHF) this is converted to a covalent ternary complex. These complexes are stable to boiling in Na-dodecylsulfate, and ^{32}P labelling of FdUMP allows tracing of the complexes in cell extracts and detection by immunoprecipitation. Applying this method, Honess and coworkers (15) detected FdUMP-binding activity in extracts from lytically owl monkey kidney cells. Complexes could be precipitated by antisera against viral proteins of about 30K. These could be discriminated from cellular thymidylate synthase by their migration velocity in gels, as the 19 N-terminal amino acids of the cellular TS-protein are missing in the viral enzyme. Formation of ternary complexes with ^{32}P-FdUMP and 5,10-MeTHF could also be obtained, if poly(A)$^+$ RNA from infected cells was translated *in vitro* with rabbit reticulocyte lysates. This was taken as the direct demonstration of thymidylate synthase activity exerted by the 33.5 K *H. saimiri* polypeptide.

4. Homologous genes in other herpesviruses.

H. ateles is a virus related with *H. saimiri* in genome structure and oncogenic properties (16). Hybridization of labelled TS DNA probes with *Herpesvirus* (*H.*) *ateles* DNA indicated the presence of an equivalent gene in colinear organization. This was confirmed by sequencing of the TS translational reading frame of *H. ateles* which has 83% nucleotide sequence homology with *H. saimiri*. *Herpesvirus* (*H.*) *aotus* type 2 is another virus of New World primates with similar genomic organization (17); extracts of *H. aotus* type 2 infected cells did also reveal thymidylate synthase activity (Honess, personal communication). Epstein-Barr virus, a more distantly related human lymphotropic herpesvirus ("gamma-herpesvirus") does not reveal nucleotide sequence homology within the entire length of its genome in a computer search. Also, no TS activity was found in cells productively infected by herpes simplex virus, pseudorabies virus, and human cytomegalovirus. In contrast, the sequence of the varizella zoster virus genome ("alpha-herpesvirus") contains a highly conserved homologous to the TS gene sequence of H. saimiri, and enzyme-substrate complexes with virus-specific TS polypeptide are readily detected in virus-infected cells (15).

DISCUSSION

Herpesvirus (H.) saimiri, an oncogenic agent of New World primates, encodes a thymidylate synthase (TS) of 33.5 K that has striking amino acid sequence homologies with TS proteins from human cells and various prokaryotic organisms. The TS gene of H. saimiri is efficiently transcribed late in virus replication; TS mRNA is the most abundant late transcript of infected cells. The 2.5 kb mRNA is a non-spliced transcript with a remarkably long 5'-untranslated sequence (1207 nt) that is manyfold interrupted by sequential ATG and translational stop codons.

Herpesviruses with high GC content of their protein-coding DNA such as herpes simplex virus, pseudorabies virus, human cytomegalovirus, and Epstein-Barr virus appear not to possess a TS gene. Varizella zoster virus, on the other hand, which has a low GC genome does encode a TS protein sharing amino acid sequence homology with the thymidylate synthase of H. saimiri and human cells; and a TS sequence was also found in the close relatives of H. saimiri, in H. ateles, and H. aotus type 2. Thus, TS genes are only found in viruses with low GC genomes, and there is no correlation with the taxonomic subgroups of herpesviruses. This raises the question concerning the evolutionary origin of TS in herpesvirus genomes. It could be hypothesized that an ancestral progenitor herpesvirus had a functional viral TS gene. The gene may have been independently in different herpesvirus subgroups, parallel to the evolution of some of their members to high GC content. On the other hand, it may also be plausible that a few herpesviruses acquired their TS genes independently from the host genome; the majority of herpesviruses in all subgroups seem not to have the gene. The idea of gene acquisi-

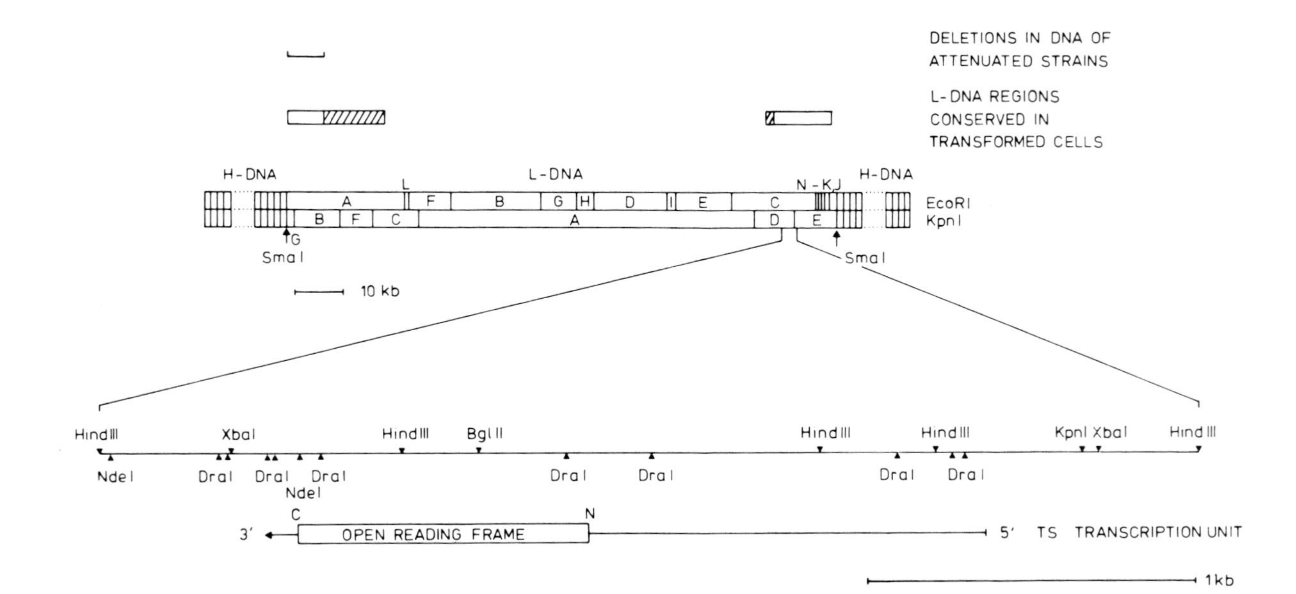

FIGURE 1. Structure of the *H. saimiri* genome and organization of the thymidylate synthase gene.

tion may also be supported by the discoordinate gene expression of the TS gene in virus replication and the loss of 19 N-terminal amino acids in the *H. saimiri* TS. The bizarre position of the translational reading frame in TS mRNA with an extremely long 5'-untranslated sequence could be due to the transposition of a cellular TS coding sequence into an unrelated viral sequence environment. The H. saimiri TS DNA is a late, non-structural gene out of control. It may be the first traceable example of cellular gene acquisition by a DNA virus.

ACKNOWLEDGMENTS

The work was supported by Deutsche Forschungsgemeinschaft.

REFERENCES

1. Fleckenstein B, Desrosiers RC (1982). *Herpesvirus saimiri* and *herpesvirus ateles*. In Roizman B (ed): "The Herpesviruses, Comprehensive Virology", New York: Plenum Publishing Corp., p 253.
2. Desrosiers RC, Fleckenstein B (1983). Oncogenic Transformation by *Herpesvirus saimiri*. In Klein G. (ed): "Advances in Viral Oncology 3", New York: Raven Press, p 307.
3. Schirm S, Weber F, Schaffner W, Fleckenstein B (1985). A transcription enhancer in the *Herpesvirus saimiri* genome. EMBO J 4:2669.
4. Werner FJ, Bornkamm GW, Fleckenstein B (1977). Episomal viral DNA in a *Herpesvirus saimiri*-transformed lymphoid cell line. J Virol 22/3:794.
5. Kaschka-Dierich C, Werner FJ, Bauer I, Fleckenstein B (1982). Structure of non-integrated, circular *Herpesvirus saimiri* and *Herpesvirus ateles* genomes in tumor cell lines and *in vitro*-transformed cells. J Virol 44/1:295.
6. Bornkamm GW, Delius H, Fleckenstein B, Werner FJ, Mulder C (1976). Structure of *Herpesvirus saimiri* Genomes: Arrangement of Heavy and Light Sequences in the M Genome. J Virol 19/1:154.
7. Knust E, Schirm S, Dietrich W, Bodemer W, Kolb E, Fleckenstein B (1983). Cloning of *Herpesvirus saimiri* DNA fragments representing the entire L-region of the genome. Gene 25:281.

8. Bankier AT, Dietrich W, Baer R, Barrell BG, Colbere-Garapin F, Fleckenstein B, Bodemer W (1985). Terminal repetitive sequences in Herpesvirus saimiri virion DNA. J Virol 55:133.
9. Desrosiers RC (1981). Herpesvirus saimiri DNA in tumor cell-deleted sequences and sequences rearrangements. J Virol 39:497.
10. Knust E, Dietrich W, Fleckenstein B, Bodemer W (1983). Virus-specific transcription in a Herpesvirus saimiri-transformed lymphoid tumor cell line. J Virol 48/2:377.
11. Maxam AM, Gilbert W (1980). Sequencing end-labelled DNA with base-specific chemical cleavages. Methods Enzymol 65:499.
12. Sanger F, Nicklen S, Coulson AR (1977). DNA sequencing with chain terminating inhibitors. Proc Natl Acad Sci 74:5463.
13. Takeishi K, Kaneda S, Ayusawa D, Shimizu K, Gotoh O, Seno T (1985). Nucleotide sequence of a functional cDNA for human thymidylate synthase. Nucl Acid Res 13:2035.
14. Bodemer W, Knust E, Angermüller S, Fleckenstein B (1984). Immediate-Early Transcription of Herpesvirus saimiri. J Virol 51/2:452.
15. Honess RW, Bodemer W, Cameron KR, Niller HH, Fleckenstein B, Randall RE (1986). The A+T-rich genome of Herpesvirus saimiri contains a highly conserved gene for thymidylate synthase. Proc.Natl. Acad.Sci. USA 83:in press.
16. Fleckenstein B, Bornkamm GW, Mulder C, Werner FJ, Daniel MD, Falk LA, Delius H (1978). Herpesvirus ateles DNA and Its Homology with Herpesvirus saimiri Nucleid Acid. J Virol 25/1:361.
17. Fuchs PG, Rüger R, Pfister H, Fleckenstein B (1985). Genome organization of Herpesvirus aotus type 2. J Virol 53:13.

Viruses and Human Cancer, pages 451–466
Published 1987 by Alan R. Liss, Inc.

Three Retroviruses Infecting Macaques at the New England Regional Primate Research Center

Ronald C. Desrosiers, Norman L. Letvin, Norval W. King, Ronald D. Hunt, Beverly J. Blake, Larry O. Arthur[1], and M.D. Daniel

New England Regional Primate Research Center
Harvard Medical School
Southboro, MA 01772

[1]Program Resources, Incorporated
NCI-Frederick Cancer Research Facility
Frederick, MD 21701

The New England Regional Primate Research Center (NERPRC) cares for approximately 1500 monkeys at any one time, nearly half of which are macaques (genus *Macaca*), an Asian primate. The NERPRC macaque colony is composed primarily of three species: *Macaca mulatta* (rhesus monkey), *Macaca fascicularis* (cynomolgus monkey), and *Macaca cyclopis* (Taiwanese rock macaque). Around 1980, NERPRC veterinarians began to notice an unusual number of deaths in the macaque colony, especially among *M. cyclopis*. This trend continued in 1981. A retrospective examination of necropsy records from 1978 through 1982 yielded somewhat startling results (1). The mortality rate of *M. cyclopis* leaped from 8.0% in 1978 and 13.3% in 1979 to fully one-third, 33.8%, of this group in 1980 and maintained nearly the same rate, 29.0%, in 1981. Increased mortality was also observed in the other two macaque species in 1980, but it was not nearly as striking as in *M. cyclopis*. Much of this increased mortality appeared to be due to a chronic wasting syndrome. This disease pattern has since been better characterized as an acquired immune deficiency syndrome of macaque monkeys (1,2); some of the salient features of this disease are summarized in Table 1.

Table 1.
FEATURES OF SPONTANEOUS AIDS-LIKE DISEASE IN NERPRC MACAQUES

Diarrhea
Wasting
Immunologic Abnormalities
Opportunistic Infections
Lymphadenopathy
Lymphoproliferative Disorders and Lymphomas
Transmissibility

The time-space clustering of deaths due to this wasting syndrome occurred on a relatively large scale in 1980-1981. Since that time, however, only sporadic cases of what appears to be the same disease have been observed at a rate of approximately six per year. Here we will discuss the potential contributions of three different retroviruses to spontaneous disease in our macaque colony, and we also describe some of the biological and molecular properties of these viruses. The three viruses are: retrovirus D/New England, simian T-lymphotropic virus type I (STLV-I), and simian T-lymphotropic virus type III (STLV-III). STLV-I and STLV-III are named after their human counterparts HTLV-I, the cause of adult T cell leukemia/lymphoma and HTLV-III/LAV, the cause of AIDS.

The original isolation of type D retrovirus from macaques was first described by Chopra and Mason in 1970 from a spontaneous mammary tumor of a rhesus monkey (3). Except for a brief report of two isolates in 1974 which were not further characterized (4), the frequent isolation of type D retrovirus as a common infectious agent of macaques has been achieved only recently. At NERPRC, type D retrovirus was first isolated from macaques with the immunodeficiency syndrome in 1983 and procedures for efficient isolation and growth of the virus were developed (5). Cells infected with retrovirus D/New England contain intracytoplasmic A particles and virus buds from the cell membrane with a complete nucleoid. Extracellular mature particles have a central nucleoid, lack prominent surface projections and, when sectioned in the appropriate plane, have a characteristic cylindrical or bullet-shaped nucleoid (5). These morphologic features are indistinguishable from other type D retroviruses and, in fact, define this group.

Similar related viruses have also been isolated from macaques at the California (6), Seattle, Washington (7) and Oregon (8) Regional Primate Research Centers.

We have now isolated type D virus from over 30 macaques of the NERPRC colony. The correlations observed after the first 17 isolations (9) have been maintained throughout subsequent isolations. The type D virus has never been isolated from peripheral blood lymphocytes (PBL) of clinically healthy macaques in spite of more than 100 attempts. Type D virus has been isolated from PBL of virtually every macaque with chronic wasting and immunodeficiency. Of animals from which D/New England was isolated, those severely ill with wasting-immunodeficiency died within the first few weeks or months following virus isolation; others, much less ill at the time of D/New England isolation, have since recovered and remained well for longer than one year (9). D/New England has not been re-isolated from PBL following recovery from illness. This suggests that these animals have recovered from primary D/New England infection and are no longer viremic.

D/New England has been molecularly cloned and its DNA has been compared with that of Mason-Pfizer monkey virus (MPMV), the prototype D retrovirus of macaques (9). MPMV DNA was cloned by Chris Barker and Eric Hunter of the University of Alabama. Comparison of these two related viruses revealed that only about 50% of restriction sites were conserved, which corresponds to sequence homology of about 92%. Similar to other retroviruses, the greatest divergence was observed in the envelope region. There is no homology between type D virus and HTLV-I, II or III.

One interesting feature of the D retroviruses isolated from the NERPRC colony is the lack of strain variability (9). Five isolates from three different species of NERPRC macaques were compared at over 30 restriction sites. Strain differences were <u>not</u> observed. This is a surprising result given the variability that generally exists among different isolates of a virus and the variability observed among type D isolates from different primate research centers. Type D virus may have been introduced into our colony as a rare event followed by its spread through the captive colony. Infection of captive macaques with type D retrovirus appears to be common; however, very little is known about infection of feral macaques. Although type D infection is prevalent at the Seattle, Washington Primate Research Center, groups of feral macaques recently imported into the colony were seronegative (10).

Despite the association of D/New England isolation with disease described above, experimental infection with D/New England induced only lymphadenopathy, neutropenia and a transient decrease in peripheral blood blastogenic responsiveness in macaques greater than 1 month of age; 2 of 6 macaques less than one month of age failed to thrive and died following D/New England infection (11). These results and the high rate of infection with D virus in our colony (see below) raised questions regarding the significance of the disease association. This prompted a search for other agents which might contribute to or be responsible for immunodeficiency disease in the macaque colony and led to the first description of HTLV-III/LAV related viruses in primates (12).

To date, STLV-III has been isolated from four macaques of the NERPRC colony, three of which were involved in a lymphoma transmission study (12). STLV-III was isolated from frozen splenocytes of rhesus monkey Mm 251-79, who died with lymphoma 26 months after inoculation with minced tissue from a spontaenous _M. mulatta_ lymphoma (13). Inoculation of other rhesus with minced tissue from Mm 251-79 resulted, not in lymphoma, but in a fatal AIDS-like disease; STLV-III was isolated from two of these macaques (12). A rhesus monkey, 142-83, from which STLV-III was also isolated, had been born at NERPRC and was not involved in any transmission study; this animal, therefore, must have been infected naturally. Diarrhea, facial rash, generalized lymphadenopathy and splenomegaly were noted grossly at death and histopathological examination of necropsy tissues revealed an extensive lymphoproliferative disease (14).

Subsequent examination of sera revealed Mm 142-83 to be seropositive to STLV-III at six weeks of age while still in the NERPRC nursery. (It should also be noted that Mm 142-83 became infected additionally later in life with type D retrovirus). Although the mother of this animal is alive and healthy 2.5 years later, she is seropositive for STLV-III at this time. This suggests that transmission of virus from mother to offspring may have occurred. These and other examples, to be described in more detail elsewhere, indicate that the macaque isolates of STLV-III are not necessarily pathogenic during natural transmission among macaques. The potential pathogenicity of the macaque isolates of STLV-III may be similar to or quite different from that of similar isolates from African primates; more detailed study will be needed to make these comparisons.

STLV-III has growth properties and T cell tropism similar to HTLV-III (12,15) and a characteristic morphology indistinguishable from HTLV-III/LAV and the other lentiviruses (12). Infected cells have no cytoplasmic type A particles, the virus buds in the manner of the type C viruses and mature, extracellular particles appear to have a rod-shaped, cylindrical nucleoid when sectioned in the appropriate plane.

Importantly, STLV-III has easily demonstrable antigenic relatedness to the human AIDS virus. This has been demonstrated using radioimmune precipitation by Kanki et al (16) and by the use of a *gag* protein radioimmune competition assay (Fig. 1). HTLV-III and STLV-III were tested in both heterologous and homologous competition radioimmunoassays for HTLV-III p24. STLV-III competed only partially in the homologous HTLV-III p24 assay, indicating that core proteins of HTLV-III and STLV-III shared immunoreactive determinants but were immunologically different because of the low level of competition. Common determinants on the major core proteins of the two viruses were demonstrated in the heterologous HTLV-III p24 assay. In this assay, serum from an African green monkey shown to be positive to STLV-III was used to precipitate iodinated HTLV-III p24. When lysed HTLV-III and STLV-III were added as competitors, both competed equally for precipitation of radiolabeled HTLV-III p24 by the African green monkey serum. Type D virus, *Macaca arctoides* endogenous type C virus (MAC-1), HTLV-I, as well as Visna and equine infectious anemia virus gave no detectable competition in either assay. STLV-III is therefore the only other virus that has been found to compete in this assay.

Attempts to determine genetic relatedness between STLV-III and HTLV-III have also been made. DNA from STLV-III infected and uninfected HUT-78 cells and DNA from HTLV-III infected H9 cells were cut with *SstI*, electrophoresed through an agarose gel, transferred to nitrocellulose and hybridized with ^{32}P-labeled 9 kilobasepair (kbp) *Sst* I fragment of the BH10 clone of HTLV-III (Fig. 2). The BH10 clone was provided by F. Wong-Staal and R.C. Gallo (NCI, NIH). Strong hybridization to viral DNA in HTLV-III-infected cells was readily detected. However, even when 3x more DNA from STLV-III-infected cells was used and hybridization was performed under low stringency conditions, significant cross-hybridization to DNA from STLV-III-infected cells was not detected. Long overexposures revealed a weakly-hybridizing fragment of 6 kbp, but

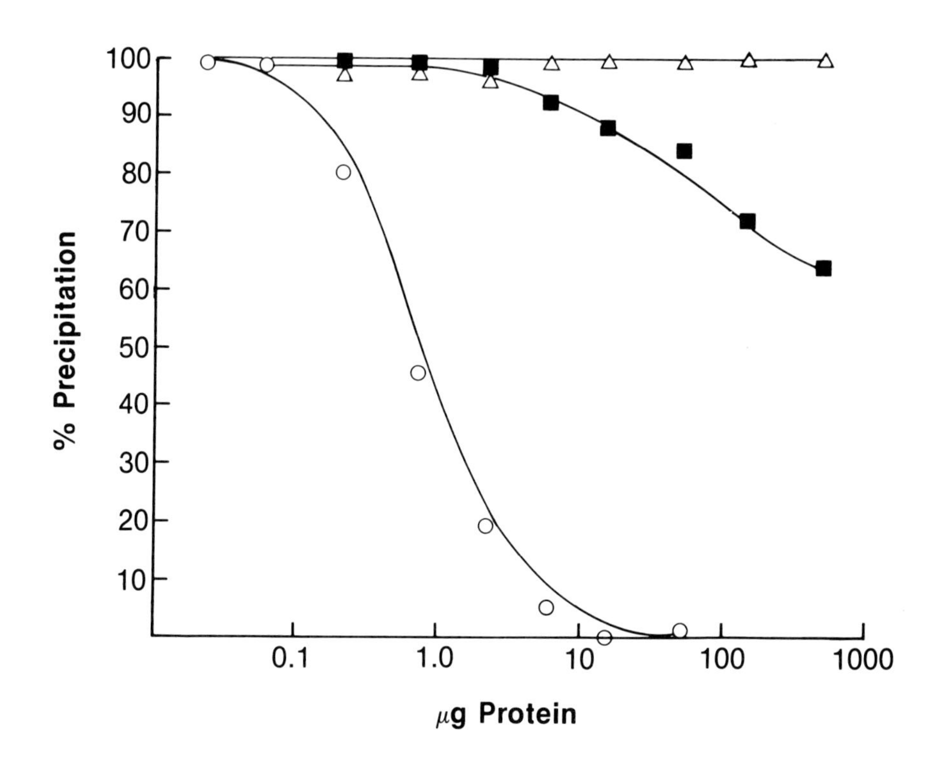

FIGURE 1a. Homologous competition immunoassay for HTLV-III p24. Unlabeled detergent-disrupted purified viruses were tested at serial three-fold dilutions for ability to compete with binding of anti-HTLV-III antisera to ^{125}I-HTLV-III p24. Viruses used in the assay were HTLV-III ( ), STLV-III ( ), and controls ( ): MAC-1, HTLV-I, SAIDS/D California, and uninfected H9 cells.

this was also observed in uninfected cells; this probably reflects some weak homology of the viral sequence to a host cell DNA sequence. These experiments have been repeated many times, using Hirt supernatant, replicative intermediate DNA as well as total cell DNA, with hybridization at moderate as well as low stringency, with the same result each time - no detectable cross-hybridizing sequences. Use of plasmid subclones derived from the gag and pol regions for the hybridization probe did not change the outcome of

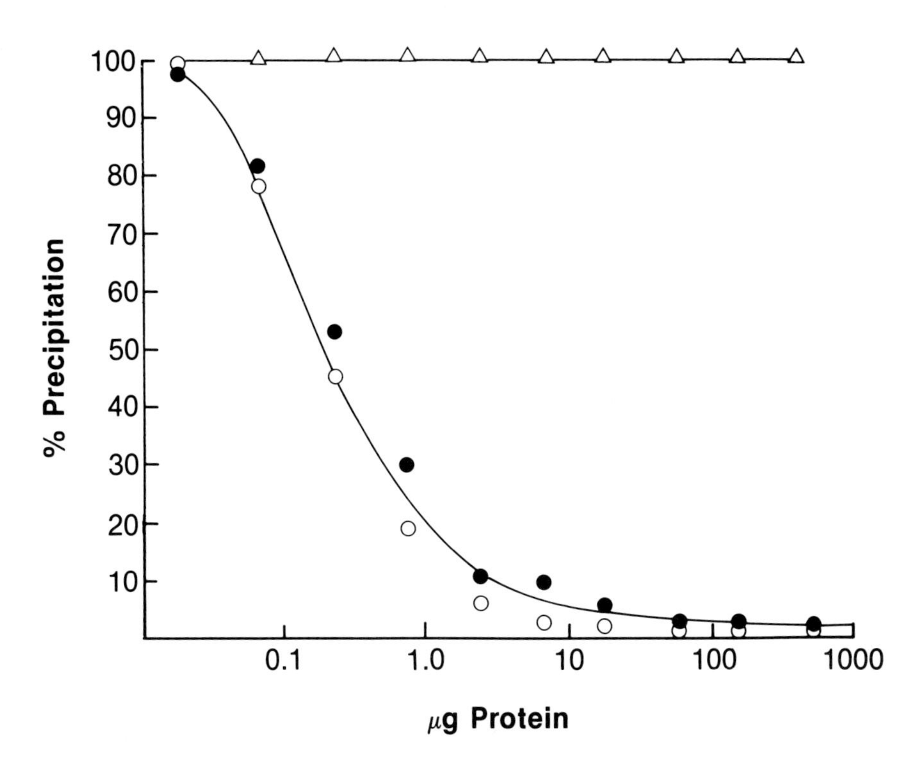

FIGURE 1b. Heterologous competition immunoassay for HTLV-III p24. Unlabeled detergent-disrupted purified viruses were tested at serial three-fold dilutions for ability to compete with binding of African green monkey serum to ^{125}I-HTLV-III p24. Viruses used in the assay were HTLV-III ( ), STLV-III ( ), and controls ( ): MAC-1, HTLV-I, SAIDS/D California, and uninfected H9 cells.

these experiments. Since the antigenic cross-reactivity of STLV-III and HTLV-III/LAV has been demonstrated by several investigators using different experimental approaches, we have to be aware of the limitations of these DNA-DNA cross-hybridization experiments. Even the low stringency conditions (hybridization in 6xSSC, 30% formamide at 37°C with washing in 2xSSC at 55°C) are only about 42° below the Tm of homologous HTLV-III - HTLV-III hybrids. This is only enough to allow for about 28% sequence mismatch. It is

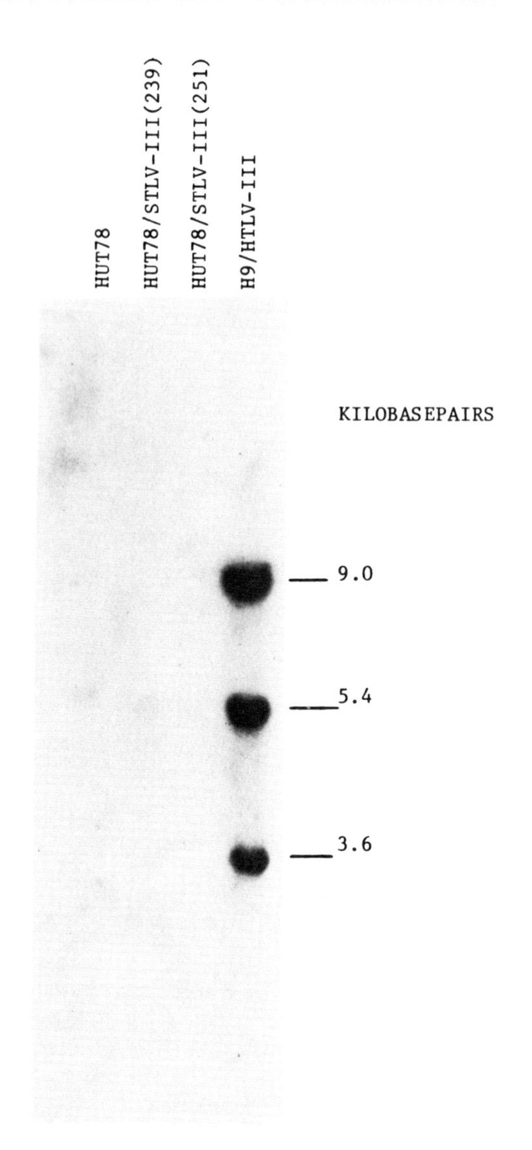

FIGURE 2. Hybridization of cloned HTLV-III DNA with DNA from cells infected with STLV-III. DNA from the indicated cell type was cut with SstI, electrophoresed through a 0.8% agarose gel and transferred to nitrocellulose by the procedure of Southern. The 9.0kbp insert of BH10 was gel purified, labeled with ^{32}P by nick translation and hybridized with the transferred DNA. Hybridization was performed in 6xSCC, 30% formamide at 37°C for 16 hours and the filter was rinsed in 2xSSC at 55°C prior to film exposure. Approximately 3 ug of H9/HTLV-III DNA was used in slot 4 while approximately 10 ug of the other DNAs were used to enhance our ability to detect cross-hybridizing sequences.

thus possible that STLV-III and HTLV-III could have 65-75% sequence homology and this would not have been detected in the hybridization depicted in Fig. 2 described above. We have tried lower hybridization stringencies without positive results, but these experiments are confounded by the high background levels produced by low stringency.

Macaques are Asian primates; the relationship of macaque STLV-III to similar isolates from African primates (mangabeys and African green monkeys) is not clear at this time. The categorization of these isolates awaits molecular cloning and sequence comparison; they may all be different isolates of the same virus or they may group according to continent or species of origin.

Probably the most important feature of the macaque-STLV-III system is the ability to induce an AIDS-like disease with high mortality in a reasonable time frame in a common laboratory primate (17). In these preliminary studies, six young rhesus monkeys received 1.7 ml of undiluted virus intravenously; virus was grown in normal human PBL in the presence of IL-2. All six animals became infected, as evidenced by the ability to repeatedly isolate STLV-III from their PBL in the weeks and months following inoculation. Four of the six animals died with a rather rapid time course, 127-160 days after inoculation. The fifth animal (Mm 74-84) died after a more prolonged period (352 days), while the sixth animal (Mm 127-83) remains healthy longer than 14 months after infection, in spite of the fact that we continue to isolate STLV-III from monthly blood samples. The clinical signs and histopathological findings at necropsy of the five animals that died are consistent with an AIDS-like disease. All five had a persistent diarrhea with chronic wasting evident in three (losses of 13%, 34% and 60% of body weight). Each of the five animals had immunolgical abnormalities, the most prominent being decreases in absolute numbers of circulating T4 helper/inducer cells. Immunologic abnormalities were manifested by the presence of opportunistic infections, which included acute adenoviral pancreatitis (in three animals), trichomoniasis, and candidiasis. Why the unusual acute adenoviral pancreatitis was observed in three of the animals is unclear at this time. Thymic atrophy with lymphocyte depletion was also observed. In addition, characteristic brain lesions described as multifocal perivascular macrophage infiltrates were present in the brains of the first four animals that died. Extensive electron microscopic examination of these brain lesions revealed

STLV-III particles in the macrophages, but not in the surrounding brain cells (17). These brain lesions are probably identical to the brain lesions described in cases of human AIDS as microglial nodules (18). The current use of the term neurotropic for HTLV-III and STLV-III may not be accurate, since no one has yet shown that these viruses are actually present in neural cells.

We have used an ELISA test to monitor serum antibody levels to STLV-III. The six macaques experimentally infected with STLV-III were examined for their antibody response at 85 days after inoculation (Fig. 3). The four macaques that died with a relatively rapid time course had little or no detectable antibodies to STLV-III at that time. The animal that died a longer time after infection (Mm 74-84) developed an intermediate antibody response, and the animal that remains healthy (Mm 127-83) had by far the strongest antibody response to STLV-III. It thus appears that the strength of the antibody response to STLV-III correlates directly with the ability of the animal to survive infection. Antibody synthesis is helper T cell-dependent; the relative strength of the antibody response may simply reflect the rate at which STLV-III is destroying T4 cells or their function.

The cause of lymphoproliferative disorders (LPD) and lymphomas in macaques with wasting - immunodeficiency is also under investigation. Tissue DNAs have been prepared from ten lymphoma - LPD cases and analyzed for the presence of viral DNA by Southern blot hybridization. Cloned probes used in these studies were pMT2 of HTLV-I, pD398 of D/New England and three different Epstein-Barr virus (EBV) DNA probes. Sequence analysis of macaque STLV-I has revealed greater than 90% sequence homology with HTLV-I (19); STLV-I proviral DNA is thus readily detected using an HTLV-I probe. Hybridization with pD398 was performed at high stringency to minimize hybridization with endogenous sequences and to allow detection of new integrated copies. Characteristics of EBV-related viruses of macaques have not been described, but the genetic relatedness of EBV to similar viruses from African green monkeys (20) and baboons (21,22), suggests that such a macaque virus should be readily detected with cloned EBV DNA probes. The results of Southern blot hybridization of five lymphoma DNAs with the ^{32}P-labeled HTLV-I probe are shown in Fig. 4. Under conditions in which DNA from the positive control HTLV-I infected cell line HUT-102 is heavily over-exposed, significant hybridization to the five macaque lymphoma DNAs was

FIGURE 3. Strength of antibody response in six macaques experimentally infected with STLV-III. STLV-III produced in Hut-78 cells was used to coat wells of ELISA plates. Unbound sites were bound with goat BSA. The indicated sera were diluted with 10T heat inactivated goat serum, 0.05% Tween-20 in PBS, and then added to coated wells of the ELISA plate. Following incubation at room temperature for one hour, the wells were washed and then incubated with alkaline phosphatase conjugated antibody to IgG. Following incubation, washing and color development, absorbance at 410 nm was monitored with a Dynatech 600 ELISA reader.

not observed. Note that two of these lymphomas (Mm 251-79 and Mm 142-83) were obtained from animals from which STLV-III was isolated. Absence of STLV-I DNA sequences was also noted in the other five lymphoma - LPD samples. Similarly, EBV DNA sequences and integrated D/New England DNA were not detected in these ten samples. Thus, as with EBV-negative AIDS-associated lymphomas, a causative virus has not been identified.

It must be emphasized that we do not yet know to what extent type D retrovirus and STLV-III contribute to spontaneous disease in the NERPRC macaque colony. One of the reasons for developing ELISA tests is to be able to screen the large numbers of serum samples necessary for an adequate epidemiologic investigation. Serum samples have been obtained recently from our entire macaque colony and determination of antibody prevalence in healthy animals and animals with disease should shed light on this issue. Preliminary results of screening sera from an initial 72 random macaques indicated that 50% were antibody positive for type D and none were STLV-III positive. Also, of the four most recent spontaneous chronic wasting cases, all were type D positive and none were STLV-III positive, both serologically and by virus isolation. A type D viral etiology for at least some of the cases of chronic wasting - immunodeficiency would be consistent with the previously described syndrome induced by experimental infection with MPMV (23,24) and by recent data on type D retrovirus infection at the California Regional Primate Research Center (P. Marx et al, personal communication). However, it should be pointed out that if type D retrovirus is a major cause of the spontaneous chronic wasting - immunodeficiency syndrome at the NERPRC, serious disease is a rare outcome of infection; the vast majority of animals survive D/New England infection and remain healthy.

Although the cause of disease in our colony is an important issue, development of a useful animal model for AIDS research is of even greater importance. Most significantly, the macaque STLV-III system provides an excellent animal model for AIDS. The virus is antigenically cross-reactive with the human AIDS virus and induces an AIDS-like disease in a common laboratory primate in a reasonable time frame. Use of this model will, hopefully, accelerate progress in understanding the genetic determinants of pathogenicity, as well as developing approaches to vaccine development and drug therapy.

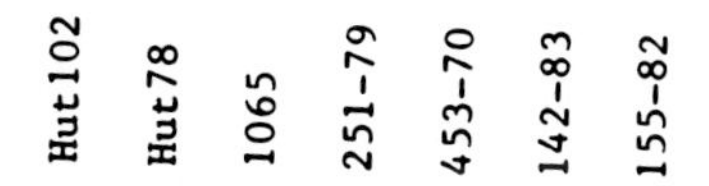

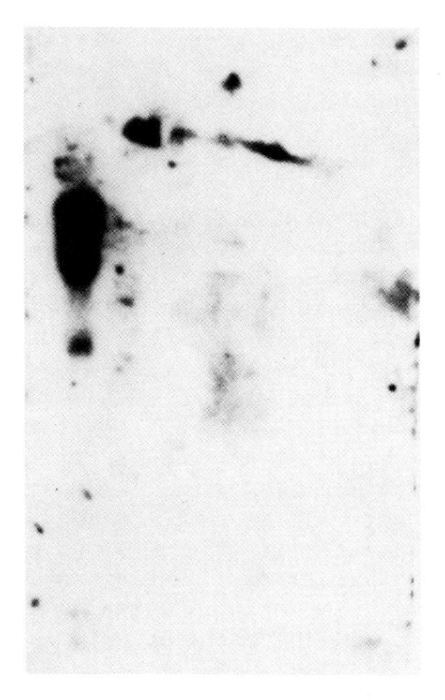

FIGURE 4. Hybridization of cloned HTLV-I DNA to macaque lymphoma DNA. DNA was prepared from the HTLV-I infected cell line Hut-102 as positive control, the uninfected cell line Hut-78 as negative control and the 5 macaque lymphoma samples shown. 5 ug of total cell DNA was cut with EcoRI, electrophoresed through an agarose gel, Southern transferred and hybridized with ^{32}P-labeled pMT2 DNA in 6xSSC, 30% formamide at 37°C. Following overnight hybridization, the filter was rinsed with 2xSSC at 55° prior to exposure to film.

Acknowledgements

This work was supported by Public Health Service grants CA 38205 CA 34949 and AI 20729 from the National Institutes of Health and by grant RR00168 from the Division of Research Resources. Research was also sponsored, at least in part, by the National Cancer Institute, DHHS, under contract N01-C0-23910 with Program Resources, Incorporated. The contents of this publication do not necessarily reflect the views or policies of the DHHS, nor does mention of trade names, commercial products, or organizations imply endorsement by the U.S. Government.

References

1. Letvin NL, Eaton KA, Aldrich WR, Sehgal PK, Blake BJ, Schlossman SF, King NW, Hunt RD (1983). Acquired immunodeficiency syndrome in a colony of macaque monkeys. Proc Natl Acad Sci USA 80:2718.
2. Letvin NL and King NW (1984). Clinical and pathologic features of an acquired immune deficiency syndrome (AIDS) in macaque monkeys. Adv Vet Sci Comp Med 28:237.
3. Chopra HC, and Mason MM (1970). A new virus in a spontaneous mammary tumor of a rhesus monkey. Cancer Res 30:2081.
4. Ahmed M, Schidlovsky G, Korol W, Vidrine G, Cicmanec JL (1974). Occurrence of Mason-Pfizer monkey virus in healthy rhesus monkeys. Cancer Res 34:3504.
5. Daniel MD, King NW, Letvin NL, Hunt RD, Sehgal PK, Desrosiers RC (1984). A new type D retrovirus isolated from macaques with an immunodeficiency syndrome. Science 223:602.
6. Marx PA, Maul DH, Osborn KG, Lerche NW, Moody P, Lowenstine LJ, Henrickson RV, Arthur LO, Gilden RV, Gravell M, London WT, Sever JL, Levy JA, Munn RJ, and Gardner MB (1984). Simian AIDS: isolation of type D retrovirus and transmission of the disease. Science 223:1083.
7. Stromberg K, Benveniste RE, Arthur LO, Rabin H, Giddens WE Jr, Ochs HD, Morton WR, Tsai C-C (1984). Characterization of exogenous type D retrovirus from a fibroma of a macaque with simian AIDS and fibromatosis. Science 224:289.
8. Marx PA, Bryant ML, Osborn KG, Maul DH, Lerche NW, Lowenstine LJ, Kluge JD, Zaiss CP, Henrickson RV,

Shiigi SM, Wilson BJ, Malley A, Olson LC, McNulty WP, Arthur LO, Gilden RV, Barker CS, Hunter E, Munn RJ, Heidecker G, Gardner MB (1985). Isolation of a new serotype of simian acquired immune deficiency syndrome type D retrovirus from Celebes black macaques (Macaca nigra) with immune deficiency and retroperitoneal fibromatosis. J Virol 56:571.

9. Desrosiers RC, Daniel MD, Butler CV, Schmidt DK, Letvin NL, Hunt RD, King NW, Barker CS, Hunter E (1985). Retrovirus D/New England and its relation to Mason-Pfizer monkey virus. J Virol 54:552.
10. Benveniste RE, Stromberg K, Morton WR, Tsai C-C, Giddens WE Jr (1985). Association of retroperitoneal fibromatosis with type D retroviruses. In Salzman L (ed): "Animal Models of Retrovirus Infection," New York: Academic Press. In press.
11. Letvin NL, Daniel MD, Sehgal PK, Chalifoux LV, King NW, Hunt RD, Aldrich WR, Holley K, Schmidt DK, and Desrosiers RC (1984). Experimental infection of rhesus monkeys with type D retrovirus. J Virol 52:683.
12. Daniel MD, Letvin NL, King NW, Kannagi M, Sehgal PK, Hunt RD, Kanki PJ, Essex M, Desrosiers RC (1985). Isolation of a T-cell tropic HTLV-III-like retrovirus from macaques. Science 228:1201.
13. Hunt RD, Blake BJ, Chalifoux LV, Sehgal PK, King NW, Letvin NL (1983). Transmission of naturally occurring lymphoma in macaque monkeys. Proc Natl Acad Sci USA 80:5085.
14. Chalifoux LV, King NW, Daniel MD, Kannagi M, Desrosiers RC, Sehgal PK, Waldron LM, Hunt RD, and Letvin NL. A lymphoproliferative syndrome in an immunodeficient rhesus monkey naturally infected with an HTLV-III-like virus (STLV-III). Lab Invest in press.
15. Kannagi M, Yetz JM, Letvin NL (1985). In vitro growth characteristics of simian T-lymphotropic virus type III. Proc Natl Acad Sci USA 82:7053.
16. Kanki PJ, McLane MF, King NW Jr, Letvin NL, Hunt RD, Sehgal P, Daniel MD, Desrosiers RC, Essex M. (1985). Serologic identification and characterization of a macaque T-lymphotropic retrovirus closely related to HTLV-III. Science 228:1199.
17. Letvin NL, Daniel MD, Sehgal PK, Desrosiers RC, Hunt RD, Waldron LM, MacKey JJ, Schmidt DK, Chalifoux LV and King NW (1985). Induction of AIDS-Like disease in

macaque monkeys with T-cell tropic retrovirus STLV-III. Science 230:71.

18. Shaw GM, Harper ME, Hahn BH, Epstein LG, Gajdusek DC, Price RW, Navia BA, Petito CK, O'Hara CJ, Groopman JE, Cho E-S, Oleske JM, Wong-Staal F, and Gallo RC (1985). HTLV-III infection in brains of children and adults with AIDS encephalopathy. Science 227:177.
19. Watanabe T, Seiki M, Tsujimoto T, Miyoshi H, Hayami M, and Yoshida M (1985). Sequence homology of the simian retrovirus genome with human T-cell leukemia virus type 1. Virology 144:59.
20. Bocker JF, Tiedemann K-H, Bornkamm GW and Zur Hausen H (1980). Characterization of an EBV-like virus from African green monkey lymphoblasts. Virology 101:291.
21. Falk L, Deinhardt F, Nonoyama M, Wolfe L, and Bergholz C (1976). Properties of a baboon lymphotropic herpes-virus related to Epstein-Barr virus. Int J Cancer 18:798.
22. Heller M, and Kieff E (1981). Colinearity between the DNAs of Epstein-Barr Virus and Herpesvirus Papio. J Virol 37:821.
23. Fine DL, Landon JC, Pienta RJ, Kubicek MT, Valerio MG, Loeb WF, and Chopra HC (1975). Responses of infant rhesus monkeys to inoculation with Mason-Pfizer monkey virus materials. J Natl Cancer Inst 54:651.
24. Fine DL (1984). Mason-Pfizer monkey virus and simian AIDS. Lancet i:335.

Viruses and Human Cancer, pages 467–478

ANTISENSE TRANSCRIPTS SYNTHESIZED BY HEPATITIS B VIRUS TRANSFECTED CELLS[1]

Peter M. Price,[2,3] Arthur Zelent,[2] Mary Ann Sells,[2] Judith K. Christman,[2,3] and George Acs[2,3]

Departments of [2]Biochemistry and [3]Pediatrics, Mount Sinai School of Medicine, New York, NY 10029

ABSTRACT A clone of mouse 3T3 cells cotransfected with a plasmid containing a tandem head-to-tail arrangement of hepatitis B virus genomes (pTHBV-1) and DNA with multiple copies of a gene coding for methotrexate resistance was found to contain multiple copies of the intact HBV tandem integrated into the DNA. This clone, C4.10, synthesizes viral surface, core, and "e" antigens in the absence of detectable viral replication. Since all known hepatitis B virus proteins are translated from mRNAs transcribed from the L or (-) strand of the viral genome and since the transcripts of the S or (+) strand apparently carry little coding potential for protein synthesis, it has been generally assumed that the viral S strand does not serve as a template for RNA synthesis. However, polyadenylated RNAs extracted from C4.10 cells contained transcripts of the S polarity DNA strand of hepatitis B virus. Evidence is presented indicating that these transcripts are initiated and terminated within the intact integrated head-to-tail tandem viral genomes. In view of indications that antisense RNAs can regulate gene expression at the translational level, it is possible that this RNA transcribed from the S strand of the hepatitis B genome may have a regulatory function. Additionally, the polyadenylated antisense RNA may code for as yet unidentified intermediates in the life cycle of HBV.

[1]This work was supported by grant number CA34818 from the National Cancer Institute.

INTRODUCTION

The genome of hepatitis B virus is a partially single stranded, circular DNA (Figure 1). The long strand is approximately 3200 bases in length and contains a nick. The short strand is of variable length but always overlaps by approximately 250 bases the nick of the long strand. The genome of HBV has been cloned in E.coli (1-4), sequenced (5-7), and introduced into mammalian cells where expression of some HBV proteins was detected (8-11).

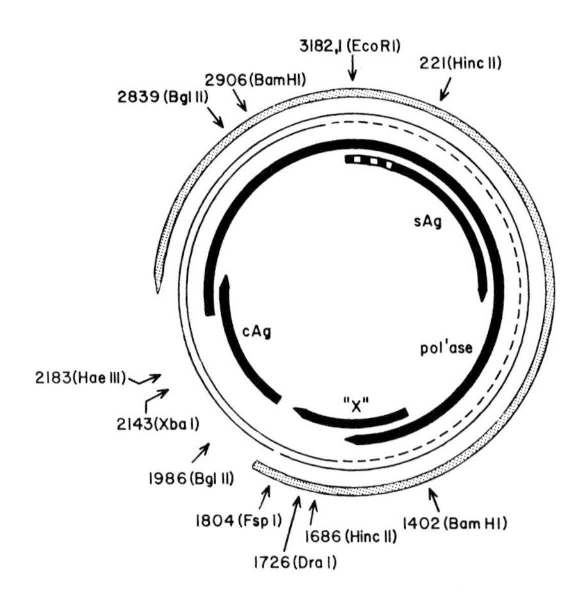

FIGURE 1. The genome of HBV. Thin lines represent the two DNA strands, long and short; solid arrows are the potential coding regions on the long strand transcript; shaded arrow indicates the short strand (antisense) transcript. Restriction enzyme sites are those used to prepare strand specific probes.

The transcript of the long (-) strand contains four potential coding regions, two of which are assigned to the hepatitis surface and core antigen genes. Another region, covering about 80% of the genome, probably codes for the viral polymerase, and the fourth region codes for an unknown or "X" protein. During infection, the (-) strand serves as template for transcription of the predominant forms of virus specific polyadenylated RNAs, a 2.4 kb envelope gene transcript and a 3.8 kb greater-than-genome length transcript originating in the region of the core gene (12). A transcript of the short (+) strand contains information for only one major potential coding region and no gene function has been assigned to it. Nevertheless, the possibility that transcripts of the short strand might play a role in the viral life cycle was suggested by the observation that RNA polymerase III is capable of transcribing a 700 base long RNA from the short (+) strand in vitro (13). In this report, we describe the detection of polyadenylated transcripts from both strands of the HBV genomes present as intact head-to-tail tandem copies integrated into the genome of a clone of mouse 3T3 cells.

METHODS

The methods used for constructing and cloning a plasmid (pTHBV-1) containing tandem copies of the HBV genome in a head-to-tail arrangement and the cotransfection of 3T3 cells with this plasmid and undigested genomic DNA from a derivative of the hamster line A29 as well as the selection procedure with methotrexate have been described (8,11).

The RNA was extracted from cells using guanidine thiocyanate (14) and the poly A(+) RNA (15) was electrophoresed on formaldehyde-agarose gels according to the method of Lehrach et al.(16). The RNA was transferred to nitrocellulose paper and hybridized to radiolabeled probes according to Maniatis et al.(17) and Melton et al.(18). S1 nuclease mapping using 5μg of poly A(+) RNA was performed according to Berk and Sharp (19) and primer extension using 10μg of poly A(+) RNA according to Broome and Gilbert (20).

Uniformly labeled strand-specific HBV probes were transcribed from HBV DNA fragments subcloned into pSP64 and pSP65 (18) and isolated from agarose gels (17).

Radiolabeled DNA probes were prepared by phosphorylation of restriction fragments with ^{32}P-γ-ATP, and digested with a second restriction enzyme in order to generate fragments in which only one strand was labeled at the 5' end (21). Probes in which only one strand was labeled at the 3' end were synthesized by treatment of restriction fragments containing protruding 5' ends with the Klenow fragment of E.coli polymerase I in the presence of

^{32}P-α-dNTPs (17), prior to digestion with a second restriction enzyme. The resulting 5' or 3' labeled DNA fragments were isolated from polyacrylamide gels (21).

Restriction endonucleases were purchased from New England Biolabs and used according to the recommendations of the manufacturer. Radioactive nucleotides were purchased from Amersham Corporation.

RESULTS

Characterization of Integrated HBV Sequences

The total number of copies of HBV DNA in C4.10 cells is approximately 80 as estimated by dot-blot hybridization (Sells et al., submitted). Hybridization of genomic DNA on Southern blots with ^{32}P-labeled nick-translated HBV DNA indicated that: 1) Episomal DNA was absent since the size

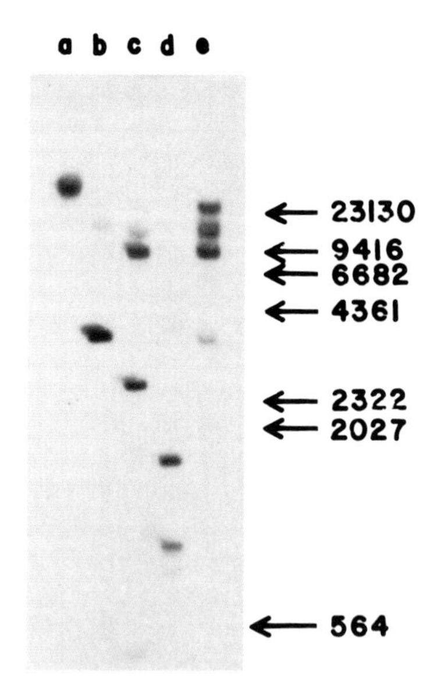

FIGURE 2. Restriction analysis of HBV-specific sequences in clone C4.10 genomic DNA (5μg). Lane a, untreated; b, EcoRI; c, Bgl II; d, BamHI; e, Hind III. Fragment size markers are indicated.

of the undigested DNA is larger than 25,000 base pairs (bp) (Figure 2, lane a); 2) The predominant HBV genome present was full length since EcoRI digestion of genomic DNA generated HBV fragments that were primarily 3182 bp (Figure 2, lane b); 3) The junction between the head-to-tail HBV dimers was preserved since the Bgl II digest of genomic DNA showed the presence of a 2329 bp fragment (Figure 2, lane c) and the BamHI digest showed the presence of a 766 bp fragment (Figure 2, lane d), both of which could only be derived from a head-to-tail HBV DNA dimer; 4) The fidelity of the HBV DNA was preserved since digestion with BamHI showed bands at 1504 and 912 bp which are internal HBV fragments and; 5) The HBV DNA has been integrated either at three different sites in the 3T3 genome or at one site as a reiterated tandem interspersed with varying sequences of carrier DNA since digestion with Hind III (Figure 2, lane e), an enzyme having no sites within the HBV, resulted in three bands hybridizable with HBV DNA.

Characterization of RNA Transcripts

The cells of clone C4.10 synthesized all of the major hepatitis B proteins, HBsAg, HBeAg, and HBcAg (Sells, et al., submitted). A radiolabeled strand-specific probe capable of detecting sense or (+) polarity RNAs (Figure 3a) hybridized with a 2400 nucleotide (nt) long RNA as well as with greater-than-genome length RNAs. Radiolabeled strand-specific probes capable of detecting (-) polarity or "short strand" transcripts, i.e. antisense RNA, hybridized to three RNA species: a major band approximately 2400 nt long, and minor bands of 2000 and 2800 nt long (Figure 3b). These bands were not detected in poly A(+) RNA from untransfected mouse 3T3 cells although three minor bands of different mobility were visualized, suggesting a low level of nonspecific interaction with the probe. Poly A(+) RNAs isolated from the PLC/PRF5 cell line (22; a human liver tumor line developed from a patient with HBV infection, containing multiple copies of integrated HBV DNA, some of which are rearranged) also contain several weakly hybridizing bands detectable with the antisense-specific probes (Figure 3c). In the absence of "control" cells for the PLC/PRF5 cell line, we cannot determine whether these are really virus specific transcripts. Poly A(-) RNAs from all three cell lines (C4.10, 3T3, and PLC/PRF5) were negative with respect to both (-) and (+) polarity transcripts (not shown).

S1 nuclease analysis of the 5' end of the C4.10 HBV

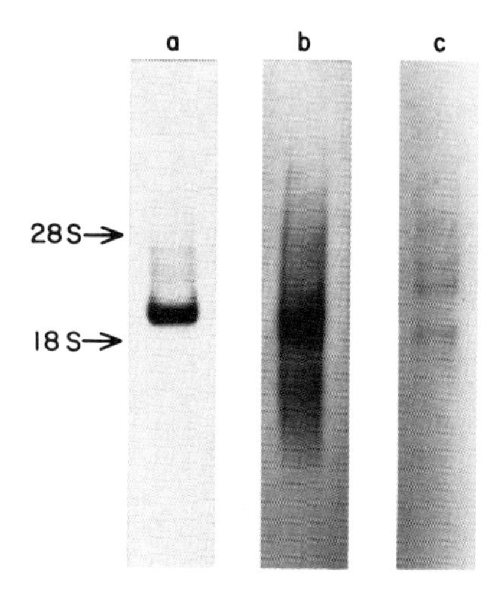

FIGURE 3. Northern analysis of polyadenylated RNA transcripts from 4.10 and PLC/PRF5 cells, a, probed for (+) polarity RNA with a fragment containing bases 2839 to 221; b, (-) polarity RNA, bases 2839 to 221 and 1686 to 1986; c, as lane b, RNA from PLC/PRF5 cells. See Fig. 1 for derivation of the fragments.

RNA using a probe corresponding to bases 1402 to 2183 (Figure 4, lane a) labeled on the 5' end of the (+) strand at base 1402 resulted in bands at 462, 490, and 515 ±20 bases. Primer extension analysis using a probe corresponding to bases 1726 to 1804 (Figure 4, lane b) labeled on the 5' end of the (+) strand at 1726, resulted in a band of 135 ±2 bases.

The analysis of the 3' end of the C4.10 HBV RNA using S1 nuclease and a probe corresponding to bases 2143 to 2906 (Figure 4, lane c) labeled on the 3' end of the (+) strand at 2906 resulted in a band at 525 ±20 bases.

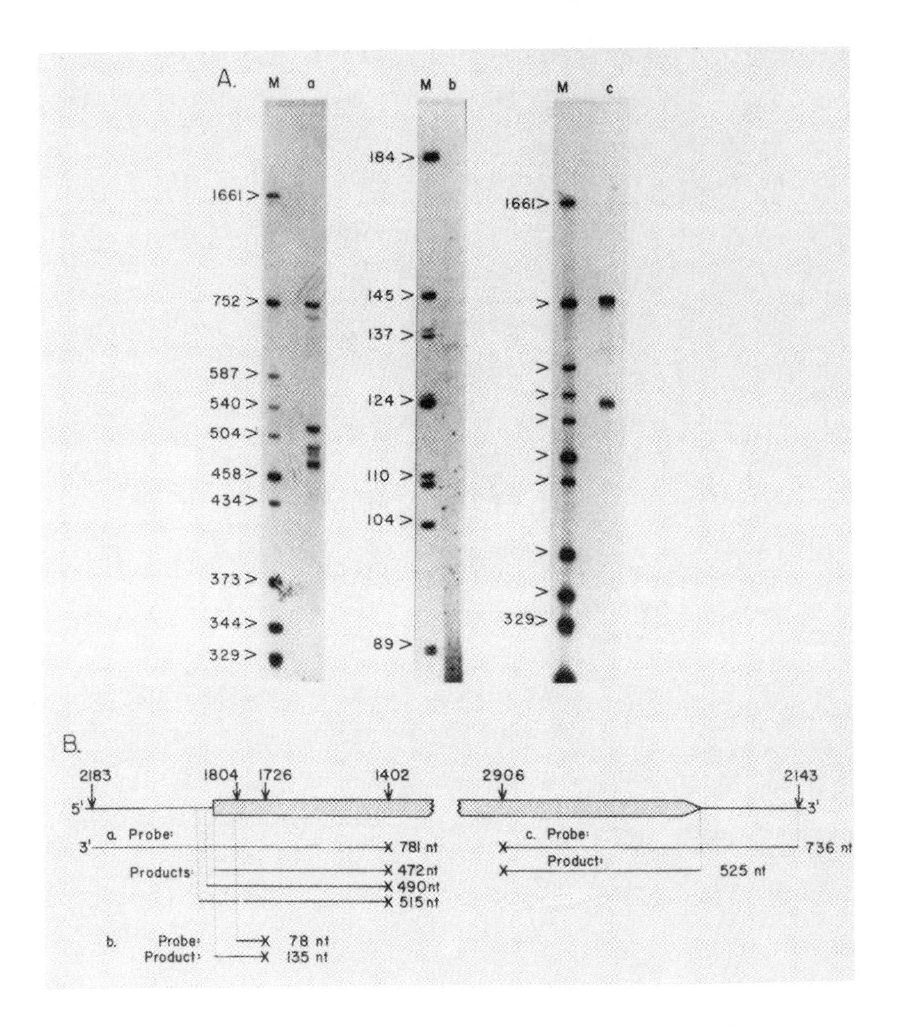

FIGURE 4. A. S1 and primer extension analysis of antisense RNA. M, markers; a, S1 analysis of 5' end using a probe corresponding to bases 1402 to 2183 labeled on the 5' end of the (+) strand at base 1402; b, primer extension using a primer of bases 1726 to 1804 labeled on the 5' end of the (+) strand at 1726; c, S1 analysis of 3' end using probe of bases 2143 to 2906 labeled on the 3' end of the (+) strand at base 2906.
B. Schematic localization of the transcriptional signals for antisense RNA on the HBV genome. The labeled 5' or 3' end is indicated with an asterisk.

DISCUSSION

We have characterized HBV RNA transcripts synthesized by C4.10 cells, a clone of 3T3 cells carrying multiple integrated copies of the HBV genome introduced in the form of a head-to-tail dimer. Using uniformly labeled strand-specific RNA probes, these RNAs consist of sense or (+) polarity RNAs of 2400 nt and greater than 3200 nt long species, and of antisense or (-) polarity RNAs of 2000, 2400, and 2800 nt long species.

Since (-) or antisense HBV RNAs in cells containing integrated HBV DNA could result from initiation of transcripts in non-HBV sequences, the antisense 4.10 cell HBV RNA was further characterized by mapping its 5' and 3' ends. Hybridization of blots of 4.10 cell poly A(+) RNAs with (-) strand specific RNA probes corresponding to various regions of the HBV genome suggested that the antisense transcripts covered almost the whole genome with the exception of a region between bases 2000 and 2300 (data not shown). S1 nuclease protection of a DNA probe by antisense RNA yielded DNA bands indicative of transcript discontinuities at bases 1917, 1892, and 1864$\pm$20. Primer extension analysis of the 5' end of the RNA indicated an initiation site of RNA synthesis at base 1861 $\pm$2, confirming one of the sites found by S1 nuclease analysis. In addition, there is a TATA-like sequence (bases 1917 to 1921) fifty nt upstream from this site which might serve as a promoter. Although this result does not rule out the possibility that the other discontinuities mapped by S1 nuclease protection are splice acceptors or derive from transcripts originating outside HBV sequences that traverse a junction with a minor fraction of partial HBV genomes, it clearly demonstrates that initiation of antisense RNA can occur at a site within the HBV genome.

Analysis of the 3' end of the antisense transcript with S1 nuclease resulted in a band corresponding to a termination site at base 2381 $\pm$20. A poly A (+) HBV transcript starting at base 1861 and ending at base 2381 would be predicted to be approximately 2800 bp long, the size of the largest transcript detected by Northern blot analysis. No known polyadenylation sites are located close to the termination site at base 2381. However, 100 nt upstream, at base 2493, is the only "consensus" sequence for polyadenylation in the HBV genome, a sequence conserved in the genomes of three other HBV subtypes (6,7,23) and in the woodchuck and ground squirrel hepatitis viruses (24,25). Analysis of the HBV genome for areas that might be involved in the formation of secondary structures revealed that in an S strand transcript, the nucleotides at bases 2358 to 2385 and 2489 to 2516 could form a hairpin in which 21 of 28 bases are hydrogen-bonded. Within this double-stranded structure, at bases 2493 to 2497, is the polyadenylation "consensus" sequence; at bases 2371 to 2380 is a stretch of pyrimidines; and at base 2381, exactly across from the "consensus" sequence, is the termination site we have determined by S1 nuclease analysis. Whether this structure could have a role in the termination of the antisense RNA transcription remains to be determined.

Thus, we have mapped both initiation and termination sites for antisense transcripts within the HBV genome that are compatible with the size of the transcripts observed. The length of the antisense transcript, as well as the fact that it is polyadenylated suggests that it is an RNA polymerase II transcript. It remains to be determined whether this (-) polarity transcript functions during the life cycle of the virus or whether it is a curiosity of integrated genomes. However, a number of possible functions can be suggested:

1) Regulation of viral gene expression. In several systems (26,27), antisense RNA suppresses the expression of exogenous and endogenous genes, presumably through hybridization with sense mRNAs. The HBV antisense RNA which we have described here is complementary to all or part of the HBV sense transcripts and thus could potentially interfere with production of all known HBV proteins. Since the amount of antisense HBV RNA in 4.10 cells is much less than that of the 2.4 kb envelope mRNAs, we could not expect to detect any effect which it might have on production of HBsAg. However, annealing with the low level of longer than-genome length transcripts might account for the failure of 4.10 cells to assemble virions even though they produce normal core proteins and make properly initiated RNAs of the same length as the proposed viral replicative intermediates (Sells et al., submitted).

2) Message for a viral protein. The polyadenylated (-)RNA is transcribed from a region including the only major open reading frame of the S strand. It might also be processed to combine small open reading frames and thus could code for as yet unidentified viral proteins.

3) Regulation of viral replication. The initiation site of antisense HBV RNA (base 1861), is in the region of the HBV genome presumed to be the origin of replication

(28) and thus has a possibility to affect either the synthesis or stability of replicative intermediates.

REFERENCES

1. Charnay,P, Pourcel,C, Louise,A, Fritsch,A, Tiollais,P (1979). Cloning in Escherichia coli and physical structure of hepatitis B virion DNA. Proc Natl Acad Sci USA 76:2222.
2. Burrell,CJ, Mackay,P, Greenaway,PJ, Hofschneider,PH, Murray,K (1979). Expression in Escherichia coli of hepatitis B virus DNA sequences cloned in plasmid pBR322. Nature 279:43.
3. Sninsky,JJ, Siddiqui,A, Robinson,WS, Cohen,SN (1979). Cloning and endonuclease mapping of the hepatitis B viral genome. Nature 279:346.
4. Price,PM, Hirschman,SZ, Garfinkel,E (1980). DNA cloned from the ayw subtype of hepatitis B virus. J Med Virol 6:139.
5. Galibert,F, Mandart,E, Fitoussi,F, Tiollais,P, Charnay,P (1979). Nucleotide sequence of the hepatitis B virus genome (subtype ayw) cloned in E.coli. Nature 281:646.
6. Pasek,M, Goto,T, Gilbert,W, Zink,B, Schaller,H, Mackay,P, Leadbetter,G, Murray,K (1979). Hepatitis B virus genes and their expression in E.coli. Nature 282:575.
7. Valenzuela,P, Quiroga,M, Zaldivar,J, Gray,P, Rutter,WJ (1980). The nucleotide sequence of the hepatitis B viral genome and the identification of the major viral genes. In Fields,B, Jaenisch,R, Fox,CF (eds): "Animal Virus Genetics," New York: Academic Press, p 57.
8. Christman,JK, Gerber,M, Price,PM, Flordellis,C, Edelman,J, Acs,G (1982). Amplification of expression of hepatitis B surface antigen in 3T3 cells cotransfected with a dominant-acting gene and cloned viral DNA. Proc Natl Acad Sci USA 79:1815.
9. Gough,NM, Murray,K (1982). Expression of the hepatitis B virus surface, core and e antigen genes by stable rat and mouse cell lines. J Mol Biol 162:43.
10. Dubois,MF, Pourcel,C, Rousset,S, Chany,C, Tiollais,P (1980). Excretion of hepatitis B surface antigen particles from cells transformed with cloned viral DNA. Proc Natl Acad Sci USA 77:4549.
11. Price,PM, Ostrove,S, Flordellis,C, Sells,MA, Thung,S,

Gerber,M, Christman,J, Acs,G (1983). Characterization of RNA transcripts and virally coded proteins synthesized in mouse fibroblasts transfected with hepatitis B DNA. Biosci Rep 3,1017.

12. Cattaneo,R, Will,H, Schaller,H (1984). Hepatitis B virus transcription in the infected liver. EMBO J 3:2191.
13. Standring,DN, Rall,LB, Laub,O, Rutter,WJ (1983). Hepatitis B virus encodes an RNA polymerase III transcript. Molec Cell Biol 3:1774.
14. Chirgwin,JM, Przybyla,AE, MacDonald,RJ, Rutter,WJ (1979). Isolation of biologically active ribonucleic acid from sources enriched in ribonuclease. Biochemistry 18:5294.
15. Aviv,H, Leder,P (1972). Purification of biologically active globin messenger RNA by chromatography on oligothymidylic acid-cellulose. Proc Natl Acad Sci USA 69:1408.
16. Lehrach,H, Diamond,D, Wozney,JM, Boedtker,H (1977). RNA molecular weight determinations by gel electrophoresis under denaturing conditions. A critical reexamination. Biochemistry 16:4743.
17. Maniatis,T, Fritsch,EF, Sambrook,J (1982). "Molecular Cloning" New York: Cold Spring Harbor Press.
18. Melton,DA, Krieg,PA, Regagliati,MR, Maniatis,T, Zinn,K, Green,MR (1984). Efficient in vitro synthesis of biologically active RNA and RNA hybridization probes from plasmids containing a bacteriophage SP6 promoter. Nucleic Acids Res 12:7035.
19. Berk,AJ, Sharp,PA (1977). Sizing and mapping of early adenovirus mRNAs by gel electrophoresis of S1 endonuclease-digested hybrids. Cell 12:721.
20. Broome,S, Gilbert,W (1985). Rous sarcoma virus encodes a transcriptional activator. Cell 40:537.
21. Maxam,AM, Gilbert,W (1980). Sequencing end-labeled DNA with base-specific chemical changes. In Grossman L, Moldave K (eds): "Methods in Enzymology, vol 65," New York: Academic Press, p 499.
22. Alexander,JJ, Bey,EM, Geddes,EW, Lecatsas,G (1976). Establishment of a continuously growing cell line from primary carcinoma of the liver. S Afr Med J 50:2124.
23. Kobayashi,M, Koike,K (1984). Complete nucleotide sequence of hepatitis B virus DNA of subtype adr and its conserved gene organization. Gene 30:227.
24. Galibert,F, Chen,TN, Mandart,E (1982). Nucleotide sequence of a cloned woodchuck hepatitis virus genome:

Comparison with the hepatitis B virus sequence. J Virol 41:51.

25. Mandart,E, Kay,A, Galibert,F (1984). Nucleotide sequence of a cloned duck hepatitis B virus genome: comparison with woodchuck and human hepatitis virus sequences. J Virol 49:782.
26. Marx,JL (1984). New ways to "mutate" genes. Science 225:819.
27. Izant,JG, Weintraub,H (1985). Constitutive and conditional suppression of exogenous and endogenous genes by anti-sense RNA. Science 229:345.
28. Summers,J, Mason,WS (1982). Replication of the genome of a hepatitis B-like virus by reverse transcription on an RNA intermediate. Cell 29:403.

Viruses and Human Cancer, pages 479–491

THE ROLE OF THE PRECORE REGION OF THE HEPATITIS B VIRUS GENOME IN THE COMPARTMENTALIZATION OF CORE GENE PRODUCTS

Jing-hsiung Ou and William J. Rutter

Hormone Research Institute
Department of Biochemistry & Biophysics
University of California
San Francisco, CA 94143

ABSTRACT The core gene of the hepatitis B virus (HBV) genome contains two conserved in-phase initiation codons separated by about 90 nucleotides. This region ("the precore region") is conserved among various hepadna viruses. Using an SV40-derived vector, we have expressed the coding sequence of the core gene with or without the precore region in a heterologous mammalian cell line. The results show that the precore region is not required for the expression of either core antigen (HBcAg) or, a related HBV antigen, e antigen (HBeAg). But its presence results in the more efficient production of HBeAg, and is requisite for the secretion of HBeAg. In addition, the precore region is required for the association of HBcAg with cytoplasmic membranes, presumably the endoplasmic reticulum. Our results suggest that the precore region plays a role in targeting core proteins to the membrane. This may be the direct cause of HBeAg secretion and also may aid in the interaction of core and surface antigen structures.

INTRODUCTION

Hepatitis B virus (HBV), a member of hepadna viruses, causes acute or chronic hepatitis in human beings; it also may be the primary causative agent of hepatocellular carcinoma (1). The mature virus (Dane particles) contains a lipid envelope which is embedded with about 400 molecules of surface antigen (HBsAg) proteins (2). Inside the lipid envelope there is a core particle which is comprised of the viral DNA, the core antigen (HBcAg), as well as a DNA polymerase (reverse transcriptase) and a protein covalently linked to the 5' end of the DNA. During acute infection of HBV, HBsAg and the e antigen (HBeAg), an antigen closely related to HBcAg, are found in the sera (3), but the HBcAg itself is not detectable in the sera. However, during the recovery phase antibodies against HBcAg along with those against HBsAg and HBeAg can be detected. Treatment of core particles of HBV with mercaptoethanol and heat, or with mild protease digestion, results in the apparent conversion of HBcAg into HBeAg (4,5). Thus, HBeAg in the serum must somewhat be another form of or a derivative of HBcAg.

The genomic DNA of several different HBV isolates has been cloned and sequenced (6-9). The sequence of HBV DNA contains only one open reading frame for the HBsAg gene and the HBcAg gene. However, it has recently been shown that the HBsAg gene is transcribed, from two different promoters, resulting in three distinct transcripts with three distinct initiation codons (10-14). Translation of these three mRNAs produces three proteins that share a common HBsAg C-terminal domain (P24) and varying N-terminal extensions (P30, P39). In analogy with the HBsAg gene the HBcAg coding sequence contains two in-phase initiation codons. The sequence between these two initiation codons ("the precore region") is not required for the expression of HBcAg in heterologous procaryotic (7,15) or mammalian cells (16). Nevertheless, the precore region is conserved not only in different isolates of HBV but also in two other hepadna viruses, woodchuck hepatitis virus (WHV) (17) and ground squirrel hepatitis virus (GSHV) (18). The HBcAg gene of another distantly related hepadna virus, duck hepatitis B virus (DHBV), contains a less conserved precore region (19). Inspection of the precore region indicates that it is rich in hydrophobic residues and somewhat resembles a signal peptide (20). We have therefore tested the effect of the precore region on the synthesis and secretion of both HBcAg and HBeAg. When the precore region is present in the transcript, HBcAg becomes associated with the endoplasmic reticulum (ER) and HBeAg is selectively secreted.

RESULTS

SV40 Expression Vector Containing the Precore and Core Sequences

SV40-precore (PC) and core (C) recombinants were prepared according to the strategy presented in Figure 1. The PC sequence (precore fragment) starts 11 nucleotides upstream from the precore region and extends to the presurface region of the surface gene. The C sequence (core

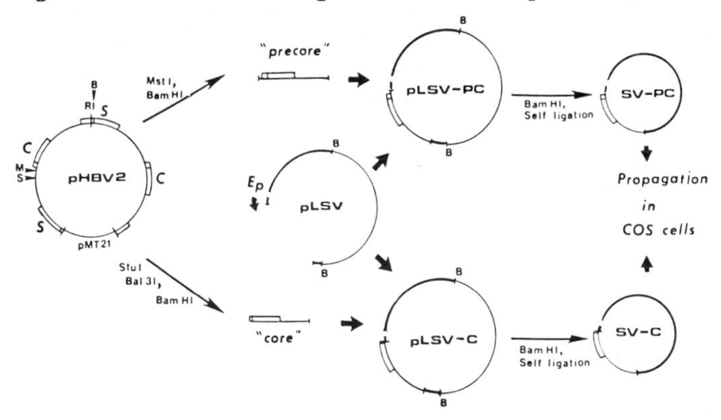

FIGURE 1. Constructions of the SV40-HBV recombinant DNA. The genomic DNA of SV40 cloned into BamHI site of pBR322 (pSV40) was digested with StuI and BclI for the removal of the early gene coding sequences. Details for the preparation of the linearized SV40 vector (pLSV) have been described (11). The plasmid pHBV2 contains two tandem copies of the genomic DNA of HBV joined at the EcoRI sites. The precore fragment was prepared by digesting pHBV2 with MstI (M) and BamHI (B). The core fragment was isolated from pHBV2 by digesting with StuI (S) followed by Bal31 exonuclease (Bethesda Research Lab) digestion. The DNA after blunt-ending with Klenow fragment of the DNA polymerase I was further digested with BamHI. Core fragments thus prepared were cloned into pLSV vector. pLSV-C which has been sequenced and was found to contain only the second ATG of the HBcAg coding sequence was isolated. pLSV-PC and pLSV-C were digested with BamHI for the removal of pBR322 sequence. The recircularized recombinant DNAs, SV-PC and SV-C, were then transfected into COS cells (11). Two weeks after transfection COS cells were lysed by three cycles of freezing and thawing. The cell lysates which contain recombinant SV40-HBV viruses were used for further infec-

tions of COS cells. Ep, early promoter of SV40; C, HBcAg gene; S, HBsAg gene. pMT21 is a pBR322 derivative constructed by H.V. Huang (29).

fragment) starts 35 nucleotides upstream from the second in-phase ATG codon and terminates at the same site as the PC fragment. The plasmid containing the precore fragment was called pLSV-PC and that containing the core fragment was called pLSV-C. In both cases the transcription is under the control of the SV40 early promoter and termination occurs at the SV40 polyA site. The pBR322 sequences of the parent vectors were removed and the recircularized recombinant DNA was transfected into COS cells which are permissive for these SV40 vectors (11,21). The viruses produced from pLSV-PC and pLSV-C transfected cells were called SV-PC and SV-C, respectively.

TABLE 1: Expression of HBcAg and HBeAg in SV-PC and SV-C Infected COS Cells[1]

CELL FRACTION	SV40 CONSTRUCT	P/N RATIO	
		HBeAg+HBcAg	HBcAg
Medium	WT	1.0	1.0
	Precore	8.4	1.4
	Core	1.5	1.1
Cytoplasm	WT	1.4	1.0
	Precore	19.4	6.0
	Core	19.8	13.5
Nucleus	WT	1.5	1.0
	Precore	7.1	1.2
	Core	3.3	2.3

[1]Growing of COS cells, transfection of DNA into cells and the infection of cells with recombinant viruses were done essentially as described (11,22). Cells were washed twice with phosphate-buffered saline (PBS) and then lysed with 1 ml PBS containing 0.5% Nonidet P-40 (NP40, Sigma) and 1mM phenylmethylsulfonylfluoride (PMSF). The nuclei were separated from cytoplasm by centrifugation at 15,000g for 1 min and then lysed in 1 ml PBS containing 1mM PMSF by three cycles of freezing and thawing. HBeAg+HBcAg activities were measured by the Abbott-HBe kit. HBcAg activities were measured by the RIA described in ref. 25. P/N = cpm of samples divided by cpm of negative controls.

<u>Expression of Precore and Core Sequences in COS Cells</u>

The expression of HBcAg and HBeAg in SV-PC and SV-C infected COS cells was detected by radioimmunoassay which discriminates HBcAg from HBeAg but does not discriminate HBeAg from HBcAg. Thus, the levels of HBeAg are qualitatively evaluated by comparison of the two values of HBcAg and HBeAg+HBcAg (Table 1). The medium of SV40 wild

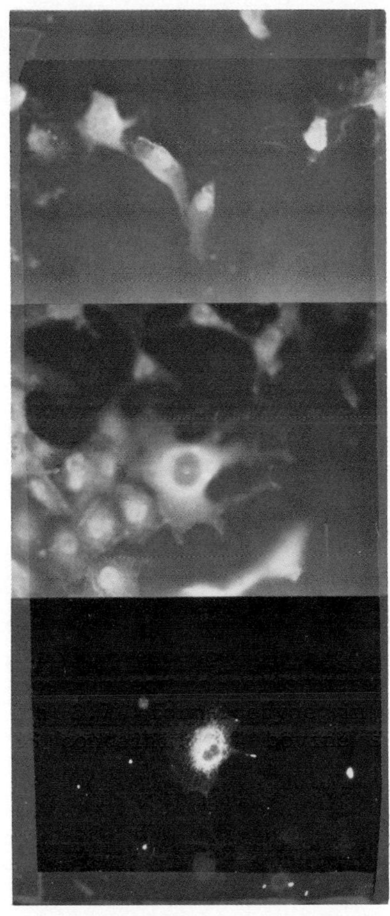

FIGURE 2. Immunofluorescence of HBcAg expressed in COS cells. 2-3days post-infection cells were washed twice with PBS, fixed with 3.7% formaldehyde in PBS and then washed with BSS (PBS containing 1% bovine serum albumin,

type (SV-WT) and SV-C infected cells contains very little HBeAg or HBcAg activity; on the other hand, the medium of SV-PC infected cells contains relatively abundant HBeAg+ HBcAg activity but little HBcAg. This indicates that the antigens accumulating in the SV-PC medium is almost exclusively HBeAg.

The cytoplasmic lysates of SV-PC and SV-C infected cells contain both HBeAg and HBcAg activities but SV-PC produces significantly greater HBeAg activity. This result confirms previous reports that the precore region is not essential for the expression of HBcAg, but it results in the more effective production of the HBeAg. The majority of the HBcAg and HBeAg expressed in COS cells is in the cytoplasm, but some HBcAg and HBeAg is also found in the crude nuclear lysates, perhaps as a rseult of contamination; here also the SV-PC nuclear lysate has higher HBcAg+ HBeAg activity but lower HBcAg activity than the SV-C nuclear lysate.

The Precore Region Facilitates the Association of the Core Gene Products with the Endoplasmic Reticulum

The antigen expressed in COS cells infected with SV-PC and SV-C was localized by immunofluorescence experiments using anti-core antibodies. As shown in Figure 2, the immunofluorescence of both cells is restricted to the cytoplasm. The nuclei are virtually unlabeled. The HBcAg immunofluorescence is weak and diffuse in the cytoplasm of SV-C infected cells (Fig. 2, middle panel). In SV-PC infected cells, however, the HBcAg fluorescence is more intense and is strikingly associated with what appears to be ER and Golgi structures (Fig. 2, bottom panel).

0.05% sodium azide and 0.02% saponin). Cells were then incubated with 75 µl of rabbit anti-cAg in BSS (50 µg/ml) for 30 min at 37°C. After two washes with BSS, the cells were incubated with 75 µl of FITC-conjugated goat anti-rabbit IgG (Miles) for 30 min at 37°C. Cells were finally washed twice with BSS, twice with PBS and mounted on a drop of 50% glycerol in PBS for microscopy. Top panel, mock-infected cells; middle panel, SV-C-infected cells; bottom panel, SV-PC-infected cells.

The Precore Region Facilitates the Secretion of the HBeAg into the Medium

As shown in Figure 3, the HBeAg+HBcAg activity increased rapidly in the medium of SV-PC infected COS cells, reaching a plateau between two to three days post-infection and declining gradually thereafter. In contrast, HBcAg activity of these samples remained at the background level throughout the incubation period. These results suggest that HBeAg accumulates almost exclusively in the medium of SV-PC infected COS cells. In contrast, the medium of SV-C infected cells showed very little HBeAg+HBcAg activity during the first stage of infection (less

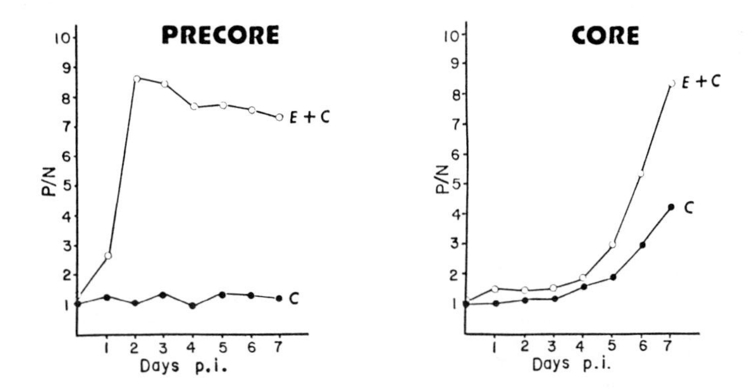

FIGURE 3. Kinetics for the accumulation of HBcAg and HBeAg in the media of SV40-HBV infected COS cells. PRECORE, SV-PC infected cells. CORE, SV-C infected cells. COS cells were infected with SV-PC or SV-C. Three hours post-infection virus was removed and cells were rinsed twice with fresh medium. 15 ml of fresh medium was then overlaid on infected cells and 1 ml of medium was collected and frozen at different time points indicated. All the samples were assayed at the same time. E+C, HBeAg and HBcAg activities measured by Abbott-HBe kit; C, HBcAg activity measured with rabbit anti-HBcAg (22).

than 4 days); on the other hand, at the later stages of infection (later than 72 hr) when cell lysis occurred HBcAg and probably also HBeAg is detected in the medium. Significantly, no HBcAg activity was detected in the SV-PC medium even at the later stages of SV-PC infection when cell lysis was occurring (Fig. 3). This may be due to its association with membranes as demonstrated in Figure 2.

DISCUSSION

Previous work has shown that the HBsAg gene is controlled by two promoters that produce three transcripts and result in three distinct proteins having a common C-terminal region, and N-terminal extensions of variable length (10-14). The core gene is less complex, but the available evidence suggests that it is composed of two separate elements which also differ at the N-terminus. Evidence that the precore region is significant is in part derived from the conservation of the amino acid sequence of the precore region of several different hepatitis B viruses. The precore region in human, woodchuck and ground squirrel hepatitis B viruess is highly conserved; it is less conserved in the duck virus, a more distantly related hepadna virus. The region is rich in hydrophobic residues (≥50%) and superficially resembles signal peptide sequences (20); however, it lacks a long continuous stretch of hydrophobic amino acids; furthermore, all known hepadna viruses contain four cysteine residues in this region. These features suggest that this domain may be involved in translocation into the endoplasmic reticulum but also may have another specialized function. The present work demonstrates that when precore sequences are in the mRNA (PC mRNA) the resulting HBcAg associated with membranes, probably the endoplasmic reticulum. Furthermore, HBeAg but not HBcAg is secreted from the cells. This supports the hypothesis that the precore sequences play a targeting function. The HBeAg is secreted only in cells transfected with the precore construct. This suggests that the precore region facilitates transposition of HBcAg into the ER. Our results suggest that whatever the form of HBeAg, it can be efficiently formed from HBcAg in the ER and subsequently secreted. Our results, therefore, may explain the observation that HBeAg but not HBcAg is found in the serum of patients during the period of rapid growth of the virus. Presumably, some HBeAg is formed and secreted during active replication of the virus. Recent studies on transcription of the core gene indicate it has multiple initiation sites as has been observed in the HBsAg gene (10-14). Enders et al. (23) and Moroy et al. (24) demonstrated that only a minor fraction of the GSHV and WHV HBcAg mRNAs initiates upstream from the precore region and can therefore code for the entire HBcAg molecule including the precore region. Similar experiments with DHBV revealed only one major transcription initiation site located in the precore region (25). The longer transcript containing precore sequences must be minor and remains undetected in these experiments or the DHBV may be exceptional among hepadna viruses and not have a functional precore region. This could be a result of the modified HBV structure combining the precore and X or B gene regions in the DHBV (19).

In normal HBV infected cells because of the predominance of the C mRNA, the function of the PC mRNA cannot easily be assessed. However, by using an artificial

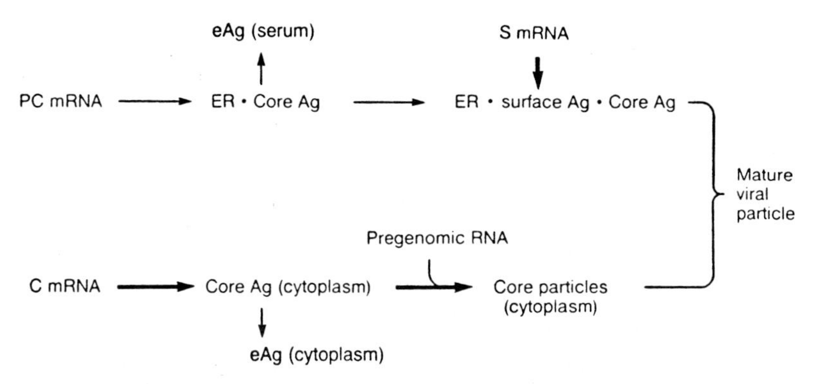

FIGURE 4. Hypothetical model for the assembly of HBV particles. cAg translated from precore (PC) mRNA bind to the ER. Some molecules are transported into the lumen, perhaps processed, and secreted as eAg, cAg translated from core (C) mRNA remains in the cytoplasm and interacts with pregenomic RNA to form core particles. Some of the cytoplasmic cAg exists in the eAg form. The core particles containing HBV RNA or DNA may then interact with ER bound cAg and sAg (P49, P30 and P24, translated from S mRNA) and become encapsulated, presumably via a budding process, with the consequent formation of mature viral particles.

promoter linked to a tailored coding sequence (PC or C). The characteristic properties of the two protein products can be discriminated and evaluated. The unique biological activities associated with the precore region as well as the conservation of its structure in the hepadna viruses argues for a viral function for the precore region. The relative paucity of PC transcripts as compared to C transcripts argues for a specialized function requiring relative small quantities of the protein product. One possibility is that it functions to produce the secreted HBeAg. HBeAg (or the core within the ER) could have an as yet undiscovered role, either at the surface of the HBV particle or independent of it. An equally tenable and more appealing (to us) hypothesis is that precore region is involved in the targeting of core particles to the ER membrane. The precore sequence might anchor HBcAg into the membrane containing HBsAg such that the C-terminal DNA binding portion of the HBcAg faces the cytoplasm. This membrane bound HBcAg may then interact with cytoplasmic core particles produced by the interaction of HBcAg (from the major C transcript) and the pregenomic RNA. In this fashion, the core-surface domains of the virus may more effectively fuse. This is consistent with the observation of Kimimura et al. (26) and Yamada et al. (27) that core particles in HBV infected hepatocytes can bud into the cisternae of the ER to form Dane particles. In addition, core particles from WHV are frequently found associated in the endoplasmic reticulum while others are found in the cytoplasm (28). In this hypothesis, the majority of the HBcAg produced from the PC transcript remains membrane bound, the transport of HBcAg within the ER lumen with the consequent formation and secretion of HBeAg may be a side reaction which may or may not have a specific viral function. Figure 4 illustrates the complementary role of C mRNA in structure formation of the core, and PC mRNA in the targeting to HBsAg rich membranes.

ACKNOWLEDGEMENTS

We thank Dr. Orgad Laub for his involvement in the early phases of these experiments, Drs. Pablo Valenzuela and Lacy Overby of Chiron Corporation (Emeryville, CA) for providing us with the rabbit anti-cAg and Dr. Peter Walter and Pablo Garcia for helpful discussions. Finally, we wish to thank Ms. Leslie Spector for preparing the manuscript. This work was supported by NIH grant AM 19744.

REFERENCES

1. Beasley RP, Hwang L-Y (1984). Epidemiology of

hepatocellular carcinoma. In Vyas GN, Dienstag JL, Hoofnagle JH (eds): "Viral Hepatitis and Liver Disease", Orlando: Grune and Stratton, p 209.
2. Tiollais P, Pourcel C, Dejean A (1985). The hepatitis B virus. Nature 317:489.
3. Magnius LO, Espmark JA (1972). New specificities in Australia antigen positive sera distinct from Le Bouvier determinants. J Immunol 109:1017.
4. Takahashi K, Akahane Y, Gotanda T, Mishiro T, Imai M, Miyakawa Y, Mayumi M (1979). Demonstration of hepatitis B e antigen in the core of Dane particles. J Immunol 122:275.
5. Mackay P, Lees J, Murray K (1981). The conversion of hepatitis B core antigen synthesized in E. coli to e antigen. J Med Virol 8: 237.
6. Galibert F, Mandart E, Fitoussi F, Tiollais P, Charnay P (1979). Nucleotide sequence of the hepatitis B virus genome (subtype ayw) cloned in E. coli. Nature 281:646.
7. Pasek M, Goto T, Gilbert W, Zink B, Schaller H, McKay P, Leadbetter G, Murray K (1979). Hepatitis B virus genes and their expression in E. coli. Nature 282: 575.
8. Valenzuela P, Quiroga M, Zaldivar J, Gray P, Rutter WJ (1980). In Fields B, Jaenisch R, Fox CF (eds): "Animal Virus Genetics". New York: Academic Press, p. 57.
9. Ono Y, Onda H, Sasada R, Igarashi K, Sugino Y, Nishioka K (1983). The complete nucleotide sequences of the cloned hepatitis B virus DNA: serotype adr and adw. Nucl Acids Res 11:1747.
10. Cattaneo R, Will H, Hernandez N, Schaller H (1983). Signals regulating hepatitis B surface antigen transcription. Nature 305:336.
11. Laub O, Rall LB, Truett M, Shaul Y, Standring DN, Valenzuela P, Rutter WJ (1983). Synthesis of hepatitis B surface antigen in mammalian cells: expression of the entire gene and the coding region. J Virol 48: 271.
12. Rall LB, Standring DN, Laub O, Rutter WJ (1983). Transcription of hepatitis B virus by RNA polymerase II. Mol Cell Biol 3:1766.
13. Standring DN, Rutter WJ, Varmus HE, Ganem D (1984). J Virol 50:563.
14. Ou J, Rutter WJ (1985). Hybrid hepatitis B virus-host transcripts in a human hepatoma cell. Proc Natl Acad Sci USA 82:83.

15. Edman JC, Hallewell RA, Valenzuela P, Goodman HM, Rutter WJ (1981). Synthesis of hepatitis B surface and core antigens in E. coli. Nature 291:503.
16. Will H, Cattaneo R, Pfaff E, Kuhn C, Roggendorf M, and Shaller H (1984). Expression of hepatitis B antigens with a simian virus 40 vector. J Virol 50:335.
17. Galibert F, Chen TN, Mandart E (1982). Nucleotide sequence of a cloned woodchuck hepatitis virus genome: comparison with the hepatitis B virus sequence. J Virol 41:51.
18. Seeger C, Ganem D, Varmus HE (1984). Nucleotide sequence of an infectious molecularly cloned genome of ground squirrel hepatitis virus. J Virol 51:367.
19. Mandart E, Kay A, Galibert F (1984). Nucleotide sequence of a cloned duck hepatitis B virus genome: comparison with woodchuck and human hepatitis B virus sequences. J Virol 49:782.
20. Sabatini DD, Kreibich G, Morimoto T, Adesnik M (1982). Mechanisms for the incorporation of proteins in membranes and organelles. J Cell Biol 92:1.
21. Gluzman Y (1981). SV40-transformed simian cells support the replication of early SV40 mutants. Cell 23:175.
22. Ou JH, Laub O, Rutter WJ (1986). Hepatitis B virus gene function: the precore region targets the core antigen to cellular membranes and causes the secretion of the e antigen. Proc Natl Acad Sci USA 83:1578.
23. Enders GH, Ganem D, Varmus HE (1985). Mapping the major transcripts of ground squirrel hepatitis virus: the presumptive template for reverse transcriptase is terminally redundant. Cell 42:297.
24. Moroy T, Etiemble J, Trepo C, Tiollais P, Buendia MA (1985). Transcription of woodchuck hepatitis virus in the chronically infected liver. EMBO J 4:1507.
25. Buscher M, Reiser W, Will H, Schaller H (1985). Transcripts and the putative RNA pregenome of duck hepatitis virus: implications for reverse transcription. Cell 40:717.
26. Kimimura T, Yoshikawa A, Ichida F, Sasaki H, (1981). Electron microscopic studies of Dane particles in hepatocytes with special reference to intracellular development of Dane particles and their relation with HBeAg in serum. Hepatology 1:392.
27. Yamada G, Sakamoto Y, Mizuno M, Nishihara T, Kobayashi T, Takahashi T, Nagashima H (1982). Electron and immunoelectron microscopic study of Dane particle

formation in chronic hepatitis B virus infection. Gastroenterology 83:348.

28. Summers J, Mason WS (1982). Replication of the genome of a hepatitis B-like virus by reverse transcription of an RNA intermediate. Cell 29:403.
29. Arias CF, Lopez S, Bell JR, Strauss JH (1984). Primary structure of the neutralization antigen of simian rotavirus as deduced from cDNA sequence. J Virol 50:657.

Viruses and Human Cancer, pages 493–507

HEPATITIS B VIRUS GENE EXPRESSION IN TRANSGENIC MICE

Francis V. Chisari, Carl A.Pinkert,[2] David R. Milich,
Alan McLachlan, Pierre Filippi,
Richard D. Palmiter,[3] and Ralph L. Brinster[2]

Scripps Clinic & Research Foundation
La Jolla, CA 92037

ABSTRACT

We have produced transgenic mice bearing a subgenomic fragment of the HBV genome containing the pre-S, S and X coding regions under the control of an exogenous promoter (pMT-PSX). HBV sequences are integrated as tandem repeats with variable copy number at 1 or 2 sites within the mouse genome. Breeding experiments reveal transmission of HBV DNA sequences in a normal mendelian fashion. Serum hepatitis B surface antigen (HBsAg) concentrations between 0.5-12.8 μg/ml are detectable in 2 of 6 pMT-PSX transgenics and their progeny. The secreted product is particulate (22 nm) and has a buoyant density of 1.20 g/cm^3. Both pre-S and HBsAg are detectable in liver and kidney as cytoplasmic granular inclusions. Multiple RNA species are present, but 2.3 and 2.5 kbp transcripts consistent with regulation by both endogenous (HBV) and exogenous (MT) promoters respectively are common to all tissues. Zinc administration to pMT-PSX mice increases tissue antigen concentration and all mice tested are immunologically tolerant to HBsAg. To

---

[1]This work was supported by Grants AI20001, CA40489, HD17321, HD09172, HD07155 and a NATO Grant for International Collaboration (675/84)
[2]Univ. of Pennsylvania, Philadelphia,PA
[3]Univ. of Washington, Seattle, WA

date, all mice are clinically and histopathologically normal which suggests that tissue injury is probably not a direct consequence of pre-S or HBs antigen expression. Studies that address the pathogenetic role of the immune response to HBV encoded antigens will be possible in this system when HBV gene expression is achieved in inbred strains of appropriate immune response phenotype.

INTRODUCTION

Since the discovery of the Australia antigen and its association with serum hepatitis much has been learned about the structure, composition, antigenic characteristics, transmission, epidemiology and disease associations of the hepatitis B virus (HBV) (reviewed in 1). In contrast, the pathogenesis of acute and chronic hepatocellular injury and the events leading to the development of primary hepatocellular carcinoma (PHC) associated with HBV infection are entirely speculative due to the failure to propagate HBV in tissue culture and the absence of an inbred animal model suitable for appropriate experimental studies. Existing models in the chimpanzee(2), the eastern woodchuck(3), the Beechey ground squirrel(4) and the Pekin duck(5), have limited usefulness for pathogenesis studies for various reasons including restricted supply, costs, unavailability of inbred strains and poorly defined immune system. Using recombinant DNA technology and egg microinjection techniques we have very recently produced a transgenic mouse model of the chronic HBV carrier state which, in man, is associated with viral DNA integration into the host genome, chronic hepatitis and the delayed development of primary hepatocellular carcinoma(6). Because of its unique properties this model has features that are complementary to the others and permit the execution of studies not otherwise possible. Prior to a description of our results we shall briefly review the field with emphasis on the areas specifically impacted by our model.

BACKGROUND

The human hepatitis B virus belongs to the recently

designated group of animal DNA viruses called the hepadnaviridae (7). It is an enveloped polymorphic virus. The infectious virion (Dane particle) is a 42 nm in diameter particle with a lipoprotein envelope, containing the hepatitis B surface antigen (HBsAg), and a nucleocapsid characterized by a distinct hepatitis B core antigen (HBcAg) containing the viral genome and DNA polymerase. The majority of particles produced during HBV infection, however, are noninfectious, 22 nm diameter spheres and filaments consisting exclusively of HBsAg.

The HBV genome is a small circular partly double stranded DNA molecule with a single stranded region of variable length (8). The long strand is of fixed length of about 3200 nucleotides and has a nick or a gap of a few nucleotides. The short strand is of variable length ranging from 50-100% of the HBV genome with its 5' end located at a constant position 260 nucleotides from the position of the nick on the long strand.

Six complete HBV sequences have been established thus far (reviewed in 9). For the 6 genomes the lengths range from 3182 to 3221 base pairs. A comparative analysis of the nucleotide sequences of these cloned genomes permits a general presentation of the genetic organization of HBV. The long strand carries virtually all of the protein coding capacity of the genome. Five large coding regions are conserved among all 6 genomes (Figure 1) and have been designated region S, pre-S, C, P and X (14). Gene S encodes the major envelope protein of HBsAg and its translation product is 226 amino acids long. HBsAg consists of 3 hydrophobic segments separated by hydrophilic regions. Immediately upstream of the S region is another coding region designated pre-S. This region codes for a polypeptide which is produced as the amino terminal component of a fusion protein with HBsAg. The pre-S

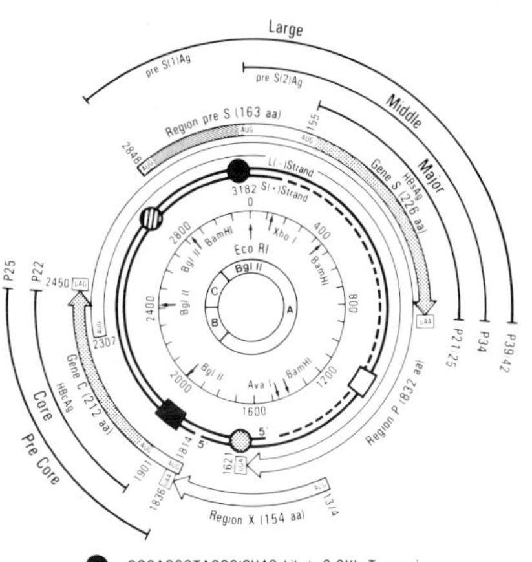

Fig. 1 HBV Genome

polypeptide appears to have the ability to bind polymerized human serum albumin and may thus constitute the polyalbumin receptor previously described to be limited to Dane particles. Region C codes for HBcAg. Although there is some variability, the gene C translation product appears to be a protein of 212-214 amino acids in length. The HBcAg sequence is characterized by an extremely hydrophobic region between residues 22-30 which might be involved in anchoring the lipid bilayer about the nucleocapsid. The carboxy terminal sequence contains several arginine residues which presumably are involved in interactions with the HBV-DNA within the nucleocapsid since they resemble protamine and other DNA binding proteins. Region P probably encodes the endogenous DNA polymerase associated with viral cores. Region X encodes a protein of approximately 154 amino acids which is not particularly hydrophobic and therefore not likely to be associated with the lipid bilayer. Antibodies reactive with the X gene product have been found in the serum of patients with chronic hepatitis and PHC (10).

Infection with HBV characteristically occurs via the blood borne or sexually transmitted routes. Following a prolonged incubation period (up to 6 months), one of several clinical manifestations may occur, including fulminant fatal hepatitis (~1%), clinically apparent acute hepatitis (~20%) or subclinical inapparent disease (~80%). Approximately 90% of patients recover completely with clearance of the virus from the circulation and resolution of hepatocellular injury. However, 10% of patients do not clear the virus and they become chronic carriers. About 1/3 of carriers have no evidence of liver disease (healthy carriers) and they have a normal life expectancy. In contrast, about 2/3 of carriers develop chronic hepatocellular injury of variable degree. The more severe form of injury, chronic active hepatitis, is clinically debilitating, may lead to the development of cirrhosis and, after 30-40 years of chronic HBV infection, leads to the development of primary hepatocellular carcinoma (PHC) at a rate of more than 200 times greater than the normal population (11). There are 200 million HBV carriers in the world today.

HBV can exist in the hepatocyte in both a freely replicating form as well as integrated in the host genome (reviewed in 12). Acute hepatocellular injury usually occurs in the context of free, episomal, viral replication and it usually resolves together with elimination of free virus and virus infected cells. In the absence of viral clearance, viral sequences may integrate into the human genome at variable intervals over a period of month to years after initial infection (occasionally more rapidly). The combined effects of prolonged carriage of integrated viral DNA and chronic hepatocellular injury and regeneration appear to predispose the infected hepatocyte to neoplastic transformation.

Unfortunately, the mechanisms responsible for hepatocellular injury in acute and chronic hepatitis are not well understood due to the lack of satisfactory tissue culture or animal models. Currently available evidence, derived almost entirely from clinical observations and correlations, suggests that while acute hepatitis may be a direct cytotoxic consequence of viral replication, very different events are probably responsible for chronic hepatocellular injury associated with viral integration in view of the healthy HBV carrier state without any associated liver disease (13). Most investigators believe that hepatocellular injury, in both acute and chronic hepatitis, is mediated by a cellular immune response to one or more viral antigens expressed at the hepatocyte membrane and that the final common pathway is antigen-specific, T cell mediated hepatocytolysis. Despite a great deal of work by many investigative teams, including our own, this hypothesis is not only unsettled, it is virtually untested since all T cell responses require the presentation of antigen in the context of autologous histocompatibility antigens. Until now, satisfactory experimental systems that fulfill these requirements have not been available (14). Thus, the very existence of a pathogenetically important cellular immune response is currently entirely hypothetical, and the specific identification of the pathogenetically relevant viral antigen(s) is even less secure.

In a similar fashion, the relationship between viral genome integration and PHC is not well understood. On the one hand, chronic HBV infection clearly leads to the development of PHC and all HBV associated PHC have been shown to contain integrated viral sequences (15,16). Additionally there is some evidence to suggest that within the viral genome there are specific sequences in the vicinity of the nick in the long strand that may be preferentially integrated in PHC (17). However since HBV is clearly not an acutely transforming virus additional co-factors must be necessary for malignant transformation to occur. It is likely that HBV genome integration is, at most, a necessary but insufficient event in the development of PHC. Whether HBV contributes to the development of PHC as a result of gene amplification, translocation, point mutation or some other mechanism is not well understood, but, in vivo, HBV alone is not a complete transforming signal.

Thus, additional signals, besides those potentially arising from integrated HBV sequences, are necessary for malignant transformation. While the nature of these signals is unclear, the one unifying feature is the coexistence of chronic liver cell injury and cellular regeneration. If chronic hepatocellular injury proves to be a consequence of virus specific immune activation then, paradoxically, the anti-viral immune response may be a cofactor in HBV induced tumorigenesis. With the development of mouse embryo microinjection technology, it became apparent that many of these questions might be addressed by creating a model in which integrated copies of the HBV genome are present, expressed, and transmitted in strains of mice that have been previously characterized in terms of their immunological responsiveness to HBV-encoded antigens (18). This report summarizes the salient features of such a model previously described in detail elsewhere (19).

RESULTS

Production and Breeding of Transgenic Lineages.

To date we have concentrated on the region within the HBV genome that encodes the viral envelope polypeptides designated hepatitis B surface antigen (HBsAg) and pre-S antigen. In early studies the subgenomic 2743 bp BglII A+C fragment, spanning nucleotides 2425 to 1986, cloned into the BamHI site of a pBR322 derivative and designated pAC (generously provided by P. Tiollais) was microinjected either after linearization at the unique HindIII site within the vector or after removal of most vector sequences by double digestion with HindIII and PstI. Only 1 of 15 transgenic mice produced with this construct expressed HBsAg, expression was weak (serum HBsAg concentration 25 ng/ml) and was not inheritable.

To improve the frequency, magnitude and transmissibility of HBV gene expression we put the HBV envelope polypeptide open reading frame under the control of the inducible metallothionein (MT) promoter. Maps of the MT vector and HBV insert are shown in Figure 2. Plasmid pMT-1 contains the mouse metallothionein I gene with its heavy metal inducible promoter located upstream of a unique BglII site. We inserted the 2329 bp BglII A fragment of HBV which is a subfragment (nucleotides 2839-1986) of the region used in our earlier studies with pAC. This places the pre-S and S genes under the control of the exogenous promoter. The recombinant construction was designated pMT-PSX since the HBV insert contains the complete open reading frames of the pre-S, S and X coding regions, although the X region is not expressed.

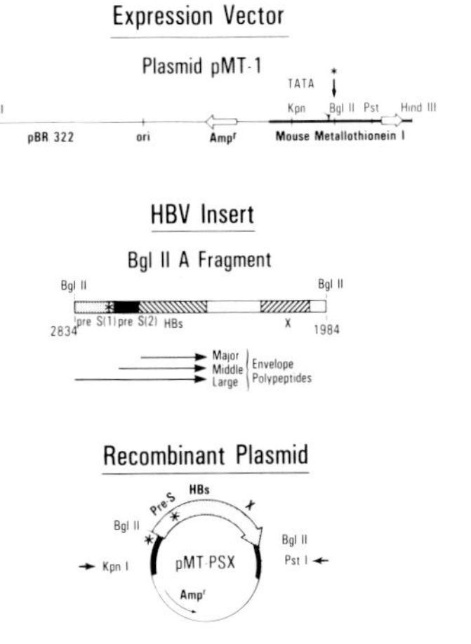

Fig. 2
Microinjected DNA

Plasmid pMT-PSX was microinjected after digestion to provide linear molecules devoid of pBR sequences. Two of six transgenics (designated TM 18-11 and TM 23-3) expressed HBsAg. TM 18-11 died after surgical liver biopsy and was not available for further study. The serum HBsAg concentration of TM 23-3 was 3.2 μg per ml. Notice that this is more than 2 orders of magnitude higher than the concentration achieved with the endogenous HBV promoter in TM 7.2(25 ng/ml). In breeding

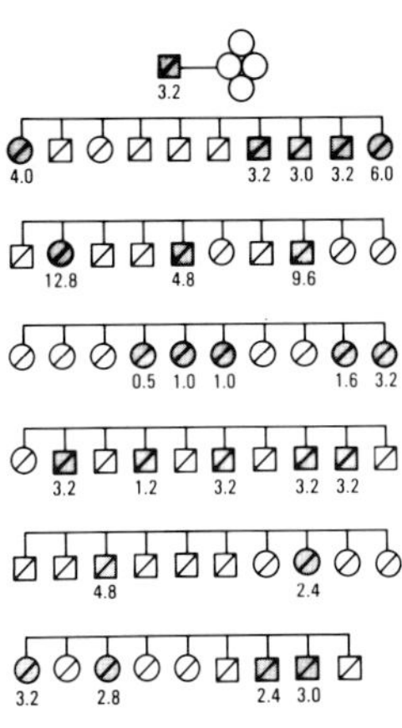

Fig. 3
Breeding TM 23-3

experiments with TM 23-3 the transgene was transmitted in a normal Mendelian fashion all of the transgenic progeny produced high serum levels of HBsAg with some reaching 10-12 μg/ml (Figure 3). This family has now been expanded by breeding and over 100 transgenic, HBsAg positive progeny are available as this is written.

Analysis of Integrated Viral Sequences

Most transgenes are usually integrated at random into a single site within the host genome as a tandem linear polymer, usually in a head to tail orientation. This was generally true in our studies as well; however we also observed head to head and tail to tail tandems in several mice. Our data are consistent with integration at random sites within the mouse genome in this model and this is analogous to the situation found in PHC in man.

Analysis of Gene Expression

Expression of the viral pre-S and S genes from the HBV promoter alone (pAC construct, TM 7-2) and the combined HBV and metallothionein promoters and exogenous regulation (pMT-PSX construct, TM 18-11 and TM 23-3) was examined. TM 7-2, containing the BglII A+C fragment produced trace quantities of serum HBsAg (25 ng/ml). Analysis of organ homogenates from this mouse revealed highest HBsAg concentration (as percent of total protein) in the stomach, followed by brain, small intestine, and liver. As less than 5 percent of transgenic offspring from TM 7-2 did not express HBsAg this lineage was not studied further. The distribution of soluble HBsAg in tissue homogenates derived from 23-3 and

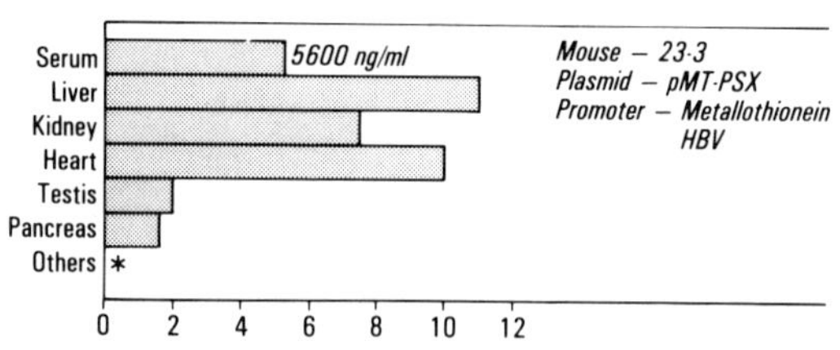

Fig. 4
Tissue Distribution of HBsAg

his progeny is summarized in Fig. 4. Note that liver and kidney HBsAg levels are high as expected based on the tissue specificity of the MT promoter. However, the unexpected high level expression HBsAg in the heart suggests that HBsAg synthesis is influenced by host factors as well. This hypothesis is supported by the pattern of antigen distribution detectable by immunofluorescence analysis of frozen liver sections from TM 23-3 and his progeny (Fig. 5). Both pre-S antigen and HBsAg were detectable as coarse granular deposits in the cytoplasm of some hepatocytes. Interestingly, antigen positive cells were restricted to the periphery of the hepatic lobule, and in this site most of the hepatocytes were positive. The important influence of cellular factors on HBsAg synthesis is apparent since many hepatocytes and all of the Kupffer cells, endothelial cells, and bile duct epithelium are antigen negative. It is notable that all the tissue of these transgenic mice and their progeny studied thus far have been entirely normal histopathologically.

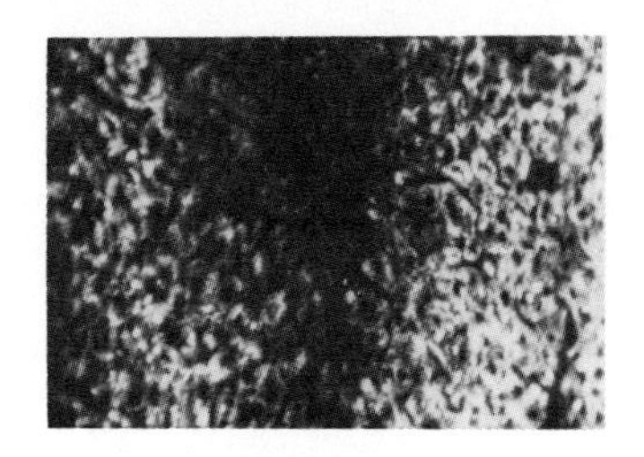

Fig. 5
TM 23-3 Liver

Expression of the integrated genes was analyzed at several levels. First, HBV-specific transcripts examined by Northern blotting of total liver RNA from TM 23-3 revealed a predominant 2.3-kb transcript (Fig. 6, 2 left lanes) consistent with promotion at or near the MT promoter and termination at the HBV polyadenylation signal. Since this transcript was inducible by zinc the MT promoter was probably functional. Several larger transcripts, some of which were zinc inducible were also observable. These inducible transcripts were

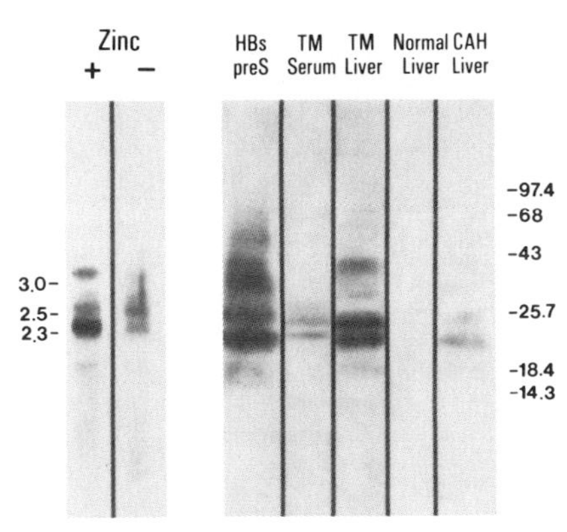

Fig. 6
Northern and Western Blots

probably due to termination downstream from the HBV polyadenylation signal. Noninducible transcripts may represent RNA's initiated from other promoters present either in adjacent mouse DNA or in the HBV sequences. HBV-specific transcripts were detectable by Northern blotting only in liver, kidney, testis, heart, and brain, suggesting that tissue HBsAg detectable in other organs may reflect serum contamination. Furthermore, only liver and kidney displayed dominant 2.3-kb transcripts, whereas in other organs either larger transcripts were dominant or all transcripts were equally represented.

The protein was also assessed by Western blotting of HBsAg-positive serum and liver homogenates (Fig. 6, 5 right lanes). Both serum and liver from TM 23-3 revealed the predicted HBsAg bands at molecular weights of 22,000 and 25,000 Daltons representing the unglycosylated and glycosylated forms of the 226 amino acid product of the HBs gene. Additional, higher molecular weight bands were observed in liver with a dominant band at about 39,000 Daltons. This is the predicted size of the large envelope protein that initiates at the first translation start codon within the pre-S region. Subsequent blots using antisera to pre-S(1) and pre-S(2) peptides confirmed the presence of

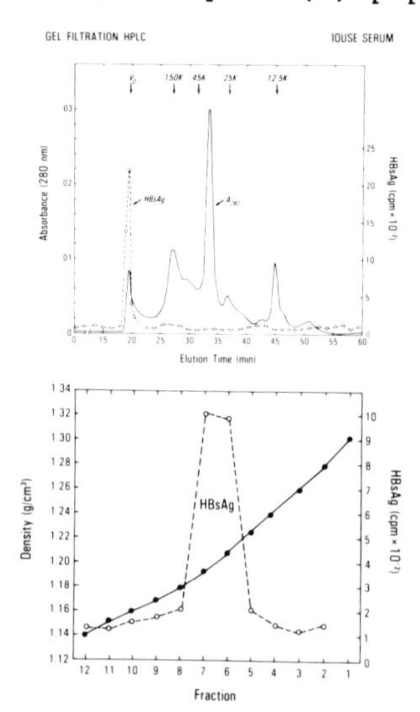

Fig. 7
Purification of HBsAg from mouse serum

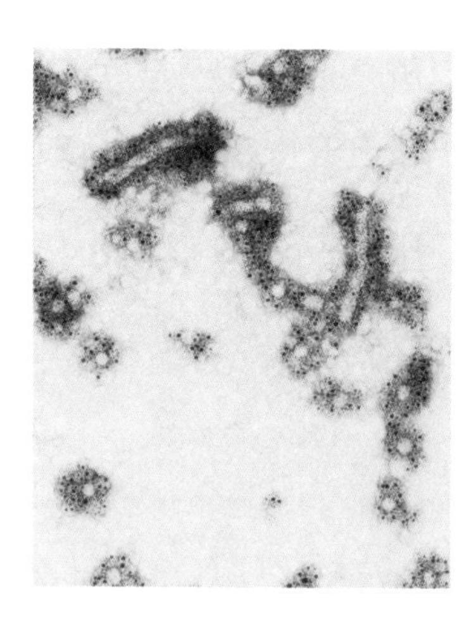

Fig. 8
Immunogold Electron Microscopy of Purified Serum HBsAg

both pre-S antigens and establish the identity of this polypeptide.

We also examined the secreted serum product. HBsAg activity migrated with an apparent molecular weight greater than 1 x 10^6 upon gel filtration by high pressure liquid chromatography (Fig. 7) consistent with the size (2.5 x 10^6 M_r) of human serum HBsAg particles from HBV-infected patients. This high molecular weight material banded at a mean density of 1.20 g/cm^3 in cesium chloride consistent with the density of human HBsAg. Finally, this purified fraction contained filamentous and spherical HBsAg particles with a mean diameter of 22 nm as shown by immunoelectron microscopy with colloidal gold labelled antibodies (Fig. 8). These particles are indistinguishable from human HBsAg particles. Thus, on the basis of molecular weight, buoyant density, diameter, morphology, and antigenicity, secreted transgenic HBsAg is indistinguishable from human HBsAg particles.

Direct Pathologic Consequences of HBV-DNA Integration.

To date there are no detectable abnormalities in fertility, litter size, sex ratio, growth, development and life span of our HBV DNA-containing transgenic mice. Further, animals randomly sacrificed at monthly intervals for nearly one year are normal at the gross and light microscopic level. Notably liver morphology is entirely normal and no hepatomas have been found despite the production by the liver of large quantities of HBs and the pre-S antigens. We conclude that HBV envelope gene expression is probably not directly cytotoxic or oncogenic.

Induction of an In Vivo Immune Response to Microinjected Gene Products.

Our transgenic mice are SJL x B6 hybrids. Since the SJL strain is immunologically nonresponsive to HBsAg at the T cell level (18) and since the B6 strain responds well to HBsAg any immunological studies in our transgenics will be variable because they are not inbred. Since 100% of our pMT-PSX transgenic mice (TM-23-3 progeny) express HBsAg one might expect them to be tolerant to this antigen. Indeed anti-HBs does not spontaneously appear and immune complexes are not found in skin or renal glomeruli as might occur in the face of a humoral immune response to this circulating antigen. Transgenic mice backcrossed twice against the responsive B6 parental strain have been immunized with purified native HBsAg particles of the same (ay) and the alternate (ad) subtype of the microinjected construct (ay). Only the alternate (d) subtype antigen elicited an antibody response to the subtype specific epitope after up to 2 injections of 4 μg of HBsAg in complete Freund's adjuvant. We conclude that our transgenics are indeed immunologically tolerant to the subtype determinant of the microinjected gene product. These observations are compatible with the hypothesis that liver cell injury in human HBV infection is not a direct consequence of viral envelope antigen expression and may be secondary to a cytotoxic immune response to hepatocyte surface membrane viral antigens. It should therefore be possible to design studies to circumvent tolerance and monitor immunologically mediated tissue injury specific for each of the viral antigens based on this transgenic mouse model when inbred strains of transgenic mice that express HBV encoded antigens are available.

CONCLUSIONS

These transgenic mice thus provide a model analogous to the stage in HBV infection when replication has ceased and the viral DNA has integrated into the host genome as occurs in the chronic carrier state and in hepatocellular carcinoma. Other data suggest that the pAC construct is able to establish high serum levels of HBsAg in transgenic mice (20). Because of the low HBsAg serum titers we achieved with this construct (pAC), the MT fusion gene may provide a useful alternative system. The virtue of the MT promoter is its strength and inducibility. Indeed, it leads to expression of high levels of HBV-encoded antigens within the liver and this makes it useful for subsequent studies of immunologically mediated hepatocellular injury in transplantation and adoptive transfer experiments. Although the transgenic mouse model lacks many features of HBV infection, it provides an opportunity to study the consequences of expression of integrated HBV DNA in genetically defined mice of predetermined immune responsiveness to HBV-encoded antigens. Such studies should provide useful information pertaining to the pathogenesis of the diseases associated with HBV in man.

ACKNOWLEDGEMENTS

We thank P. Tiollais for enabling us to work in his laboratory and for providing the plasmid pAC and the recombinant HBsAg particles; M. Trumbauer for performing the embryo microinjections; R. Neurath and R. Lerner for antipeptide reagents; B. Thornton and A. Moriarty for pre-S and X antigen analysis; M. Riggs and J. Hughes for technical assistance; C. M. Chang for performing the electron microscopy; and J. Verenini for typing the manuscript.

REFERENCES

1. Blumberg BS (1984). Background and Perspective. In Chisari FV (ed): "Advances in Hepatitis," New York: Masson Publishing Inc, p 1.
2. Barker LF, Chisari FV, McGrath, PP, Dalgard DW, Kirschstein RL, Almeida JD, Edgington TS, Sharp DG, Peterson MR (1973). Transmission of type B viral hepatitis to chimpanzees. J Infect Dis 127:648.
3. Summers J, Smolec JM, Snyder R (1978). A virus similar to human hepatitis B virus associated with hepatitis and hepatoma in woodchucks. Proc Natl Acad Sci USA 75:4533.
4. Marion PL, Oshiro LS, Regnery DC, Robinson WS, Scullard GH (1980). A virus in Beechey ground squirrels that is related to hepatitis B virus in humans. Proc Natl Acad Sci USA 77:2941.
5. Mason WS, Seal G, Summers J ((1980). Virus of Pekin ducks with structural and biological relatedness to human hepatitis B virus. J Virol 36:829.
6. Shafritz DA, Hadziyannis SJ (1984). Hepatitis B virus DNA in liver and serum, viral antigens and antibodies, virus replication, and liver disease activity in patients with persistent hepatitis B virus infection. In Chisari FV (ed): "Advances in Hepatitis,", New York: Masson Publishing Inc, p 80.
7. Robinson WS, Marion P, Feitelson M, Siddiqui A (1982). The hepadna virus groups: hepatitis B and related

viruses. In Szumness W, Alter HJ, Maynard JE (eds): "Viral Hepatitis," Philadelphia, PA: Franklin Institute Press, p 57.

8. Summers J, O'Connell A, Millman I (1975). Genome of hepatitis B virus: Restriction enzyme cleavage and structure of DNA extracted from Dane particles. Proc Natl Acad Sci USA 72:4597.

9. Tiollais P, Wain-Hobson S (1984). Molecular genetics of the hepatitis B virus. In Chisari FV (ed): "Advances in Hepatitis Research," New York: Masson Publishing Inc.

10. Moriarty AM, Alexander H, Lerner RA, Thornton GB (1984). Antibodies to peptides detect new hepatitis B antigen: Serological correlation with hepatocellular carcinoma. Science 227:429.

11. Beasley RP, Hwang LY, Lin CC, Chien CS (1981). Hepatocellular carcinoma and hepatitis B virus. A prospective study of 22,707 men in Taiwan. Lancet 2:1129.

12. Tiollais P, Pourcel C, Dejean A (1985). The hepatitis B virus. Nature 317:419.

13. Kam W, Rall LB, Smuckler EA, Schmid R, Rutter WJ (1982). Hepatitis B virus DNA in liver and serum of asymptomatic carriers. Proc Natl Acad Sci USA 79:7522.

14. Chisari FV, Milich DR, Tiollais P (1984). Hepatitis B virus infection: A model for immunologically mediated hepatocellular injury. In Keppler D, Reutter W, Bianchi L (eds): "Mechanisms of Hepatocyte Injury and Death," MTP Press, p 293.

15. Brechot C, Pourcel C, Louise A, Rain B, Tiollais P (1980). Presence of hepatitis B virus DNA sequences in cellular DNA of human hepatocellular carcinoma. Nature 286:533.

16. Shafritz DA, Kew MC (1981). Identification of integrated hepatitis B virus DNA sequences in human hepatocellular carcinomas. Hepatology 1:1.

17. Dejean A, Sonigo P, Wain-Hobson S, Tiollais P. Specific hepatitis B integration in hepatocellular carcinoma DNA through a viral II base pair direct repeat. Proc Natl Acad Sci USA, in press.

18. Milich DR, Alexander H, Chisari FV (1983). Genetic regulation of the immune response to hepatitis B surface antigen (HBsAg). III. Circumvention of nonresponsiveness in mice bearing nonresponder haplotypes. J Immunol 130:1401, 1983.

19. Chisari FV, Pinkert CA, Milich DR, Filippi P, McLachlan A, Palmiter RD, Brinster R (1985). A transgenic mouse model of the chronic hepatitis B surface antigen carrier state. Science 230:1157.
20. Babinet C, Farza H, Morello D, Hadchouel M, Pourcel C (1985). Specific expression of hepatitis B surface antigen in transgenic mice. Science 230:1160.

Index